Wind Energy Design

Wind Energy Design provides an understanding of how wind farm design impacts wind turbine performance and offers rigorous design optimization coverage. It explores the blade design features, including aerodynamics, structures, and environmental impact, that yield an efficient rotor.

Demonstrating how theoretical aspects can be applied to produce a new conceptual wind energy design, this book addresses topics on wind characteristics and modeling, rotor aerodynamics, lightweight and flexible structures, wind farm aerodynamics, wind turbine control, acoustics, energy storage, and economics. The new edition features new sections on minimizing the costs of electricity and wind farm design for maximum energy capture. It includes a new case study on deployable wind turbines for defense and disaster response missions.

This book is intended for upper-level undergraduate engineering students taking courses in Renewable Energy, Wind Power, Energy Systems, and Energy Conversion.

Instructors will be able to utilize a solutions manual and figure slides for their course.

Wind Energy Design

Second Edition

Thomas C. Corke and Robert C. Nelson

CRC Press
Taylor & Francis Group
Boca Raton London New York

CRC Press is an imprint of the
Taylor & Francis Group, an **informa** business

Front cover image: engel.ac/Shutterstock

Second edition published 2026
by CRC Press
2385 NW Executive Center Drive, Suite 320, Boca Raton FL 33431

and by CRC Press
4 Park Square, Milton Park, Abingdon, Oxon, OX14 4RN

CRC Press is an imprint of Taylor & Francis Group, LLC

First edition published by CRC Press 2018

Library of Congress Cataloging-in-Publication Data
Names: Corke, Thomas C. author | Nelson, Robert C., 1942- author
Title: Wind energy design / Thomas Corke and Robert Nelson.
Description: Second edition. | Boca Raton, FL : CRC Press, 2026. | Includes bibliographical references and index.
Identifiers: LCCN 2025014627 (print) | LCCN 2025014628 (ebook) | ISBN 9781041007524 hbk | ISBN 9781041007531 pbk | ISBN 9781003611486 ebk
Subjects: LCSH: Wind turbines–Design and construction | Wind power plants–Design and construction
Classification: LCC TJ828 .C67 2026 (print) | LCC TJ828 (ebook) | DDC 621.4/5–dc23/eng/20250701
LC record available at https://lccn.loc.gov/2025014627
LC ebook record available at https://lccn.loc.gov/2025014628

ISBN: 978-1-041-00752-4 (hbk)
ISBN: 978-1-041-00753-1 (pbk)
ISBN: 978-1-003-61148-6 (ebk)

DOI: 10.1201/9781003611486

Typeset in CMR
by codeMantra

Access the Support Material: Routledge.com/9781041007524

This 2nd Edition is dedicated to my wife Bobbie, daughters Catherine, Laura and Sarah, sons-in-law Anthony, Michael and Clinton and grand children Patrick, Allison, Madeline, Claire, and Gabriel. They make every day a joyful adventure.

In the span of time since the 1st Edition, my co-author Robert Nelson passed away. He however has left a lasting imprint on this edition and I am sure he would acknowledge the support he has received from his wife, Julie, and family.

—Thomas C. Corke

Contents

Foreword **xii**
List of Figures **xv**
List of Tables **xxv**

1 Introduction **1**
1.1 History of Wind Energy 1
1.1.1 Modern Era of Wind Energy 10
Problems 20
Further Reading 20

2 Atmospheric Boundary Layer and Wind Characteristics **22**
2.1 Origin of Wind 22
2.2 Atmospheric Boundary Layer 23
2.3 Temporal Statistics 26
2.4 Wind Speed Probability 27
2.5 Statistical Models 29
2.5.1 Weibull Distribution 30
2.5.2 Methods for Weibull Model Fits 33
2.5.3 Rayleigh Distribution 36
2.6 Energy Estimation of Wind Regimes 37
2.6.1 Weibull-Based Energy Estimation Approach 38
2.6.2 Rayleigh-Based Energy Estimation Approach 40
2.7 Wind Condition Measurement 43
2.7.1 Wind Speed Anemometers 44
Problems 49
References 50

3 Introduction to Aerodynamics **51**
3.1 Introduction 51
3.2 Airfoil Geometry 53
3.3 Dimensional Analysis 54
3.4 Airfoil Aerodynamics 59
3.5 Airfoil Geometry 61
3.6 Aerodynamic Characteristic of Three NACA Airfoils 62
3.7 Airfoil Sensitivity to Leading Edge Roughness 65
3.8 New Airfoil Designs for the Wind Power Industry 68
3.9 Summary 70

Problems . . . 71
Note . . . 72
References . . . 72

4 Aerodynamic Performance of a Wind Turbine Rotor 73
4.1 Momentum Theory . . . 73
4.2 Momentum Theory with Wake Rotation . . . 83
4.3 Blade Element Momentum (BEM) Theory . . . 87
4.4 Prandtl's Tip Loss Factor . . . 92
4.5 Solution of the BEM Equations . . . 95
4.5.1 Example BEM Equation Solution . . . 96
Problems . . . 104
References . . . 106

5 Horizontal Wind Turbine Rotor Design 107
5.1 Designing a New Wind Turbine . . . 107
5.2 Initial Blade Sizing . . . 107
5.2.1 Example Rotor Design . . . 113
Problems . . . 117
References . . . 118

6 Wind Turbine Control 119
6.1 Aerodynamic Torque Control . . . 121
6.1.1 Electrical Torque Control . . . 122
6.2 Wind Turbine Operation Strategy . . . 124
6.2.1 Fixed Speed Designs . . . 125
6.2.2 Variable Speed Designs . . . 126
6.2.3 Variable Speed Adaptive Torque Control . . . 126
6.3 Axial Induction Control . . . 129
6.4 Lift Control . . . 135
Problems . . . 139
References . . . 141

7 Structural Design 143
7.1 Aerodynamic Loads . . . 143
7.2 Gravitational Loads . . . 144
7.3 Dynamic Loading . . . 146
7.4 Control Loads . . . 147
7.5 Rotor Response to Loads . . . 148
7.6 Rotor Vibration Modes . . . 153
7.7 Design for Extreme Conditions . . . 157
Problems . . . 160
References . . . 163

8 Wind Farms 164
8.1 Wind Turbine Wake Effects . . . 165

8.2 Wind Farm Design Optimization 170
8.3 Wind Farm Design for Maximum Energy Capture 172
8.3.1 Evolutionary Genetic Algorithm 179
Problems 188
References 192

9 Wind Turbine Acoustics 194
9.1 Acoustics Fundamentals 195
9.2 Sound Pressure Measurement and Weighting 198
9.3 dB Math 199
9.4 Low Frequency and Infrasound 199
9.5 Wind Turbine Sound Sources 201
9.6 Sound Propagation 205
9.7 Background Sound 210
9.8 Noise Standards 211
9.9 Wind Turbine Project Noise Assessment 211
9.9.1 Ambient Background Levels 211
9.9.2 Source Sound Levels 212
9.9.3 Sound Propagation Model 212
9.9.4 Comparison of Calculated Sound Levels with Baseline Sound Levels 212
9.10 Wind Turbine Sound Mitigation Techniques 212
Problems 213
References 214

10 Wind Energy Storage 216
10.1 Electro-Chemical Energy Storage 217
10.1.1 Lead-Acid Batteries 218
10.1.2 Nickel-Based Batteries 219
10.1.3 Lithium-Based Batteries 219
10.1.4 Additional Electro-Chemical Storage Technologies . . 220
10.1.5 Sodium Sulfur Batteries 221
10.1.6 Redox Flow Battery 221
10.1.7 Metal-Air Battery 222
10.2 Supercapacitor Storage 222
10.3 Hydrogen Storage 224
10.4 Mechanical Energy Storage Systems 226
10.4.1 Pumped Storage Hydroelectricity 226
10.4.2 Compressed Air Storage 227
10.4.3 Flywheel Storage 229
10.5 CAES Case Study 232
10.5.1 Cost Function 235
10.5.2 Net Benefit 238
10.6 Battery Case Study 239
10.7 Hydro-Electric Storage Case Study 240

10.8 Buoyant Hydraulic Energy Storage Case Study 241
Problems 245
References 247

11 Economics **248**
11.1 Cost of Energy, *COE* 248
11.2 Component Estimate Formulas 251
11.2.1 Rotor Blade Mass 251
11.2.2 Rotor Blade Cost 251
11.2.3 Rotor Hub Cost 253
11.2.4 Pitch Mechanism and Bearings Cost 253
11.2.5 Spinner Nose Cone Cost 253
11.2.6 Low-Speed Shaft Cost 254
11.2.7 Main Bearings Cost 254
11.2.8 Gearbox Cost 254
11.2.9 Mechanical Brake/High-Speed Coupling Cost 255
11.2.10 Electric Generator Cost 255
11.2.11 Variable-Speed Electronics Cost 256
11.2.12 Yaw Drive and Bearing Cost 256
11.2.13 Mainframe Cost 256
11.2.14 Electrical Connections Cost 257
11.2.15 Hydraulic and Cooling Systems Cost 257
11.2.16 Nacelle Cover Cost 257
11.2.17 Control, Safety System, Condition Monitoring Cost 257
11.2.18 Tower Cost 257
11.2.19 Transportation Cost 258
11.2.20 Roads, Civil Work Cost 259
11.2.21 Assembly and Installation Cost 259
11.2.22 Electrical Interface/Connections Cost 259
11.2.23 Engineering and Permit Cost 260
11.2.24 Levelized Replacement Cost 260
11.2.25 Operations and Maintenance Cost 260
11.2.26 Land Lease Cost 260
11.3 Example Cost Breakdown 261
11.4 Optimized Economics 263
11.5 Summary 265
Problems 266
References 267

12 Design Summary and Trade Study **269**
12.1 Design Power 271
12.2 Design Structure 271
12.3 Design Economics 273
Problems 279

13 New Concepts **280**
13.1 Vertical Axis Wind Turbine 280
13.2 Wind Focusing Concepts 283
13.2.1 Shrouded Rotors 283
13.3 Bladeless Wind Turbine Concepts 287
13.3.1 Airborne Wind Turbine Concepts 289
13.4 Other Concepts 289
13.4.1 Lateral Axis Wind Turbine 290
13.4.2 Tree-Shaped Wind Turbine 290
13.4.3 Wind Turbine Phone Charger 291
13.4.4 Miniature Wind Turbine 291
13.4.5 Wind Powered Street Border Lights 291
References 292

14 Capstone Design - Deployable Wind Turbines for Disaster Response Missions **294**
14.1 Wind Turbine Design 295
14.1.1 Site Wind Conditions 296
14.1.2 Rotor Design 297
14.2 Wind Turbine Tower and Base Designs 302
14.2.1 Tower 303
14.2.2 Base 305
14.3 Wind Turbine Arrangement 307
14.3.1 Total Power 307
14.3.2 Sound Levels 311
14.4 Battery Storage 313
14.5 Estimated Cost 315
14.6 Summary 315
Problems 316
Further Reading 317

Appendix **319**

Index **333**

Foreword

This book is intended to be a text for a senior-level Engineering course dealing with the conceptual design of a wind energy system. It is based on the experience gained in teaching "capstone" design classes in Aerospace Engineering for the past 30 years. The emphasis here is being toward wind energy.

The approach is to demonstrate how the theoretical aspects, drawn from topics on wind characteristics and modeling, rotor aerodynamics, lightweight and flexible structures, wind farm aerodynamics, wind turbine control, acoustics, energy storage, and economics, can be applied to produce a new conceptual wind energy design. This book not only cites theoretical expressions wherever possible but also emphasizes the interplay of different aspects of the design, which often require compromises. As necessary, it draws on historical information to provide needed input parameters, especially at an early stage of the design process. In addition, historical wind energy systems are used to provide checks on design elements to determine if they deviate too far from historical norms.

The conceptual design process of a wind energy system is divided into 11 steps. These are covered in Chapters 4–12. This book emphasizes the use of interactive computational approaches for iterative and/or repetitive calculations. Sample calculations covering each step of the design are provided in each chapter, except Chapters 1 and 13. In addition, there are individual problems at the end of each chapter in which the students are asked to document different degrees of dependence of the design characteristics on changing input conditions. Some of these problems are "open ended" and require interpretation and discussion.

A wind energy system's design summary and trade study are presented in Chapter 12. This incorporates aerodynamic design and structural design, wind characteristic modeling and siting, design power and annual expected power, and economics including purchase and power on station costs, and ultimately the cost of electricity. Provided software allows students to explore the impact of various initial design parameters on the design outcomes.

New to the 2nd Edition is an emphasis on "design optimization". This is first illustrated in Chapter 8 on the design of wind farms to maximize energy capture. Specifically, it utilizes a genetic algorithm to determine the optimum arrangement of wind turbines that minimizes wake losses for the full range of wind directions and wind speeds. It includes the full power curves for the wind turbines. The detailed steps involved in the genetic algorithm are presented so that they can easily be programmed by the students. The formulation to solve the integral form of the total wind turbine power, whose maximum is the objective function for the optimization, is presented, so that it can also be implemented by the students. A full example of optimized wind farms with two to seven wind turbines is presented. Overall, the optimized design approach provides a powerful tool for examining the aspects of wind farm design, such as minimum turbine spacing, site selection, and distributed capture concepts proposed to enhance energy capture.

The design optimization approach using the genetic algorithm is also applied in Chapter 11 to determine wind turbine designs with the objective of minimizing the cost of electricity (COE). This utilizes the correlation between various cost factors and the turbine rotor radius. It then seeks to answer the question on what is more cost-effective in a wind farm, a larger number of smaller wind turbines or a fewer number of larger wind farms. This also utilizes the optimization approach presented in Chapter 8.

To tie all of the concepts and further illustrate design optimization, the 2nd Edition includes a new chapter (Chapter 14) that is meant as a Capstone Design project that is a common conclusion to a design course. Here it involves the design of "Deployable Wind Turbines for Defense and Disaster Response Missions". The overall project involves the design of a mobile horizontal axis wind turbine that can be erected singly or in a group to provide electric power to a remote site where no electric infrastructure exists. The designs are to adhere to transport packaging space constraints, and assembly and anchoring of the wind turbines are important considerations. Compared to conventional wind turbine applications, this application places a premium on rapid shipping and installation, short-duration operation (days to months), and quick teardown upon mission completion. A full design that includes an arrangement of six such wind turbines in a compound area, which includes power estimates that include wake effects and sound level predictions for a range of wind directions, is presented as an example. An economic analysis is also included. Accompanying problem sets elaborate on the possible designs and optimization comparisons. This chapter can be used either as a demonstration of the methods of wind energy *system* design or as an open-ended design project.

The learning objectives in the 2nd Edition are the same, namely: (1) to understand how to characterize the properties of the wind resource from which the power is to be extracted; (2) to understand how to predict the performance of a horizontal axis wind turbine using blade element momentum (BEM) theory; (3) to understand the blade design features including aerodynamics, structures, and environmental impact that yield an efficient rotor; (4) to understand how wind farm design impacts wind turbine performance; (5) to understand the aspects of active control to improve off-design turbine performance, (6) to understand the impact electric power storage can have on the wind energy system; and (6) to understand the economic issues related to the wind energy system. To this, the 2nd Edition adds (7) rigorous design optimization and (8) a detailed and open-ended Capstone Design that involves all of the elements (1)–(7).

This 2nd Edition book can be used in either of two ways. First, it can be used to develop a *complete* conceptual design of a new wind energy system. This is the way that we personally teach this material. Starting at the beginning, the students develop a complete design (similar to the case study) in a step-by-step fashion. This is accomplished over one semester (15 weeks).

Second, this book can be used to consider the individual aspects of a wind energy system without developing a complete design. This approach makes the best use of the problem sets at the end of each chapter. The effect of different input parameters can be easily investigated, and optimums can be sought. We know of instructors who prefer this approach.

The following is a list of chapters:

Chapter 1: Introduction

Chapter 2: Atmospheric Boundary Layer and Wind Characteristics

Chapter 3: Introduction to Aerodynamics

Chapter 4: Aerodynamic Performance of a Wind Turbine Rotor

Chapter 5: Horizontal Wind Turbine Rotor Design

Chapter 6: Wind Turbine Control

Chapter 7: Structural Design

Chapter 8: Wind Farms

Chapter 9: Wind Turbine Acoustics

Chapter 10: Wind Energy Storage

Chapter 11: Wind Energy Economics

Chapter 12: Design Summary and Trade Study

Chapter 13: New Concepts

Chapter 14: Capstone Design - Deployable Wind Turbines for Defense and Disaster Response Missions

Appendix

For a complete conceptual design, the chapters are intended to be followed in chronological order. A conscious attempt has been made to include within each chapter all of the supplementary material that is needed to develop that aspect of the design. This minimizes the need to search for formulas or graphs in other chapters or references.

Chapter 12 summarizes the case study which runs throughout the text and discusses the role of a trade study on a complete design. This is illustrated with the case study design and in the problems at the end of the chapter. Chapter 13 presents new concepts for wind energy. Some of these are topical, which leads to a discussion on the motivation and practicality of the concepts.

Thomas C. Corke
March 2025

List of Figures

1.1 Photograph of circa 900 AD vertical axis wind turbine used for grinding grain that is located in modern Afghanistan. . . 2
1.2 Schematic drawing of the early vertical axis wind turbine is shown in Figure 1.1. 2
1.3 Photograph of an early Chinese vertical axis wind turbine utilizing cloth sails. 3
1.4 Sketch of an early horizontal axis wind turbine utilizing cloth sails and driving a water ladder pump. 3
1.5 Medieval illustration of a sunk post mill. 4
1.6 Photograph of post mill presumably been build in 1683 at Essern, District of Nienburg Germany, and drawing showing post mill internal design. 5
1.7 Photograph of a Smock mill that was common to the Netherlands. 6
1.8 Circa 1750 photograph of John Smeaton who developed early mathematical models to predict windmill efficiency. 7
1.9 Photograph of the Blyth vertical axis wind turbine that was the first to produce electricity. 7
1.10 Scientific American page showing different views of the Brush horizontal axis wind turbine built in 1887. 8
1.11 Photograph of Paul la Cour (1846–1908) and his 1897 horizontal axis wind turbine that produced electricity used in the production of hydrogen gas. 9
1.12 Photograph of the Johannes Juul designed Gedser wind turbine built in 1956 and located in Vester Egesborg, Denmark. 9
1.13 Photograph of the Smith-Putman wind turbine built in 1941 and located in Castleton, Vermont. 10
1.14 Photograph of an early Darrieus wind turbine in the field (a) and rotor airfoil section shape for Alcoa aluminum extrusion (b). 11
1.15 Photograph of the largest built Darrieus wind turbine located in Cap-Chat, Quebec Canada. 12
1.16 Photograph of the Giromill vertical wind turbine. 13

1.17 Illustration of a Savonius wind turbine design, and the difference in the drag coefficient between concave and convex surfaces that is the basis for this wind turbine design. 14
1.18 Photograph of a wind turbine design that combines Darrieus and Savonius concepts. 15
1.19 Plots of coefficients of power versus rotor tip-speed-ratio for different vertical and horizontal axis wind turbine designs. 15
1.20 Photograph of General Electric 2.5 MW wind turbines making up a 240 turbine wind farm. 16
1.21 Photograph of the General Electric Haliade off-shore wind turbine which has a rated power of 6 MW, and a rotor diameter of 151 m. 16
1.22 Illustration of the internal components in the nacelle of a modern HAWT. 17
1.23 Trend in average HAWT rotor diameters since 1980. 17
2.1 Mechanism of wind generation through global temperature gradients. 23
2.2 Effect of Coriolis force on the wind between pressure isobars. 23
2.3 Schematic of geostropic wind in the Northern hemisphere that results from a steady state balance of Coriolis force and pressure isobars. 24
2.4 Schematic of atmospheric boundary layer profiles for small and large surface roughness. 24
2.5 Hypothetical power curve for wind turbine with a rated power of 250 kW. 27
2.6 Probability distribution of wind speeds at the Notre Dame White Field wind turbine site, and a best-fit Rayleigh distribution. 29
2.7 Sample Weibull distributions for atmospheric boundary layer data at different sites. 30
2.8 Weibull distributions fit for the data in Table 2.3. $k = 2.0$ and $c = 6.68\,\text{m/s}$. 35
2.9 Photograph of Robinson 1846 cup anemometer. 45
2.10 Example of a propeller anemometer that is designed to point into the wind. 46
2.11 Schematic drawing of a Pitot-static probe anemometer. 47
2.12 Photograph of a three-component sonic anemometer located on the meteorological tower of the University of Notre Dame research wind turbines. 48
3.1 Sketch of a wind turbine showing the different blade section shapes across the blade span. 52
3.2 Aerodynamic forces and moment acting on an airfoil. 52
3.3 Geometry defining a symmetric and cambered airfoil section shapes. 54

3.4 Sample lift coefficient versus angle of attack for a thick symmetric airfoil section. 60
3.5 Drag coefficient versus angle of attack for the same airfoil section that produced the lift coefficient versus angle of attack shown in Figure 3.4. 60
3.6 Aerodynamic characteristics of a NACA-0006 airfoil section. 62
3.7 Aerodynamic characteristics of a NACA-0012 airfoil section. 63
3.8 Aerodynamic characteristics of a NACA-4412 airfoil section. 63
3.9 Effect of chord Reynolds number on the lift-to-drag ratio versus angle of attack of a NACA-4412 airfoil section. 65
3.10 Effect of leading edge roughness on the lift-to-drag ratio versus angle of attack of a NACA-4412 airfoil section. 66
3.11 NREL thin-airfoil family for use in medium sized wind turbine blades. 67
3.12 NREL thick-airfoil family for use in medium sized wind turbine blades. 68
3.13 NREL thick-airfoil family for use in large sized wind turbine blades. 69
3.14 NREL thick-airfoil family for use in large sized wind turbine blades. 70
4.1 Flow field of a wind turbine and actuator disc. 74
4.2 Variation of the velocity and dynamic pressure through the stream-tube. 75
4.3 Variation of the static and total pressure along the steam-tube. 75
4.4 Cylindrical control volume surrounding the stream-tube. . . 76
4.5 Variation of the rotor thrust and power coefficients, C_T and C_P, with the axial induction factor, a. 80
4.6 Thrust coefficient as a function of axial induction factor, a, indicating a valid range for momentum theory. 81
4.7 Schematic of the induced rotation of the flow downstream of the rotating actuator disc. 84
4.8 Example of the variation in chord and geometric twist along the radial distance of a wind turbine rotor blade. 88
4.9 Illustration of the aerodynamic forces acting on a wind turbine blade section at a distance r from the axis of rotation. . . . 89
4.10 Illustration of rotor tip vortices from a three-bladed wind turbine rotor. 92
4.11 Photograph of the cross-section of the tip vortices from a two-bladed wind turbine that was visualized in a wind tunnel experiment. 93
4.12 Prandtl tip loss factor along the span of a wind turbine rotor. 94

4.13 Example of a wind turbine blade divided into ten sections for BEM analysis. 94
4.14 Flow chart for the iterative procedure used in solving the BEM equations. 96
4.15 Photograph of the University of Notre Dame Research Wind Turbines and Meteorological tower. 97
4.16 Blade chord distribution for the University of Notre Dame Research Wind Turbines. 98
4.17 Blade twist distribution for the University of Notre Dame Research Wind Turbines. 99
4.18 Spanwise distribution of the rotor blade angles ϕ and θ_T for the University of Notre Dame Research Wind Turbines. . . . 99
4.19 Spanwise distribution of the induction factors, a and a' for the University of Notre Dame Research Wind Turbines. 100
4.20 Spanwise distribution of the lift-to-drag ratio for the University of Notre Dame Research Wind Turbines. 100
4.21 Spanwise distribution of the Prandtl loss coefficient for the University of Notre Dame Research Wind Turbines. 101
4.22 Spanwise distribution of the differential thrust for the University of Notre Dame Research Wind Turbines. 101
4.23 Spanwise distribution of the differential torque for the University of Notre Dame Research Wind Turbines. 102
4.24 Spanwise distribution of the differential power for the University of Notre Dame Research Wind Turbines. 102
4.25 Power curve for the University of Notre Dame Research Wind Turbines. 103
5.1 Power coefficient as a function of the rotor tip speed ratio. 109
5.2 Comparison of theoretical wind turbine rated power versus rotor diameter for two C_p values and that of modern wind turbines. 110
5.3 Power coefficient as a function of tip-speed-ratio for different numbers of rotor blades based on inviscid simulation. 112
5.4 Relative wind angle, $\phi(r)$, and blade twist angle, $\theta_T(r)$ along the rotor radius for a Betz optimum design. 114
5.5 Radial distribution of the local rotor chord length of a rotor for a Betz optimum design. 114
5.6 Example of a modification to the Betz optimum chord distribution to reduce the weight of the rotor. 115
5.7 Transformation from Betz optimum chord distribution to a realistic tapered distribution that is a close approximation to the optimum for $r/R \geq 0.5$. (a) Belz design and (b) tapered design. 116
5.8 Overlay of Betz optimum chord distribution and a realistic tapered distribution from Figure 5.7. 116

6.1 Schematic of the wind turbine functional control elements. . 119
6.2 Section view of typical components of a wind turbine that are involved in its monitoring and control. 120
6.3 Schematic of a wind turbine closed-loop control system. . . . 121
6.4 Example of the relation between the rotor tip-speed ratio and rotor pitch angle on the coefficient of power for a 600 kW two-bladed horizontal wind turbine. 122
6.5 Schematic drawing of a four-pole synchronous machine along with the sinusoidal waveform of the induced electromotive force (emf) which has units of volts, that is produced by the rotation of the rotor. 123
6.6 Power curve for a stall regulated wind turbine with variable speed design. 127
6.7 Example of control trajectory to seek the optimum tip-speed ratio for the wind turbine performance shown in Figure 6.4 with $\theta_{cp} = -1°$. 128
6.8 Generic power curve for a wind turbine illustrating optimum (Betz) and actual performance in Region II. 129
6.9 Plot of $A = a(1-a)^2$ versus a showing that the maximum occurs at $a = 1/3$. 132
6.10 Plot of ratio of the "not ideal" (NI) to the "ideal" (I) values of $a(1-a)^2$ versus a. 132
6.11 Plot of percent improvement obtained by optimizing the axial induction factor. 133
6.12 Plot of the rotor radial distribution of the axial induction factor for three tip speed ratios of an existing current-generation wind turbine. 133
6.13 Plot of the rotor radial distribution of the lift coefficient for three tip speed ratios of an existing current-generation wind turbine. 134
6.14 Plot of the rotor radial distribution of the lift coefficient for which the axial induction factor is the ideal 1/3 for three tip speed ratios of an existing current-generation wind turbine. . 135
6.15 Plot of the rotor radial distribution of the change needed in the lift coefficient to achieve the ideal 1/3 axial induction factor for three tip speed ratios of an existing current-generation wind turbine. 135
6.16 Comparison of the performance of different active lift control approaches. 136
6.17 Airfoil section illustrating positive camber. 137
6.18 Lift as a function of angle of attack (a) and drag polar (b) for a zero camber airfoil (solid curve) and with a plane trailing edge flap with downward deflection (dashed curve). 137
6.19 Illustration of spanwise segmented flaps as it might apply to a wind turbine rotor. 138

6.20 Illustration of a Gurney flap for lift control. 138
6.21 Illustration of multiple spanwise Gurney flaps for spanwise varying lift control. 139
6.22 Illustration of multiple spanwise plasma Gurney flaps for spanwise varying lift control. 140
7.1 Force vectors based on BEM analysis (left) and illustration of 3-D lift and drag force distribution resulting in maximum shear forces and bending moments at the rotor root. 144
7.2 Illustration of gravitational and centrifugal loads acting on a spinning wind turbine rotor. 145
7.3 Illustration of types of coned or "flapping" rotor conditions of the horizontal axis wind turbine. 146
7.4 Illustration of the gyroscopic restoring moment produced by the yawed motion of the rotor. 147
7.5 Section view of a HAWT rotor illustrating the internal structure. 148
7.6 Illustration of shear force and bending moment on a small spanwise element of the loaded rotor. 149
7.7 Spanwise element of rotor blade used in beam analysis to determine principle bending axis. 150
7.8 Deflection amplitude distribution for the first bending (flapping) eigenmode, u^{1f}, of a cantilevered beam that is representative of a HAWT rotor blade. 155
7.9 Deflection amplitude distribution for the first edgewise bending eigenmode, u^{1e}, of a cantilevered beam that is representative of a HAWT rotor blade. 156
7.10 Deflection amplitude distribution for the second flapping eigenmode, u^{2f}, of a cantilevered beam that is representative of a HAWT rotor blade. 157
7.11 Simplified internal structure of a HAWT rotor designed to resist bending moments extreme wind loads. 159
8.1 Photographs of modern onshore and offshore wind farms. . . 165
8.2 Photograph showing the wakes from wind turbines made visible by low level fog over an an offshore wind farm. 166
8.3 Schematic drawing of wind turbine wake model. 167
8.4 Velocity on the wake centerline of an upstream ideal, $a = 1/3$, wind turbine based on the wake model equations. 168
8.5 Rule of thumb pattern of wind turbines in a wind farm. The predominant wind direction is from bottom to top. 171
8.6 Impact of site area and number of wind turbines on wind farm efficiency. 171
8.7 Schematic illustration of the half cone of a wind turbine wake. 174

8.8 Schematic illustration of a wind turbine inside the cone of an upstream wind turbine for a wind vector angle, θ, and the resulting wind direction parameter, β. 175
8.9 Example of a wind direction probability (wind rose) map. 177
8.10 Example of roulette wheel with pie slices that correspond to the cumulative probability values, Cum(i), for all the solutions. 182
8.11 Three examples of arrangements of two wind turbines that maximize the captured power based on the wind direction probability given in Table 8.5. Wind $\theta = 75°$, where $\theta = 0°$ is along x-axis. 187
8.12 Three examples of arrangements of three wind turbines that maximize the captured power based on the wind direction probability given in Table 8.5. Wind $\theta = 75°$, where $\theta = 0°$ is along x-axis. 188
8.13 Examples of arrangements of from two to seven wind turbines that maximize the captured power based on the wind direction probability given in Table 8.5. Wind $\theta = 75°$, where $\theta = 0°$ is along x-axis. 189
8.14 Wind farm expected power as a function of the number of wind turbines with optimized locations based on the wind direction probability. 190
9.1 Schematic examples of wind turbine sound sources, propagation paths and receivers. 195
9.2 Schematic representation of a sound pressure wave. 196
9.3 Examples of sound pressure levels that occur in different activities. 197
9.4 Frequency response curves for A, B, and C weighting scales. 198
9.5 Perception threshold of the human ear for low frequency sound. 200
9.6 Example of the type of interaction that occurs, when the rotor plane cuts through the unsteady wake vortex street produced by the tower, resulting in "bursts" of sound observed in the time traces from a microphone. 202
9.7 Mechanisms for sound generation due to the air flow over the turbine rotor. 204
9.8 Sound level power scaling for different aerodynamic sound source mechanisms on the turbine rotor. 204
9.9 Sound pressure level azimuthal radiation pattern for a wind turbine. 205
9.10 Trends in sound pressure levels as a function of rotor diameter for different generations of wind turbines. 206

9.11 Example of the effect of wind on the propagation of low frequency rotational harmonic noise from a large-scale HAWT. 207
9.12 Example of the effects of wind-induced refraction on acoustic rays radiating from an elevated source. 207
9.13 Sound pressure as a function of distance the wind turbine example problem with $\alpha = 0.005\,\text{dB/m}$. 210
10.1 Example of a two week period of system loads, system loads minus wind generation, and wind generation. 216
10.2 Wind turbine energy storage optimization flow chart. 217
10.3 Illustration of an electro-chemical storage battery cell. . . . 218
10.4 Specific power versus specific energy for types of electro-chemical storage batteries. 220
10.5 Schematic drawing of a flow battery. 222
10.6 Schematic of a super capacitor. 223
10.7 Illustration of the elements in the use of electricity for hydrogen production and possible storage. 225
10.8 Illustration of pumped storage hydroelectric power plant. . . 227
10.9 Illustration of compressed air storage power plant. 228
10.10 Components of a basic compressed air storage power plant. . 229
10.11 Recuperated cycle representation of a compressed air storage power plant. 230
10.12 Illustration of a flywheel energy storage system. 230
10.13 Summary comparison of different electric power storage systems with regard to power rating and discharge rate. 232
10.14 Thermodynamic representation of a CAES power plant. . . 234
10.15 Example of the electric power demand and corresponding consumer price of electricity over a 24 hours period. 236
10.16 Charging and discharging price functions that correspond to the price function shown in Figure 10.15. 236
10.17 Result of optimization based on a range of heat price for a CAES power plant. 239
10.18 Schematic of a hydro-electric storage configuration. 240
10.19 Example of a floating off-shore platform supporting a wind turbine. 242
10.20 Principle behind the floating off-shore platform concept used for wind energy storage. 242
10.21 Example of a sequence of floating position based on the amount of water contained in an internal compartment of the floating structure. 243
10.22 Schematic representation of the buoyant energy storage. . . 243
10.23 Relation between the projected area of the floating structure and the immersion depth for a given stored power level. . . . 244
11.1 Wind turbine rotor blade mass correlation with rotor radius. 252

11.2 Wind turbine rotor blade cost, labor cost, and baseline and advanced material cost correlations with rotor radius. 252
11.3 Wind turbine tower mass correlation with the product of the rotor swept area and hub height. 258
12.1 Rotor blade cross-section illustrating internal structure to resist bending. 270
12.2 Rotor blade load and bending coordinate system. 270
12.3 Effects of wind speed on the maximum total power for different rotor radii. 271
12.4 Rotor z-component deflection at different radial locations for $R = 51$ m at three wind speeds. 272
12.5 Rotor y-component deflection at different radial locations for $R = 51$ m at three wind speeds. 272
12.6 Rotor maximum deflection as a function of wind speed for $R = 51$ m. 273
12.7 Rotor maximum deflection as a function of rotor radius for a wind speed of 14 m/s. 274
12.8 Wind turbine cost as a function of the rotor radius. 275
12.9 Balance of station cost as a function of the rotor radius. . . 276
12.10 Capacity factor as a function of the rotor hub height. 276
12.11 Annual energy production as a function of the rotor radius. 277
12.12 Cost of electricity as a function of the rotor radius. 277
13.1 Example of a modern vertical axis wind turbine design. . . . 281
13.2 Concept for highway electric power generation using overhead horizontally oriented helical VAWTs. 281
13.3 Concept for highway electric power generation using a series of small VAWTs lining the roadway median strip. 282
13.4 Photograph of a pilot test of a concept for wind farms made up of small VAWTs. 282
13.5 Schematic drawing of a shrouded horizontal wind turbine. . 283
13.6 Ducted horizontal wind tunnel undergoing wind tunnel tests at Clarkson University. 285
13.7 Artificial hill concept to accelerate ground wind around wind turbines. 286
13.8 Photograph of SheerWind 2.2 MW INVELOX wind capture system under construction in China. 286
13.9 Building design that incorporates wind energy. 287
13.10 Illustration of an oscillating wind disk blade-less wind turbine concept. 287
13.11 Flexible wind stalks blade-less wind turbine concept. 288
13.12 Altaeros energies 35 ft lighter-than-air flying wind turbine. . 288
13.13 "Sky Serpent" tethered flying wind turbines. 289
13.14 Lateral axis wind turbine design. 290
13.15 Ultra-tiny micro wind turbine design developed at the University of Texas Arlington. 290

13.16 Road lighting concept using helical VAWTs on light poles. . 291
14.1 Example of deployable wind energy system concept that makes use of the storage container for the wind turbine base. . . . 296
14.2 Airfoil section lift-to-drag ratio as a function of angle of attack used to find the angle of attack of $(C_l/C_d)_{\max}$. 299
14.3 Relative wind angle, $\phi(r)$ and blade twist angle, $\theta(r)$ along the rotor radius for a Betz design. 299
14.4 Radial distribution of the local chord length for the rotor for a Betz optimum design (solid curve) and modification to reduce the rotor weight (dashed line). 300
14.5 Effect of the constant rotor pitch, θ_{CP} on the total wind turbine power. 301
14.6 Radial distribution of the induction factors, $a(r)$ and $a'(r)$ for the wind turbine. 301
14.7 Radial distribution of the differential thrust normal to the rotor plane, $T_n(r)$ for the wind turbine. 302
14.8 Radial distribution of the differential power, $dP(r)$, for the wind turbine. 302
14.9 Tower hollow tube general cross-section. 303
14.10 Maximum stress in tower structure normalized by the ultimate material stress. Horizontal line marks $\sigma_{\text{tot}}/\sigma_u = 0.5$. 305
14.11 Deployable wind energy system concept including an outrigger of width W and length, L, on the storage container that is used to balance moment "b". 307
14.12 Restoring moment "b" as a function of the outrigger length, L. 308
14.13 Hypothetical mobile wind farm for generating power at a disaster response site. 309
14.14 To-scale representation of the wake from Turbine A on Turbine D for wind from the North direction. 311
14.15 To-scale representation of the wake from Turbine C on Turbine D for wind from the North-East direction. 312

List of Tables

1.1 Worlds Largest Wind Turbines in 2017 14
1.2 Wind Energy by the Numbers 18
2.1 Classes of Surface Roughness for Atmospheric Boundary Layers 25
2.2 Sample Frequency Distribution of Monthly Wind Velocity . 29
2.3 Sample Wind Velocity Frequency Distribution 34
2.4 Monthly Average Wind Speed Data 43
3.1 Summary of Effects of Airfoil Geometry on Aerodynamic Characteristics 65
3.2 Estimated Annual Energy Improvements from NREL Airfoil Series 69
4.1 Properties of the Actuator Disk 74
4.2 Equations Used in the Solution of the BEM Equations in Analysis of a Wind Turbine Rotor 95
4.3 Characteristics of the University of Notre Dame Wind Turbines 97
4.4 Rotor Geometry of the University of Notre Dame Wind Turbines 98
5.1 Power Train Efficiencies for Modern Wind Turbines at Rated Power Conditions 108
5.2 Summary of Equations for Estimating the Blade Chord and Twist Angle as a Function of the Local Rotor Radius 111
8.1 Functional Evaluation of Initial Population 180
8.2 Chromosomes in Mating Pool after RW Selection Process . . 182
8.3 Chromosomes in Mating Pool before and after Crossover Process 184
8.4 Replacement Population for the Next Iteration 185
8.5 Wind Direction Probability Used in Optimized Wind Farm Design 186
9.1 Wind Turbine Aerodynamic Sound Mechanisms 203
9.2 ISO 1996-1971 Recommendations for Community Noise Limits 211
10.1 Capital Costs of Installed Storage 233
10.2 Efficiency and Hours at Full Power of Installed Storage . . . 233
11.1 Component Cost Breakdown for a Land-Based 1,500 kW (rated) Wind Turbine with a Rotor Diameter of 70 m and a Hub Height of 65 m 262

11.2 Functional Evaluation of Initial Population 264
11.3 Results from Consecutive Runs of the Economic Genetic Algorithm in Which Wind Turbine Radii and Hub Heights Are Sought That Minimize the Cost of Electricity 265
11.4 Ranges of COEs for Land-Based and Off-shore wind Turbine Installations 265
12.1 Mass and Cost (2002$) Breakdown for Wind Turbine Case Study with $R = 51$ m and $V_{\text{cut-in}} = 4$ m/s 278
12.2 Economic Summary for Wind Turbine Case Study with $R = 51$ m and $V_{\text{cut-in}} = 4$ m/s 279
14.1 Dimensions of a Standard 20 ft Shipping Container 295
14.2 Calculations for Wind Turbine Design Power 298
14.3 Computed Weights of Key Wind Turbine Components . . . 303
14.4 Wind Farm Power Calculation Input 310
14.5 Wind Farm Turbine Coordinates 311
14.6 Wind Farm Power 311
14.7 Wind Turbine Generated Sound Levels at the Compound Center 313
14.8 Component Cost Breakdown for a Land-Based Mobile Wind Turbine with a Rotor Diameter of 12 m and a Hub Height of 17.5 m 314
14.9 Dimensions of a Standard 40 ft Shipping Container 314

1

Introduction

The use of the energy in the wind has played a long and important role in the history of human civilization. The first known application of wind energy dates back 5,000 years to Egypt, where sails were used to propel boats. The first true windmill, a wind-powered machine with vanes attached to an axle to produce circular motion, can be traced back to the Persians around 1700 BC. By the 10th century AD, windmills were being used to grind grain in the area now known as eastern Iran and Afghanistan. The western world began using windmills much later, with the earliest written references dating from the 12th century. These too were used for milling grain. A few hundred years later, windmills were used to pump water and reclaim much of Holland from the sea. The first "modern" wind turbines appeared in the mid-20th century, with a focus on producing electricity. This chapter provides a background on the historical development of wind turbines leading to present wind turbine designs.

1.1 History of Wind Energy

Transforming the kinetic energy in the wind into useful mechanical power dates back to antiquity. It is difficult to determine a precise historical date for the earliest use of wind energy. There is, however, evidence that the Persians used vertical axis wind machines as early as 1700 BC in the region of modern Iraq, Iran, and Afghanistan. The remains of one of these ancient wind turbines are shown in the photograph in Figure 1.1. A schematic of the design is shown in Figure 1.2. This wind turbine was used to grind grain. It employed a number of modern concepts. For example, it used a ventury to accelerate the wind. The ventury also shielded one-half of the turbine rotor, thereby reducing the drag on the advancing rotor blades. This design also prescribed a single rotation direction. Presumably, the wind turbine would have been aligned with a prevailing wind direction since the structure apparently was not able to rotate.

The first evidence of wind turbines in China dates back to 1219 AD. Likely owing to their well-developed sailing vessels, these early wind turbines utilized cloth sails rather than rigid wooden rotors. An example of an early vertical

DOI: 10.1201/9781003611486-1

FIGURE 1.1
Photograph of circa 900 AD vertical axis wind turbine used for grinding grain that is located in modern Afghanistan.
Source: unknown.

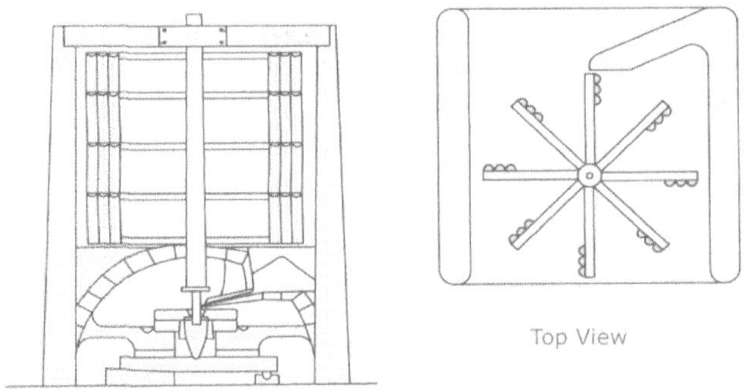

FIGURE 1.2
Schematic drawing of the early vertical axis wind turbine is shown in Figure 1.1.
Source: https://commons.wikimedia.org/wiki/User:Kaboldy, CC By-SA 3.0 (https://creativecommons.org/licenses/by-sa/3.0/). Shown without changes.

axis wind turbine design is shown in the photograph in Figure 1.3. An example of a horizontal axis wind turbine design that also utilized cloth sails as rotor blades is sketched in Figure 1.4. This was intended to pump water from a

FIGURE 1.3
Photograph of an early Chinese vertical axis wind turbine utilizing cloth sails.
Source: https://commons.wikimedia.org/wiki/User:Carl_von_Canstein, Commons: GNU Free Documentation License, version 1.2. Shown without changes.

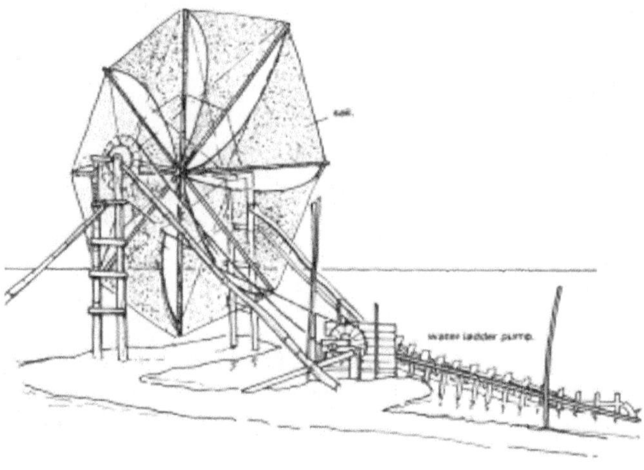

FIGURE 1.4
Sketch of an early horizontal axis wind turbine utilizing cloth sails and driving a water ladder pump.
Source: Food and Agriculture Organization of the United Nations, Irrigation and Drainage Paper 43, P.L. Fraenkel, 1986, ISBN 92-5-102515-0. Reproduced with permission.

reservoir. Horizontal axis wind turbines were widely used in the southeastern region of China from the 14th to 17th centuries AD.

The first historical reference to horizontal wind turbines in Europe dates back to the late 12th century AD. This innovation appears to have originated in Yorkshire, England and was possibly motivated by Roman water wheels which rotated on a horizontal axis. The post mill, a horizontal axis windmill, first appeared at the end of the 13th century in Canterbury, England. Figure 1.5 shows a medieval illustration of a sunk post mill. This design incorporated rotor blades attached to a wooden cog-and-ring gear set that translated the horizontal shaft rotation of the turbine into a vertical shaft rotation of a grind stone. A photograph of a 17th century post mill wind turbine that contains all of the earlier features is shown in Figure 1.6.

A derivative of the post mill design that was mainly found in the Netherlands is referred to as the "Smock" mill because of its resemblance to an article of cloths called a smocking. The photograph in Figure 1.7 provides an example of this unique shape. The smock mill improved upon the post mill design by being able to rotate the roof cap that held the wind shaft and sails. As a result, it could be aligned with the wind direction. The base of the smock mill that housed the milling equipment was fixed in place. This allowed the base to be taller, placing the wind shaft at a higher elevation than that of the post mill.

FIGURE 1.5
Medieval illustration of a sunk post mill.

Source: https://fr.wikipedia.org/wiki/Utilisateur:Poussin_jeance, CC By-SA 3.0 (https://creativecommons.org/licenses/by-sa/3.0/). Shown without changes.

FIGURE 1.6
Photograph of post mill presumably been build in 1683 at Essern, District of Nienburg Germany, and drawing showing post mill internal design.
Source photograph: https://commons.wikimedia.org/wiki/User:Heinz-Josef_Lucking, CC By-SA 3.0 de (https://creativecommons.org/licenses/by-sa/3.0/de/deed.en). Shown without changes. Source drawing: https://commons.wikimedia.org/wiki/File:Encyclopedie_volume_1-040.png, PD-US - published in the U.S. before 1923 and public domain in the U.S.

One of the early pioneers in wind energy was John Smeaton who was one of the first scientists to develop mathematical models to predict windmill efficiency. He also showed that wind turbine blades had to be twisted to obtain the best efficiency. A circa-1750 photograph of John Smeaton is shown in Figure 1.8.

The first vertical axis wind turbine used to produce electricity was produced by Scottish Professor James Blyth. The wind turbine rotor had a 17 m. diameter and a hub height of 18 m. It generated 12 kW of electric power. A photograph of his wind turbine is shown in Figure 1.9.

Charles Brush was another pioneer in the use of wind turbines to produce electricity. He was one of the founders of the American electrical industry. He invented an efficient DC dynamo that was used in a public electric power grid as well as an efficient method for manufacturing lead-acid batteries. In 1887, Charles Brush built the first automated operating wind turbine for electricity generation. A photograph of this wind turbine is shown in Figure 1.10. The wind turbine rotor was 15.2 m. in diameter and had 144 blades made of cedar.

FIGURE 1.7
Photograph of a Smock mill that was common to the Netherlands.
Source: https://commons.wikimedia.org/wiki/User:Cnyborg, CC By-SA 3.0 (https://creativecommons.org/licenses/by-sa/3.0/). Shown without change.

It had a power rating of 12 kW and operated for 20 years. Brush Electric merged with Edison Electric to form the General Electric Corporation (GE).

One of the pioneers of modern aerodynamics who built wind turbines in Denmark in the late 1800s was Paul la Cour. He was originally trained as a meteorologist, which gave him an appreciation for wind characteristics. His knowledge of the intermittent nature of the wind made him particularly concerned with the storage of wind-generated energy. He subsequently used the electricity from his wind turbines for electrolysis to produce hydrogen gas that was stored and used for gas lights. Figure 1.11 shows an 1897 photograph of one of his test wind turbines at the Askov Folk High School in Askov, Denmark. He had to replace the windows in the adjoining buildings several times when the stored hydrogen exploded. In 1905, Paul la Cour founded the Society of Wind Electricians.

Johannes Juul was one of the first students of la Cour, attending a 1904 course on "Wind Electricians". Juul was a pioneering developer of the world's first AC electric generator wind turbine, which was built in 1956 in Vester Egesborg, Denmark. The so-named Gedser wind turbine was rated at 200 kW.

FIGURE 1.8
Circa 1750 photograph of John Smeaton who developed early mathematical models to predict windmill efficiency.

Source: https://en.wikipedia.org/wiki/User:Magnus_Manske, PD-US - published in the U.S. before 1923 and public domain in the U.S.

FIGURE 1.9
Photograph of the Blyth vertical axis wind turbine that was the first to produce electricity.

Source: unknown. PD-US - published in the U.S. before 1923 and public domain in the U.S.

It employed many concepts that are standard on modern wind turbines including rotor stall control and emergency braking. Figure 1.12 shows a photograph of the Gedser wind turbine. It operated for 11 years without maintenance and

FIGURE 1.10
Scientific American page showing different views of the Brush horizontal axis wind turbine built in 1887.

Source: (1890-12-20). "Mr. Brush's Windmill Dynamo". Scientific American 63: p. 54. Author: Anonymous. PD-US - published in the U.S. before 1923 and public domain in the U.S.

was refurbished in 1975 at the request of NASA to provide data for the U.S. wind energy program.

The Smith-Putman wind turbine, shown in the photograph in Figure 1.13, was built in 1941 and located in Castleton, Vermont. It was the first megawatt-scale wind turbine connected to an electrical distribution system. It was designed by Palmer Cosslett Putnam and manufactured by the S. Morgan Smith Company. The turbine had two blades forming a rotor diameter of 175 ft. The rotors were located on the downwind side of a 120 ft steel lattice tower. Each blade was approximately 8 ft wide and 66 ft long and weighed eight tons. The blades were built on steel spars and covered with a stainless steel skin. The blade spars were hinged at their root attachment to the hub, allowing them to assume a slight cone shape. The wind turbine operated for

FIGURE 1.11
Photograph of Paul la Cour (1846–1908) and his 1897 horizontal axis wind turbine that produced electricity used in the production of hydrogen gas.
Courtesy of the Paul la Cour Museum, Askov.

FIGURE 1.12
Photograph of the Johannes Juul designed Gedser wind turbine built in 1956 and located in Vester Egesborg, Denmark.
Source: Status of Wind-Energy Conversion, R. Thomas and J. Savino, NASA TM X-71523, 1973 (Figure 3).

FIGURE 1.13
Photograph of the Smith-Putman wind turbine built in 1941 and located in Castleton, Vermont.

Source: "Wind Energy Conversion Systems - Workshop Proceedings", Ed. J. Savino, June 11-13, Washington, D.C. NSF/RA/N-73-006.

only 1,100 hours before a rotor blade failed at a known weak point, which had not been reinforced due to war-time material shortages. It remained the largest wind turbine ever built until 1979.

1.1.1 Modern Era of Wind Energy

The modern era of wind turbine design reflects the aerodynamics insights that also influenced the development of modern aircraft. The designs of the early 20th century included both vertical and horizontal axis wind turbines, generally aimed at generating electricity. Vertical axis wind turbines (VAWTs) employed either aerodynamic lift or drag to extract energy from the wind. In 1931, a French engineer named Georges Jean Marie Darrieus patented the Darrieus wind turbine. The photograph in Figure 1.14 shows an example of an early Darrieus design. It consists of a two curved rotors with an airfoil section shape. The driving force that moves the rotors is aerodynamic lift. The vertical axis had the benefit of locating the electric generator on the ground. However, the large loads at the base often required the use of guy wires for support. The largest Darrieus wind turbine ever built is the Eole turbine, located in Cap-Chat, Quebec Canada, as shown in the photograph

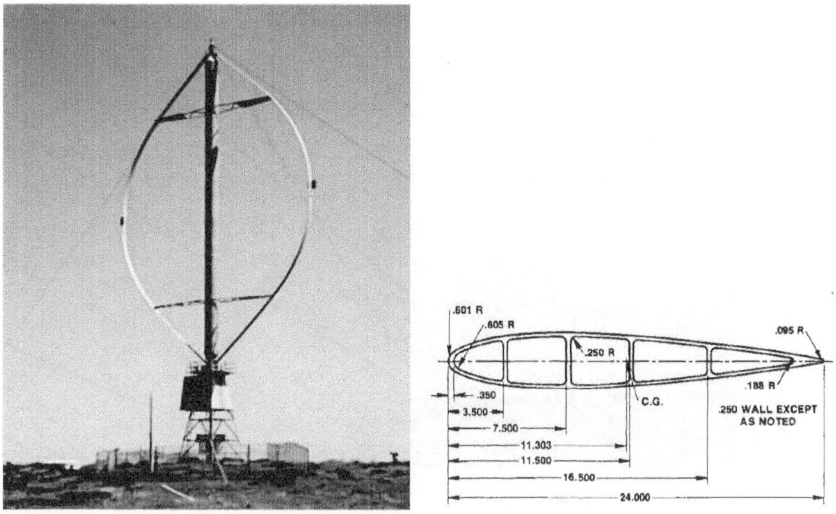

(a) (b)

FIGURE 1.14
Photograph of an early Darrieus wind turbine in the field (a) and rotor airfoil section shape for Alcoa aluminum extrusion (b).

Source: photograph https://en.wikipedia.org/wiki/User:aarchiba who grants anyone the right to use this work for any purpose, without any conditions, unless such conditions are required by law. Source: drawing NASA SEE 19800008204, 1995. Unclassified; Publicly available; Unlimited.

in Figure 1.15. The turbine is 100 m. tall and 60 m. wide. It is used only occasionally due to structural fatigue issues.

The Darrieus wind turbines have been prone to structural failures. An Alcoa 12.8 m. diameter machine collapsed at their Pennsylvania facility on March 21, 1980, when the central torque tube began vibrating and ultimately buckled. In April 1981, a 25 m. diameter machine came apart east of Los Angeles due to a failure in the software that regulated the turbine rotational speed. As a result of these incidents, the Alcoa Corporation closed down its wind turbine operation.

Another VAWT design that utilizes lift on the rotors is the Giromill as shown in the photograph in Figure 1.16. Although this is considered a Darrieus wind turbine, and covered under Darrieus's patent, it uses straight rotor blades rather than the joined curved blades. The advantage of the straight blade sections is that they allow the blade pitch angles to be adjusted. While it is cheaper and easier to build than a standard Darrieus turbine, it is not as efficient and requires strong winds (or a motor) to get it to start rotating. However, an adaptation that results from the ability to pitch the straight blades can aid in starting the rotation. In low winds, the blades are pitched

FIGURE 1.15
Photograph of the largest built Darrieus wind turbine located in Cap-Chat, Quebec Canada.

Source: https://commons.wikimedia.org/wiki/User:Guillom, CC By-SA 3.0 (https://creativecommons.org/licenses/by-sa/3.0/). Shown without changes.

flat against the wind, generating drag forces that start the turbine rotation. As the rotational speed increases, the blades are pitched back to a lower angle of attack so that they generate lift, which is the normal operating condition.

An example of a VAWT that relies completely on aerodynamic drag is the design attributed to Savonius. It was invented by the Finnish engineer Sigurd Johannes Savonius in 1922. It generally consists of open cylindrical surfaces that are attached to a vertical rotating shaft, such as illustrated in Figure 1.17. The aerodynamic drag, represented by the drag coefficient, $C_D = D/(1/2)\rho V^2 S$ where S is the frontal area, is about a factor of two lower when the wind approaches the convex side of the cylinder compared to the concave side. This is illustrated in the bottom part of Figure 1.17. The differential drag between the two orientations to the wind direction causes the Savonius turbine to spin. A wind turbine design that melds the Darrieus and Savonius concepts is shown in the photograph in Figure 1.18. This conceivably uses the drag-based Savonius turbine to address the weakness of the Darrieus wind turbine to start rotating at low wind speeds.

As mentioned, an advantage of the VAWT is that the electric generator and other related components are located on the ground where they are easily accessible. In addition, the two principal designs, Darrieus and Savonius, do not need to be aligned with the wind direction. However, they are not as effective in extracting energy from the wind as horizontal axis wind turbines (HAWT). This is illustrated in Figure 1.19 which show plots of the coefficients

FIGURE 1.16
Photograph of the Giromill vertical wind turbine.

Source: https://commons.wikimedia.org/wiki/User:Stahlkocher, CC By-SA 3.0 (https://creativecommons.org/licenses/by-sa/3.0/). Shown without changes.

of power, C_P, as a function of the rotor tip-speed-ratio, λ for various vertical and horizontal axis wind turbine designs. The coefficient of power is the ratio of the energy extracted from the wind to the available energy in the wind. The tip-speed-ratio is the ratio of the velocity of the tip of the rotor to the velocity of the wind. There is generally an optimum tip-speed-ratio that is most effective in extracting energy from the wind. We observe this in all of the cases shown in Figure 1.19. For a Savonius design the optimum tip-speed-ratio is about 0.75 at which $C_P \simeq 0.3$. In contrast, the Darrieus design has a higher optimum tip-speed-ratio of approximately 6 at which $C_P \simeq 0.35$. As a general observation, the optimum tip-speed-ratios of drag-based wind turbines are lower than that of lift-based turbines, and their power coefficients are lower as well.

Horizontal axis wind turbines are lift-based designs. Like the Darrieus that relies on lift, their optimum tip-speed-ratios generally increase as the number of rotor blades decreases. With modern three-bladed wind HAWT turbine shown in Figure 1.19), the optimum tip-speed ratio is approximately 6. More importantly, the coefficient of power of HAWT designs is higher than that of VAWT designs. In Figure 1.19, the maximum $C_P \simeq 0.46$. The theoretical maximum HAWT C_P is 0.593, which was first published in 1919 by the German physicist Albert Betz.

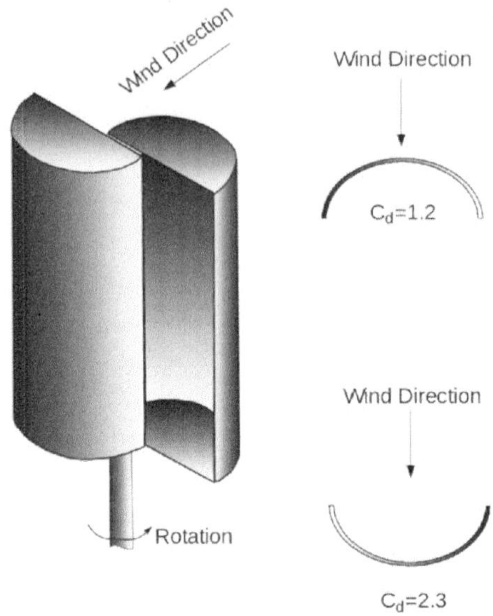

FIGURE 1.17
Illustration of a Savonius wind turbine design, and the difference in the drag coefficient between concave and convex surfaces that is the basis for this wind turbine design.

TABLE 1.1
Worlds Largest Wind Turbines in 2017

Model	Rotor Diameter (m)	Rated Power (MW)
MHI Vestas V164	164	9.5
Adwen AD-80	180	7.5
Siemens SWT-8.0-154	120–154	6–8
Enercon E-126	127	7.5
Ming Yang SCD 6	140	6
Senvion 6.2M152	126	6.15
GE Haliade	151	6

As a result of the larger coefficient of power offered by a HAWT, the emphasis of modern wind turbines has been toward horizontal axis machines. Figure 1.20 shows a photograph of a wind farm made up of 240 General Electric 2.5 MW wind turbines. These wind turbines are quite representative of a modern HAWT. The GE 2.5 MW turbine has a rotor diameter of 104 m. and a hub height of 94 m. The hub height can be varied depending on the wind conditions at the site.

As of 2017, the world's largest wind turbine is the MHI Vestas V164 with a rotor diameter of 164 m and a rated power of 9.5 MW. A list of the seven largest wind turbines in 2017 is given in Table 1.1. Some of these are land

FIGURE 1.18
Photograph of a wind turbine design that combines Darrieus and Savonius concepts.

Source: https://en.wikipedia.org/wiki/User:Fred_Hsu, CC By-SA 3.0 (https://creativecommons.org/licenses/by-sa/3.0/). Shown without changes.

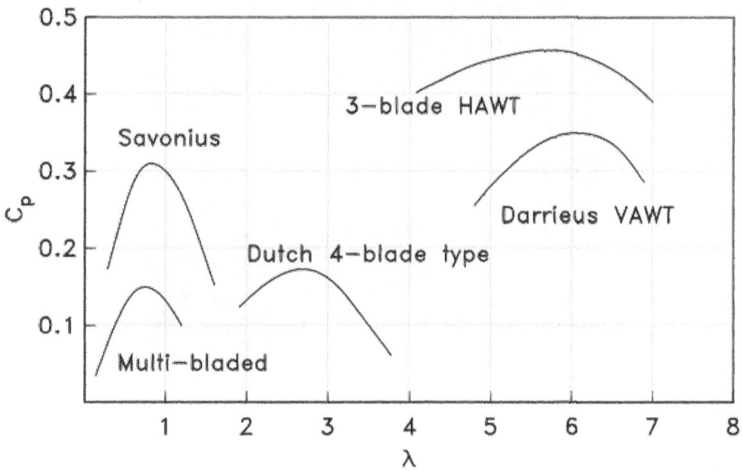

FIGURE 1.19
Plots of coefficients of power versus rotor tip-speed-ratio for different vertical and horizontal axis wind turbine designs.

FIGURE 1.20
Photograph of General Electric 2.5 MW wind turbines making up a 240 turbine wind farm.
Source: GE Renewable Energy.

FIGURE 1.21
Photograph of the General Electric Haliade off-shore wind turbine which has a rated power of 6 MW, and a rotor diameter of 151 m.
Source: GE Renewable Energy.

based. Others are water based such as the GE Haliade 6 MW pictured in Figure 1.21. This wind turbine has a direct drive generator that eliminates the sometimes troublesome gear box, and it also has a helipad.

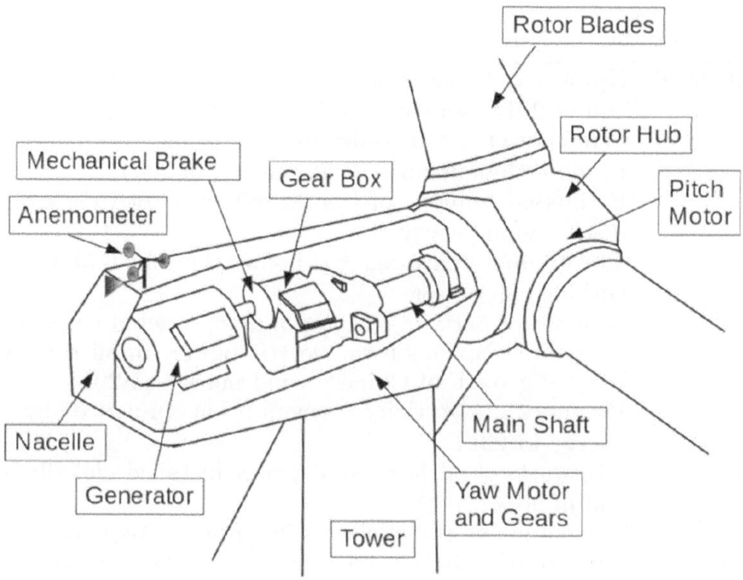

FIGURE 1.22
Illustration of the internal components in the nacelle of a modern HAWT.

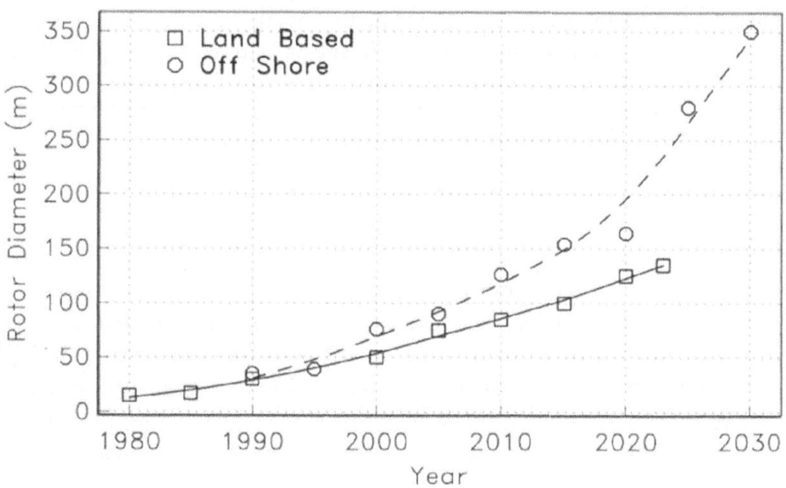

FIGURE 1.23
Trend in average HAWT rotor diameters since 1980.

With a HAWT, the electric generator and related components are located at the hub height in an enclosed nacelle. A cutaway schematic of the general interior of the nacelle of a horizontal wind turbine is shown in Figure 1.22.

TABLE 1.2
Wind Energy by the Numbers

110,000,000	Number of Chinese homes powered by wind energy at the end of 2014 (source: The Economist, 1 August 2015).
51.5	Wind power GW installed in 2014, bringing global capacity to more than 369.6 GW at the end of 2014.
608	Estimated millions of tonnes reduction of CO_2 emissions due to wind energy.
5,500	The number of average EU homes that one 6 MW offshore turbine can power.
8,000	Number of parts making up a modern wind turbine
59.6%	Amount of Spain's total electric power supplied by wind.
45%	Annual growth of Chinese wind market in 2014.
39.1%	Denmark's electricity consumption provided by wind energy in 2014.
8,759	Gigawatts of offshore wind power installed globally at the end of 2014.
3%	Percentage of global electricity supplied by wind power.
17–19%	Amount of global electricity that could be supplied by wind power in 2030.
8	Rated power (MW) of largest wind turbine, built by Vestas, with a rotor diameter of 164 m.
387	Millions of cubic meters of water saved by wind energy in the EU through wind energy (source: EWEA).
2,000	Liters of water that can be saved per MW of wind energy against other energy sources (source: US Department of Energy).

Source: Global Wind Energy Council.

The components generally include a gear box that steps up the rotation rate of the rotor to drive the generator, a yaw drive to rotate the wind turbine rotor to face the wind direction, and a main frame on which the generator, gearbox and related components mount. As will be discussed in Chapter 6, there are two approaches to control the aerodynamic lift on the rotor to maintain a constant rated power over a range of wind speeds. The first and oldest approach is called "stall-regulated" in which the rotor is designed so that above a certain wind speed, the rotor begins to lose lift and increase drag in a process aerodynamicists refer to as "stall". This approach is now only used on smaller HAWT with rated power less than 150–250 kW. The second approach that is universally used for larger wind turbines is called "pitch-regulated" whereby the blade pitch is varied equally on all of the rotor blades. This is the approach that is listed in the illustration in Figure 1.22. Pitch control is only used above the rated wind speed as a means of maintaining a constant rated power.

Besides having a larger coefficient of power, horizontal axis wind turbines have other advantages over vertical axis wind turbines. Principally, the tall tower allows the wind turbine to reach stronger and more uniform winds that occur at higher elevations above the ground. The ability to pitch the rotor blades in pitch-regulated versions further improves the performance of the wind turbine. Another advantage is that HAWT machines are generally self-starting. The principal *disadvantage* of a HAWT is the tower location of the electric generator which makes maintenance more difficult and expensive.

Wind energy is playing an ever increasing worldwide role as a renewable energy source. Countries such as Spain, Germany, and Denmark are close to meeting their goal of generating 30% of their electric power need from wind energy. Although the United States currently generates only about 5% of its electricity from wind power, with respect to the total installed capacity, it is the largest in the world, recently surpassing Germany. Table 1.2 provides an example of the worldwide scope of wind energy.

With this growing demand for wind energy, it is important to note that the present technology is far from optimized. Because of its intermittent nature, wind energy presents significant new challenges before becoming a completely reliable utility. For example on average, modern wind turbines at high-quality sites operate at only 35% of their capacity. They operate at full capacity less than 10% of the time. This is in part, a result of the variability of the wind as well as a function of economic factors that affect the aerodynamic design of the wind turbine. Although it is possible to increase wind energy capture by increasing the rotor diameter, this approach has economic limits. Figure 1.23 shows the trend in HAWT rotor diameter from 1980 to 2017. This shows a progressive increase in rotor diameter. Chapter 11 discusses the correlation between the rotor diameter and the cost of purchasing and maintaining a HAWT.

Wind energy's intermittent and unpredictable nature makes it more difficult than traditional power generation technologies to tie to a distribution grid. This also makes energy storage a key element. Predictive models for the wind conditions at a site are presented in Chapter 2. Methods for electric energy storage are presented in Chapter 10.

Current wind turbine technology has enabled wind energy to become a viable power source in the world's energy market. However, further advancements in aerodynamic design and control have the potential to make wind turbines more efficient, environmentally friendly, and to increase their useful operation life. The aerodynamic performance of horizontal axis wind turbines is presented in Chapter 4. One important aspect of wind turbines involving acoustics is presented in Chapter 9. The impact of design decisions on all these aspects of the wind energy system is investigated in Chapter 12.

Finally, new concepts for wind energy capture are appearing at a high rate. Some of these are viable, whereas others are not. A number of these are presented in Chapter 13.

A major challenge of this century will be to provide enough energy, water, and food without harming the environment and depleting these resources for future generations. A renewable energy source, such as wind, is poised to play an important role in the world's energy future. The next generation wind turbines must improve their efficiency, lower the acquisition cost, improve reliability, and have a cost of electricity that is competitive with fossil fuel electric power plants.

Problems

1. Plot a time-line of the power generated from wind turbines from the earliest designs to the present.
   (a) Are there any notable jumps in capabilities? Note them.
   (b) What might you attribute to any large changes in wind turbine capabilities that occurred over time?
2. List at least three pros and cons to horizontal and vertical wind turbines.
   (a) Of these, for the two types of wind turbine, which one explains the dominance of horizontal wind turbines as the wind farm standard?
   (b) In what applications or locations might vertical wind turbines have a preference.?
3. The yearly trend of the increase in the horizontal wind turbine rotor diameter shown in Figure 1.23 has been dominated by off-shore applications.
   (a) List at least three reasons which this has occurred.
   (b) List three reasons that might ultimately limit further increases in rotor diameters,

Further Reading

1. V. Torrey, *Wind-Catchers: American Windmills of Yesterday and Tomorrow*, Stephen Green Press, Vermont, 1976.

2. R. Righter, *Wind Energy in America*, University of Oklahoma Press, Oklahoma, 1996.
3. DoE Jobs and Economic Impacts Model, "20% wind energy by 2030", July, 2008.
4. Global Energy Council, "Global wind energy report annual market update 2015", www.gwec.net.
5. U.S. Department of Energy, "20% wind energy by 2030", July, 2008, http:www.osti.gov/bridge.
6. History of Wind Energy, www.energy.gov/eere/wind/history-wind-energy.
7. American Wind Energy Association Market Report, January, 2008, *osti.gov.*
8. Intergovernmental Panel on Climate Change, Brussels, Belgium, Published for the Intergovernmental Panel on Climate Change, Cambridge University Press, ISBN 978 0521 88010-7, 2007.
9. Regional Greenhouse Gas Initiative, Center for Climate and Energy Solutions, 2006, *c2es.org.*
10. U.S. Geological Survey, "Estimated use of water in the United States in 2000", 2005.
11. Electric Power Research Institute, "Water and sustainability", Vol. 3, Report 1006786, 2002.
12. Western Governors Association, "Water needs and strategies for a suitable future", p. 4, 2006.
13. Energy Information Administration, *Electric Power Annual*, EIA Table 2.6, Washington, DC, 2006.
14. FAO Corporate Document Repository, Natural Resources Management and Environment Department, *http://www.fao.org/docrep/010/ah810e/AH810E10.htm.*
15. https://en.wikipedia.org/wiki/Wind_turbine.
16. Wind Turbine Information: A Guide for the UK in 2025, *www.renewableenergyhub.co.uk.*
17. www.greenenergyohio.org.
18. P. L. Fraenkel, Irrigation and Drainage Paper 43, Food and Agriculture Organization of the United Nations, 1986, ISBN 92-5-102515-0.

2

Atmospheric Boundary Layer and Wind Characteristics

2.1 Origin of Wind

The proper design of a wind turbine for a site requires an accurate characterization of the wind at the site where it will operate. This requires an understanding of the sources of wind and of the turbulent atmospheric boundary layer.

Wind speeds are characterized by their velocity distribution over time, $V(t)$. Later, we will characterize this temporal variation through statistical analysis that will lead to statistical probability models.

The wind is generated by pressure gradients resulting from non-uniform heating of the Earth's surface by the Sun. Approximately 2% of the total solar radiation reaching the Earth's surface is converted to wind. On a global scale, hot air is generated in the equatorial regions. This air rises until it cools at higher altitudes and reaches buoyant equilibrium with the surrounding air. In the northern latitudes, with less solar heating, the air at higher altitudes cools further and is therefore less buoyant. It descends to the ground where it is then diverted along the ground until it reaches a warmer location, where it then becomes more buoyant and the cycle repeats. This cycle is illustrated in Figure 2.1.

In general, by virtue of different air densities with temperature, the colder regions are high pressure regions, and the warmer regions are low pressure regions. The air, like any fluid, moves from a region of higher pressure to one of lower pressure. That air movement is what we refer to as "wind". The strength (velocity) of the wind increases with the pressure difference.

The Earth's rotation has an effect on the wind. In particular, it causes an acceleration of the air mass that results in a Coriolis force

$$f_c \sim [(\text{earth's angular velocity})\sin(\text{latitude})] \cdot \text{air velocity}. \qquad (2.1)$$

This results in a curving of the wind path as it flows from high pressure and low pressure regions (isobars). This is illustrated in Figure 2.2.

At steady state, the Coriolis force balances the pressure gradient, leaving a resulting wind path that is parallel to the pressure isobars. This is referred to as the geostrophic wind. This is illustrated in Figure 2.3, which shows the

DOI: 10.1201/9781003611486-2

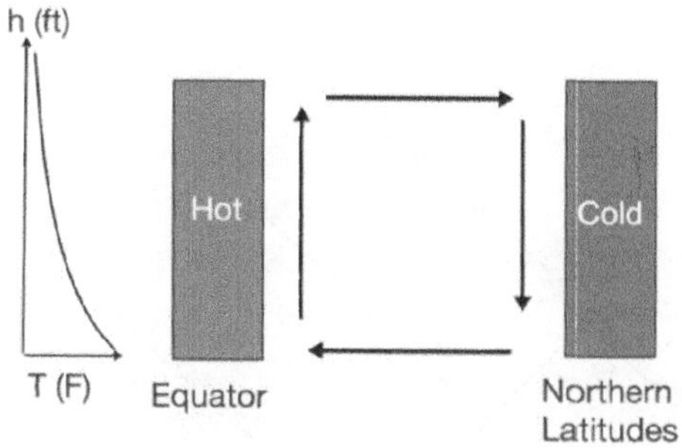

FIGURE 2.1
Mechanism of wind generation through global temperature gradients.

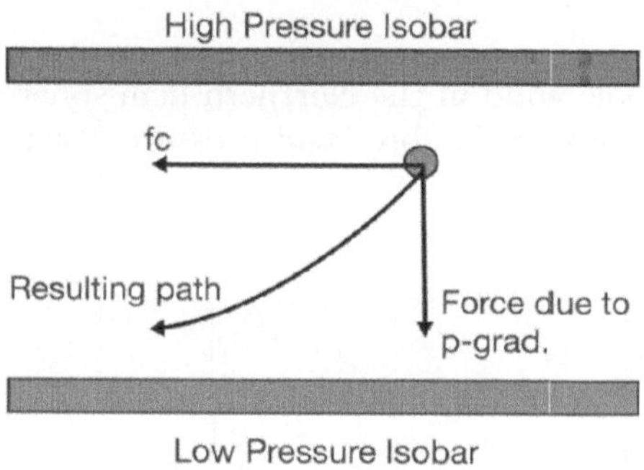

FIGURE 2.2
Effect of Coriolis force on the wind between pressure isobars.

geostrophic wind in the Northern hemisphere. The predominant geostrophic wind direction in the Northern hemisphere is from the West.

2.2 Atmospheric Boundary Layer

The flow of air (a viscous fluid) over a surface is retarded by the frictional resistance with the surface. A measure of that resistance is the coefficient of friction, $C_f = \tau_w/q$ where τ_w is the local surface shear stress, and $q = 1/2\rho V^2$ is the local dynamic pressure. The result is a boundary layer in which the minimum velocity (ideally zero) is at the surface, and the maximum velocity (ideally $V_{\text{geostropic}} = V_G$) is at the edge of the boundary layer.

The height or "thickness" of the boundary layer, δ, is affected by the coefficient of friction at the surface. This depends on the "surface roughness". The surface roughness also affects the shape of the boundary layer, which is

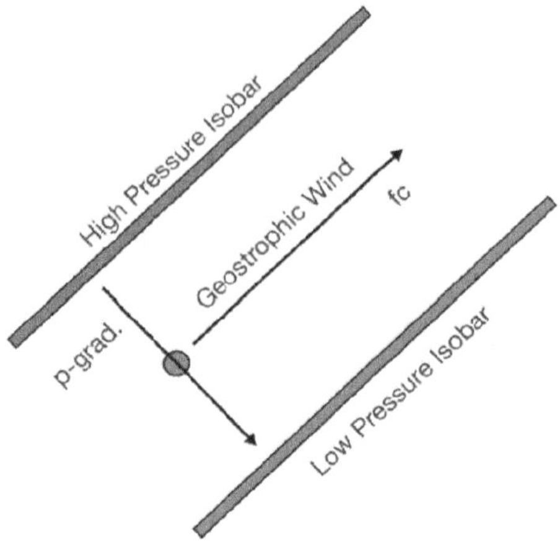

FIGURE 2.3
Schematic of geostropic wind in the Northern hemisphere that results from a steady state balance of Coriolis force and pressure isobars.

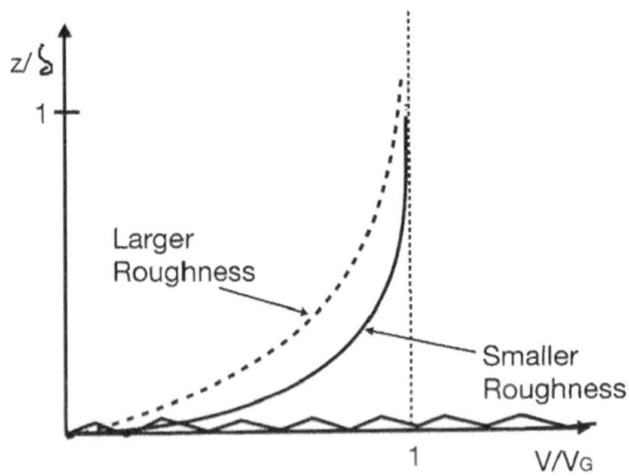

FIGURE 2.4
Schematic of atmospheric boundary layer profiles for small and large surface roughness.

defined by the change in velocity with height, $V(z)$. In atmospheric boundary layers, $dV(z)/dz$ is referred to as the "lapse rate", which is affected by surface roughness. Figure 2.4 illustrates the influence of surface roughness.

In atmospheric boundary layers, the surface roughness is represented by a category or "class". Table 2.1 lists the roughness categories, the representative roughness height, z_0, and the effect it has on the boundary layer thickness.

Wind data is available at meteorological stations around the U.S. and the world. Most airports can also provide local wind data. It is generally

TABLE 2.1
Classes of Surface Roughness for Atmospheric Boundary Layers

Category	Description	$\sim \delta$ (m)	z_0 (m)
1	Exposed sites in windy areas, exposed coast lines, deserts, etc.	270	0.005
2	Exposed sites in less windy areas, open inland country with hedges and buildings, less exposed coasts.	330	0.025–0.1
3	Well wooded inland country, built-up areas.	425	1–2

compiled at an elevation, z, of 10 m. This is the recommendation of the World Meteorological Organization (WMO).

A model for the change in the wind velocity with altitude in an atmospheric boundary layer is

$$V(z) = V(10)\frac{\ln(z/z_0)}{\ln(10/z_0)}, \tag{2.2}$$

where $z = 10\,\text{m}$ is the reference height where the velocity measurement was taken, $V(10)$ is the time-averaged wind velocity measured at the $z = 10$ m reference height, and z_0 is the roughness height at the location where the velocity measurement was taken.

The impact of the wind speed variation with elevation on a wind turbine power generation is significant. For example, if at a site $V(10) = 7\,\text{m/s}$ and $V(40) = 9.1\,\text{m/s}$, the ratio of velocities is $V(40)/V(10) = 1.3$. However, the power generated by a wind turbine scales as V^3. Therefore the ratio of power generated is $(V(40)/V(10))^3 = 2.2$. Thus in terms of sizing a wind turbine to produce a certain amount of power, knowing the wind speed at the site, at the elevation of the wind turbine rotor hub, is critically important.

In some cases, data may be available from a reference location at a certain elevation and roughness type that is different from the proposed wind turbine site. Therefore, it is necessary to project the known wind speed conditions to those at the proposed site. To do this, it is assumed that there is a height in the atmospheric boundary layer above which the roughness height does not matter. The literature suggests that this is above 60 m. Therefore assuming the log profile of the atmospheric boundary layer, at a reference location where the wind speed and roughness height are known, the wind velocity at an elevation of 60 m is given by

$$V(60) = V(10)\frac{\ln(60/z_{0_1})}{\ln(10/z_{0_1})}. \tag{2.3}$$

At the second location, where you wish to project the wind speed at an elevation of 60 m is

$$V(60) = V(z)\frac{\ln(60/z_{0_2})}{\ln(z/z_{0_2})}. \tag{2.4}$$

where z_{0_2} is the roughness height at the second location. Dividing the two expressions, one obtains a relation for the velocity at any elevation at the second site, namely

$$V(z) = V(10)\frac{\ln(60/z_{0_1})\ln(z/z_{0_2})}{\ln(60/z_{0_2})\ln(10/z_{0_1})}. \quad (2.5)$$

2.3 Temporal Statistics

The previous description of the atmospheric boundary layer was based on a steady (time averaged) viewpoint. Thus it refers to the mean wind and power. However, the atmospheric boundary is turbulent. As a result, the wind velocity and direction at any elevation vary with time, namely $V = V(z,t)$. The time scales can be relatively short, $\mathcal{O}$1–5 seconds, diurnal (24 hours periods), or seasonal (12 month periods). This extremely large range of time scales has a significant impact on wind energy power predictions and application.

The temporal variation of the wind velocity naturally leads to the use of statistical measures. The lowest (first) order statistic is the time average (mean) that is defined as

$$V_m = \frac{1}{N}\sum_{i=1}^{N} V_i \quad \text{where} \quad V_i = V_1, V_2, V_3, \ldots, V_n, \quad (2.6)$$

where V_i is a time series of velocity values. The number of the velocity values needs to be sufficiently large so that adding additional samples does not affect the average value. The time interval over which the samples are taken needs to encompass the time scale of interest.

It is important to note that since the wind turbine power scales as V^3, the average power

$$P_m \sim \frac{1}{N}\sum_{i=1}^{N} V_i^3 \neq V_m^3. \quad (2.7)$$

Based on this, a "power component", time-averaged wind speed is defined, whereby

$$V_{m_p} = \left[\frac{1}{N}\sum_{i=1}^{N} V_i^3\right]^{1/3}. \quad (2.8)$$

In this case, $P \sim V_{m_p}^3$.

Example 2.1

Consider the following set of time-varying velocity measurements: V_i=(4.3, 4.7, 8.3, 6.2, 5.9, 9.3).

For the velocity time series, $V_m = 6.45\,\text{m/s}$.

If the power were computed as $P \sim V_m^3$ then $P \sim 268.4$.

If however the power is computed correctly as $P_m \sim \frac{1}{N}\sum_{i=1}^{N} V_i^3 \neq V_m^3$ then $P \sim 333.9$.

As a result, the incorrect approach underestimates the power generation by approximately 24%.

2.4 Wind Speed Probability

Wind turbines at two different sites, with the same average wind speeds, may yield different energy output due to differences in the temporal velocity *distribution*. For example, consider a wind turbine with a rated power of 250 kW that has the following characteristics as illustrated in Figure 2.5, namely $V_{\text{cut-in}} = 4\,\text{m/s}$, $V_{\text{rated}} = 15\,\text{m/s}$, and $V_{\text{cut-out}} = 25\,\text{m/s}$. The wind turbine is considered to be located at Site A. The wind speed at Site A is constant at 15 m/s for the full 24 hours period. At a different site, Site B under consideration, the wind speed is 30 m/s for the first 12 hours, and 0 m/s for the last 12 hours of the 24 hours period. How would the power generated by the wind turbine compare over the 24 hours period at the two sites?

At Site A, for the 24 hours period the velocity is constant at 15 m/s. Therefore over this 24 hours period, the wind turbine is producing its rated

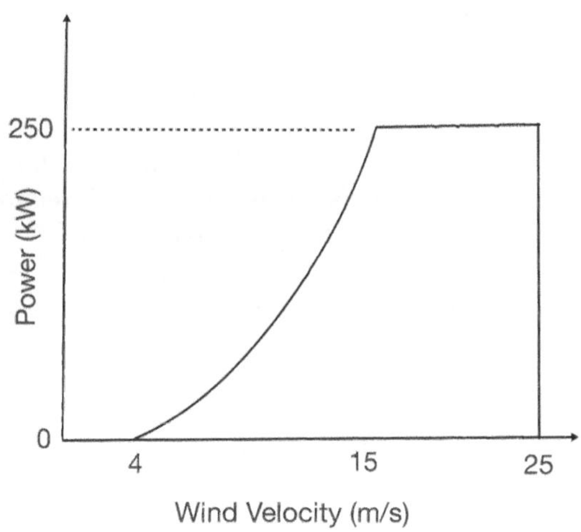

FIGURE 2.5
Hypothetical power curve for wind turbine with a rated power of 250 kW.

power of 250 kW. The power generated over the 24 hours period is then 250 kW times 24 hours, or 6,000 kW-h.

At Site B, during the first 12 hours, the wind speed is 30 m/s which exceeds the cut-out wind speed, $V_{\text{cut-out}}$, so that the wind turbine will not produce any power during this 12 hours period. During the second 12 hours period, the wind speed is 0 m/s, which is below the cut-in speed of the wind turbine, $V_{\text{cut-in}}$. As a result, the wind turbine will not produce any power during the second part of the 24 hours period as well. As a result at Site B, the total amount of power produced over the 24 hours period is 0 kW-h.

This rather simple example illustrates (in the extreme) the impact that the wind speed variation can have on a wind turbine's power generation. Therefore, it is important to quantify the variation that occurs in the wind speed over time. One such statistical measure is the "standard deviation" or second statistical moment which is defined as

$$\sigma_i = \left[\frac{1}{N}\sum_{i=1}^{N}(V_i - V_m)^2\right]^{1/2}, \tag{2.9}$$

where V_i is again the time series of velocity values which have a mean value of V_m. In this definition, σ_i is a measure of the average deviation of a velocity data point from the mean of the velocity data set.

The previous definition is somewhat inconvenient to calculate because the mean quantity, V_m needs to be computed first before determining the standard deviation. It is however easy to show that

$$\sigma_i = \left[\frac{1}{N}\sum_{i=1}^{N}V_i^2 - \left(\frac{1}{N}\sum_{i=1}^{N}V_i\right)^2\right]^{1/2}, \tag{2.10}$$

which is more convenient to compute since the sum of the V_i and V_i^2 can be accumulated together and subtracted at the end.

Wind data is most often grouped in the form of a frequency distribution such as shown in Table 2.2. This shows the number of hours per month in which the wind speed is within a specified range. With regard to such a wind velocity frequency, the power-weighted time average is

$$V_{m_p} = \left[\frac{\sum_{i=1}^{N} f_i V_i^3}{\sum_{i=1}^{N} f_i}\right]^{1/3}. \tag{2.11}$$

and the standard deviation is

$$\sigma_v = \left[\frac{\sum_{i=1}^{N} f_i \left(V_i - V_{m_p}\right)}{\sum_{i=1}^{N} f_i}\right]^{1/2}. \tag{2.12}$$

For the frequency data in the Table 2.2, $V_{m_p} = 8.34$ m/s and $\sigma_v = 0.81$ m/s.

It is important to note that V_{m_p} is *not the most probable* velocity. It generally does not occur unless the "skewness" (3rd statistical moment) is zero. This occurs only if the frequency distribution is Gaussian (random).

TABLE 2.2
Sample Frequency Distribution of Monthly Wind Velocity

Velocity (m/s)	Hours/month	Cumulative Hours
0–1	13	13
1–2	37	50
2–3	50	100
3–4	62	162
4–5	78	240
5–6	87	327
6–7	90	417
7–8	78	495
8–9	65	560
9–10	54	614
10–11	40	654
11–12	30	684
12–13	22	706
13–14	14	720
14–15	9	729
15–16	6	735
16–17	5	740
17–18	4	744

2.5 Statistical Models

In order to predict the power generated on a yearly basis, statistical models of the wind velocity frequency of occurrence are needed. It has been found

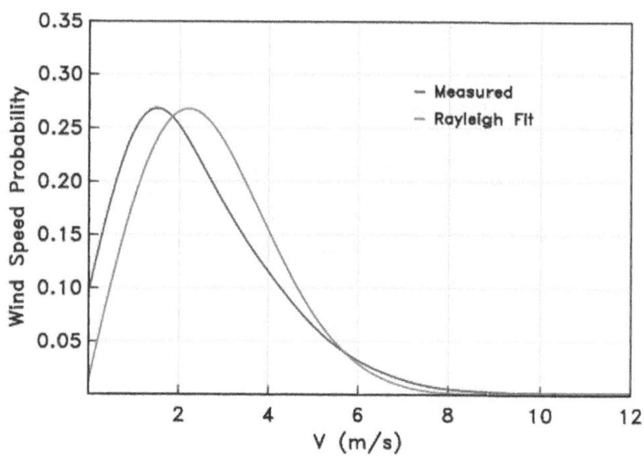

FIGURE 2.6
Probability distribution of wind speeds at the Notre Dame White Field wind turbine site, and a best-fit Rayleigh distribution.

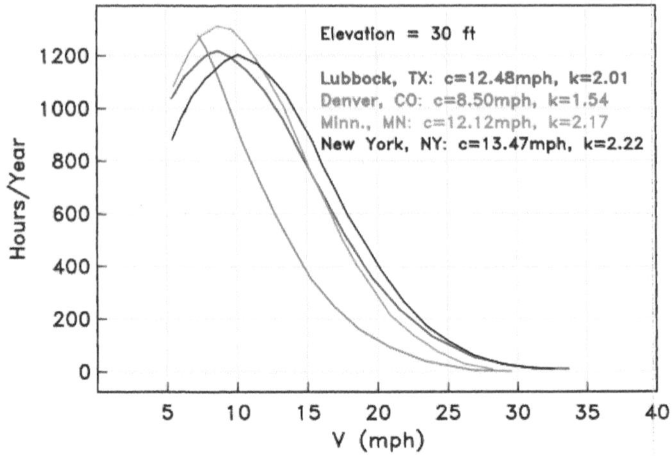

FIGURE 2.7
Sample Weibull distributions for atmospheric boundary layer data at different sites.

that Weibull and Rayleigh ($k = 2$) distributions can be used to describe wind variations with acceptable accuracy [2]. Figure 2.6 shows the probability distribution of wind speeds at the University of Notre Dame White Field wind turbine site, and an accompanying best-fit Rayleigh distribution. The advantage of using well-known analytic distributions like these is that the probability functions are already formulated.

2.5.1 Weibull Distribution

In the Weibull distribution, the probability in a years time of a wind speed, $V \geq V_p$, where V_p is an arbitrary wind speed is given as

$$p(V \geq V_p) = \exp\left[-(V_p/c)^k\right]. \tag{2.13}$$

For this, the number of hours in a year in which $V \geq V_p$

$$H(V \geq V_p) = (365)(24)\exp\left[-(V_p/c)^k\right]. \tag{2.14}$$

In these statistical representations, c and k are Weibull coefficients that depend on the elevation and location. In general, frequency data would be accumulated for a particular site and the wind turbine hub-height elevation that is being considered. The data would then be fit to a Weibull distribution to find the best c and k, An example of Weibull distributions with different coefficients is shown in Figure 2.7

Suggested corrections to Weibull coefficients k and c to account for different elevations, z, are

$$k = k_{\text{ref}} \frac{[1 - 0.088 \ln(z_{\text{ref}}/10)]}{[1 - 0.088 \ln(z/10)]}, \tag{2.15}$$

$$c = c_{\text{ref}} \left[\frac{z}{z_{\text{ref}}} \right]^n, \tag{2.16}$$

$$n = \frac{[0.37 - 0.088 \ln(c_{\text{ref}})]}{[1 - 0.088 \ln(z_{\text{ref}}/10)]}. \tag{2.17}$$

Justus et al. [1] suggested that $n = 0.23$ is a good representation of the atmospheric data.

The cumulative distribution is the integral of the probability density function, namely

$$\mathcal{P}(V) = \int_0^\infty p(V)dV = 1 - \exp\left[-(V/c)^k\right]. \tag{2.18}$$

The average wind speed is then

$$V_m = \int_0^\infty V p(V) dV \tag{2.19}$$

$$= \int_0^\infty V \frac{k}{c} \left(\frac{V}{c}\right)^{k-1} \exp\left[-(V/c)^k\right] dV \tag{2.20}$$

$$= k \int_0^\infty \left(\frac{V}{c}\right)^k \exp\left[-(V/c)^k\right] dV.$$

Letting $x = (V/c)^k$ and $dV = (a/k)x^{(\frac{1}{k}-1)}dx$, and substituting into Equation 2.19,

$$V_m = c \int_0^\infty e^{-x} x^{1/k} dx, \tag{2.21}$$

which we note the similarity to the Gamma function

$$\Gamma_m = c \int_0^\infty e^{-x} x^{n-1} dx, \tag{2.22}$$

therefore

$$V_m = c\Gamma\left(1 + \frac{1}{k}\right). \tag{2.23}$$

Note that Gamma function calculators are readily available on the internet.

The standard deviation of the wind speed, σ_v is found from

$$\sigma_V = \left(\mu_2' - V_m^2\right)^{1/2}, \tag{2.24}$$

where μ_2' is the second statistical moment of the data set that is defined as

$$\mu_2' = \int_0^\infty V^2 p(V) dV. \tag{2.25}$$

In this case, substituting $x = (V/c)^k$ and $dV = (a/k)x^{(\frac{1}{k}} - 1)dx$, one obtains

$$\begin{aligned} \mu_2' &= c^2 \int_0^\infty e^{-x} x^{2/k} dx \\ &= c^2 \Gamma\left(1 + \frac{2}{k}\right). \end{aligned} \tag{2.26}$$

Therefore, the standard deviation of the wind speeds can be written in terms of the Gamma function, namely

$$\sigma_V = c \left[\Gamma\left(1 + \frac{2}{k}\right) - \Gamma^2\left(1 + \frac{1}{k}\right)\right]^{1/2}. \tag{2.27}$$

The cumulative distribution function, $\mathcal{P}(V)$, can be used to estimate the time over which the wind speed is between some interval, V_1 and V_2. Therefore

$$\begin{aligned} \mathcal{P}(V_1 < V < V_2) &= p(V_2) - p(V_1) \\ &= \exp\left[-(V_1/c)^k\right] - \exp\left[-(V_2/c)^k\right]. \end{aligned} \tag{2.28}$$

This can also be used to estimate the time over which the wind speed exceeds a value, namely

$$\begin{aligned} \mathcal{P}(V > V_x) &= 1 - \left[1 - \exp\left[-(V_x/c)^k\right]\right] \\ &= \exp\left[-(V_x/c)^k\right]. \end{aligned} \tag{2.29}$$

Example 2.2

A wind turbine with a cut-in velocity of 4 m/s and a cut-out velocity of 25 m/s is installed at a site where the Weibull coefficients are $k = 2.4$ and $c = 9.8$ m/s. How many hours in a 24 hours period will the wind turbine generate power?

The number of hours that the wind turbine will operate is based on the probability that the wind speed falls between cut-in and cut-out values. Based on Equation 2.28,

$$\begin{aligned} \mathcal{P}(V_4 < V < V_{25}) &= p(V_{25}) - p(V_4) \\ &= \exp\left[-(4/9.8)^{2.4}\right] - \exp\left[-(25/9.8)^{2.4}\right] \\ &= 0.890 - 7.75 \times 10^{-5} \\ &= 0.890. \end{aligned} \tag{2.30}$$

Therefore, the number of hours in a 24 hours period where the wind speed is between 4 and 25 m/s is: $H = (24)(0.89) = 21.36$ hours. The amount of power generated by the wind turbine will be the product of the rated power (in kW) and the time during the 24 hours period where the velocity is between cut-in and cut-out, with the power having units of kW-h.

2.5.2 Methods for Weibull Model Fits

The methods for estimating the best k and c for a Weibull distribution include:

1. Graphical method,
2. Standard deviation method,
3. Moment method,
4. Maximum likelihood method, and
5. Energy pattern factor method.

2.5.2.1 Weibull Graphical Method

For a Weibull distribution, the cumulative distribution probability is

$$\mathcal{P}(V) = 1 - \exp\left[-(V/c)^k\right], \tag{2.31}$$

or,

$$1 - \mathcal{P}(V) = \exp\left[-(V/c)^k\right], \tag{2.32}$$

so that taking the natural log of both sides of the equality leads to a linear relation, namely

$$\underbrace{\ln\left[-\ln[1-\mathcal{P}(V)]\right]}_{y} = \underbrace{k\ln(V_i)}_{Ax} - \underbrace{k\ln(c)}_{B}. \tag{2.33}$$

Therefore by plotting $\ln\left[-\ln[1-\mathcal{P}(V)]\right]$ versus $\ln(V_i)$ for the velocity samples V_i, $i = 1, N$, the slope of the best fit straight line represents the Weibull coefficient, k, and the y-intercept represents $-k\ln(c)$, from which the Weibull scale factor, c can be found. Alternatively, one can perform a least-square curve fit of the linear function to find the slope and intercept.

A sample set of wind velocity frequency data is given in Table 2.3. The first column corresponds to wind speeds (km/h) at a site. The frequency of occurrence (Hours/month) that each wind speed occurs is given next to each wind speed in the second column. The probability of occurrence of a given wind speed, $p(V)$. is given in the third column. The probability, $p(V)$, equals the hours/month of a given wind speed (from column 2) divided by the total hours/month given by the sum of all the rows in column 2. Finally, the cumulative probability, $\mathcal{P}(V)$, in column 4 is the running sum of $p(V)$.

Figure 2.8 shows a plot of the data in Table 2.3 in the format of Equation 2.33. A best drawn straight line through the points provides the two Weibull coefficients, k and c.

2.5.2.2 Weibull Standard Deviation Method

For a Weibull distribution, one can show that the square of the ratio of the standard deviation, σ_V and mean velocity, V_m are given as

$$\left(\frac{\sigma_V}{V_m}\right)^2 = \frac{\Gamma\left(1+\frac{2}{k}\right)}{\Gamma^2\left(1+\frac{1}{k}\right)} - 1. \tag{2.34}$$

TABLE 2.3
Sample Wind Velocity Frequency Distribution

V(km/h)	Hours/month	$p(V)$	$\mathcal{P}(V)$
0	1.44	0.002	0.002
2	3.60	0.005	0.007
4	5.76	0.008	0.015
6	10.08	0.014	0.029
8	18.00	0.025	0.054
10	26.64	0.037	0.091
12	34.56	0.048	0.139
14	36.72	0.051	0.190
16	41.04	0.057	0.247
18	36.72	0.051	0.298
20	49.68	0.069	0.367
22	50.40	0.07	0.437
24	52.56	0.073	0.510
26	53.28	0.074	0.584
28	51.84	0.072	0.656
30	47.52	0.066	0.722
32	41.76	0.058	0.780
34	38.88	0.054	0.834
36	29.52	0.041	0.875
38	23.76	0.033	0.908
40	20.16	0.028	0.936
42	15.12	0.021	0.957
44	12.24	0.017	0.974
46	7.92	0.011	0.985
48	5.76	0.008	0.993
50	2.88	0.004	0.997
52	1.44	0.002	0.999
54	0.72	0.001	1
56	0	0	1
58	0	0	1
60	0	0	1

For this formulation, σ_V and V_m are calculated as an initial step. To satisfy the equation, the right-hand side of the equality must equal the left-hand side, namely $(\sigma_V/V_m)^2$. An iterative approach is then used to determine the value of k that satisfies the equality. Thus values of k are put into the equation, then the Gamma function is calculated, and the result is checked to determine if the equality is satisfied. If not, a new value of k is tried. The iterative process continues until the chosen value of k satisfies the equality. Once k is found, then

$$c = \frac{V_m}{\Gamma\left(1 + \frac{1}{k}\right)}. \tag{2.35}$$

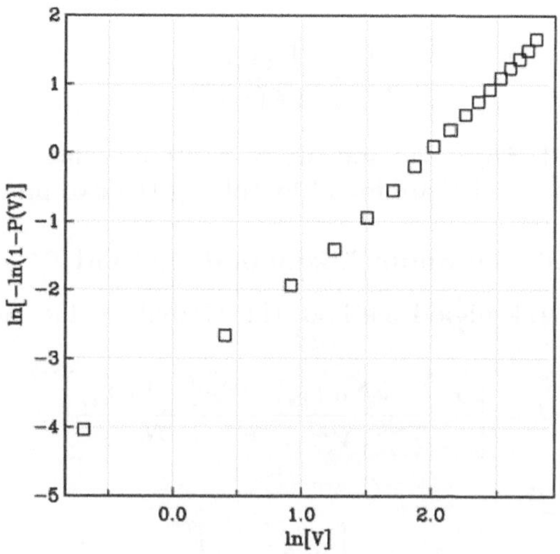

FIGURE 2.8
Weibull distributions fit for the data in Table 2.3. $k = 2.0$ and $c = 6.68\,\text{m/s}$.

A simpler approach whereby

$$k \simeq \left(\frac{\sigma_V}{V_m}\right)^{-1.090}, \tag{2.36}$$

and

$$c \simeq \frac{2V_m}{\sqrt{\pi}}, \tag{2.37}$$

can provide good approximate values of the Weibull coefficients for a set of wind time series data.

2.5.2.3 Weibull Moment Method

The Moment method is another approach to estimate the Weibull coefficients, k and c. The method is based on the a general formula for the nth statistical moment of a Weibull distribution

$$M_n = c^n \Gamma\left(1 + \frac{n}{k}\right). \tag{2.38}$$

If M_1 and M_2 are the first and second statistical moments, equal to the time mean, V_m in Equation 2.8, and the standard deviation, σ_i, given by Equation 2.9, respectively, then

$$c = \frac{M_2}{M_1} \frac{\Gamma\left(1 + \frac{1}{k}\right)}{\Gamma\left(1 + \frac{2}{k}\right)}, \tag{2.39}$$

and similarly,

$$\frac{M_2}{M_1^2} = \frac{\Gamma\left(1+\frac{2}{k}\right)}{\Gamma^2\left(1+\frac{1}{k}\right)}. \tag{2.40}$$

In this method, $M_1 = V_m$ and $M_2 = \sigma_i$ are calculated on the wind data beforehand. Then c and k are found by solving the two previous equations.

2.5.2.4 Weibull Maximum Likelihood Method

In the Maximum Likelihood method, the Weibull coefficient, k, is estimated as

$$k = \left[\frac{\sum_{i=1}^{N} V_i^k \ln(V_i)}{\sum_{i=1}^{N} V_i^k} - \frac{\sum_{i=1}^{N} \ln(V_i)}{N}\right]^{-1}, \tag{2.41}$$

and

$$c = \left[\frac{1}{N}\sum_{i=1}^{N} V_i^k\right]^{/k}. \tag{2.42}$$

We note that Equation 2.41 is a transcendental equation in the unknown, k. As such it needs to be solved iteratively.

2.5.2.5 Weibull Energy Pattern Method

The Energy Pattern method is based on the energy pattern factor, E_{PF}, which is the ratio of the total power available in the wind and the power corresponding to the cube of the mean wind speed, namely

$$E_{\mathrm{PF}} = \frac{\frac{1}{N}\sum_{i=1}^{N} V_i^3}{\left[\frac{1}{N}\sum_{i=1}^{N} V_i\right]^3}. \tag{2.43}$$

Having found the energy pattern from the wind velocity data at a given site, the approximate value of k is found from

$$k = 3.957 E_{\mathrm{PF}}^{-0.898}. \tag{2.44}$$

The value for c can be found using any of the previous methods.

2.5.3 Rayleigh Distribution

The Rayleigh distribution is a *special case* of the Weibull distribution in which $k = 2$. For $k = 2$,

$$V_m = c\Gamma\left(3/2\right), \tag{2.45}$$

or

$$c = 2\frac{V_m}{\sqrt{\pi}}, \tag{2.46}$$

which we note was used as a simplification to the Weibull standard deviation method for determining the unknown coefficients with the Standard Deviation Method.

In terms of the probability functions, substituting c into the Weibull expression for $p(V)$, one obtains

$$p(V) = \frac{\pi}{2}\frac{V}{V_m^2}\exp\left[-\frac{\pi}{4}\left(\frac{V}{V_m}\right)^2\right], \tag{2.47}$$

of which then

$$\mathcal{P}(V) = 1 - \exp\left[-\frac{\pi}{4}\left(\frac{V}{V_m}\right)^2\right], \tag{2.48}$$

so that

$$\mathcal{P}(V_1 < V < V_2) = \exp\left[-\frac{\pi}{4}\left(\frac{V_1}{V_m}\right)^2\right] - \exp\left[-\frac{\pi}{4}\left(\frac{V_2}{V_m}\right)^2\right], \tag{2.49}$$

and

$$\mathcal{P}(V > V_x) = 1 - \left[1 - \exp\left[-\frac{\pi}{4}\left(\frac{V_x}{V_m}\right)^2\right]\right] = \exp\left[-\frac{\pi}{4}\left(\frac{V_x}{V_m}\right)^2\right]. \tag{2.50}$$

2.6 Energy Estimation of Wind Regimes

The ultimate estimate to be made in selecting a site for a wind turbine or wind farm is the *energy* that is available in the wind at the site. This involves calculating the wind energy density, E_D, for a wind turbine unit rotor area and unit time. The wind energy density is a function of the wind speed and temporal distribution at the site. In assessing this, other parameters of interest are the most frequent wind velocity, $V_{F_{\max}}$, and the wind velocity contributing the maximum energy, $V_{E_{\max}}$, at the site. The most frequent wind velocity, $V_{F_{\max}}$, corresponds to the maximum of the probability distribution, $p(V)$. As a result, as the power generated scales with the cube of the wind velocity, the maximum energy usually corresponds to velocities that are higher than the most frequent ones.

Horizontal wind turbines are usually designed to operate most efficiently at its design power wind speed, V_d. Therefore, it is advantageous if V_d and $V_{E_{\max}}$ at the site are made to be as close as possible. Once $V_{E_{\max}}$ is estimated for a site, it is then possible to match the characteristics of the wind turbine to operate most efficiently under that condition. The following sections present statistical approaches for estimating $V_{E_{\max}}$ based on Weibull and Rayleigh wind speed distributions.

2.6.1 Weibull-Based Energy Estimation Approach

The power that is available in a wind stream of velocity V over a unit rotor area is

$$P_V = \frac{1}{2}\rho_a V^3. \tag{2.51}$$

For a given velocity, V, the *unit* amount of time that velocity is present is $1 \times p(V)$. Therefore, the energy per unit time is $P_V p(V)$. The total energy for all possible wind velocities at a site is therefore

$$E_D = \int_0^\infty P_V p(V) dV. \tag{2.52}$$

Substituting for P_V, and $p(V)$ for a Weibull distribution, and simplifying one obtains

$$E_D = \frac{\rho_a k}{2c^k} \int_0^\infty V^{(k+2)} \exp\left[-(V/c)\right]^k dV. \tag{2.53}$$

Making a change in variables where

$$x = \left(\frac{V}{C}\right)^k, \tag{2.54}$$

the expression for E_D becomes

$$E_D = \frac{\rho_a c^3}{2} \int_0^\infty x^{3/k} e^{-x} dx. \tag{2.55}$$

As before, the integral has the form of a standard Gamma function so that

$$E_D = \frac{\rho_a c^3}{2} \Gamma\left(\frac{3}{k} + 1\right). \tag{2.56}$$

Applying the general reduction formula for a Gamma function given by

$$\Gamma(n) = (n-1)\Gamma(n-1), \tag{2.57}$$

one obtains the following form for the energy density

$$E_D = \frac{\rho_a c^3}{2} \frac{3}{k} \Gamma\left(\frac{3}{k}\right). \tag{2.58}$$

With E_D known for a site, the energy that is available over a period of time, T, is

$$E_T = E_D T = \frac{\rho_a c^3 T}{2} \frac{3}{k} \Gamma\left(\frac{3}{k}\right). \tag{2.59}$$

To calculate the energy that is available over a 24 hours period, $T = 24$.

An expression for the most frequent wind speed, V_F, starts with the probability distribution, $p(V)$, for a Weibull velocity distribution namely

$$p(V) = \frac{k}{c^k} V^{k-1} \exp\left[-(V/c)^k\right]. \tag{2.60}$$

The most frequent wind speed is then the maximum of the probability function. This is found as the condition where

$$\frac{dp(V)}{dV} = 0, \tag{2.61}$$

which gives the following

$$\frac{k}{c^k} \exp\left[-(V/c)^k\right] \left[-\frac{k}{c^k} V^{2(k-1)} + (k-1)V^{(k-2)}\right] = 0. \tag{2.62}$$

Solving this expression for V gives

$$V = c\left(\frac{k-1}{k}\right)^{1/k}. \tag{2.63}$$

To demonstrate that this is a maximum, we note that

$$\frac{dp(V)}{dV} > 0 \text{ in the interval } \left[0,\, c\left(\frac{k-1}{k}\right)^{1/k}\right], \tag{2.64}$$

and

$$\frac{dp(V)}{dV} < 0 \text{ in the interval } \left[c\left(\frac{k-1}{k}\right)^{1/k},\, \infty\right]. \tag{2.65}$$

Therefore, this verifies that V in Equation 2.63 is a maximum that represents the most frequent wind velocity in a Weibull distribution, namely,

$$V_{F_{\max}} = c\left(\frac{k-1}{k}\right)^{1/k}. \tag{2.66}$$

In order to determine the wind velocity that results in the maximum energy, we again start with the energy per unit time produced by a given velocity which is

$$E_V = P_V p(V). \tag{2.67}$$

Again substituting for P_V, and $p(V)$ for a Weibull distribution, and simplifying one obtains

$$E_V = \frac{\rho_a V^3}{2} \frac{k}{c} \left[\frac{V}{c}\right]^{k-1} \exp\left[-(V/c)^k\right]. \tag{2.68}$$

Introducing a change in variables where

$$B = \frac{\rho_a}{2} \frac{k}{c^k}, \tag{2.69}$$

the expression for E_V becomes

$$E_V = BV^{(k+2)} \exp\left[-(V/c)^k\right]. \tag{2.70}$$

We then seek the conditions on V that maximize E_V by setting

$$\frac{dE_V}{dV} = 0, \tag{2.71}$$

which gives the following expression

$$B\left[-\exp\left[-(V/c)^k\right] V^{(k+2)} \frac{k}{c^k} V^{(k-1)} + \exp\left[-(V/c)^k\right] (k+2) V^{(k+1)}\right] = 0. \tag{2.72}$$

Solving this expression for V gives

$$V = \frac{c(k+2)^{1/k}}{k^{1/k}}. \tag{2.73}$$

In this case to demonstrate that this is a maximum, we note that E_V increases in the interval

$$\left[0, \frac{c(k+2)^{1/k}}{k^{1/k}}\right], \tag{2.74}$$

and decreases in the interval

$$\left[\frac{c(k+2)^{1/k}}{k^{1/k}}, \infty\right]. \tag{2.75}$$

Therefore this verifies that V in Equation 2.73 represents the wind velocity in a Weibull distribution that maximizes energy, or

$$V_{E_{\max}} = \frac{c(k+2)^{1/k}}{k^{1/k}}. \tag{2.76}$$

2.6.2 Rayleigh-Based Energy Estimation Approach

When considering a Rayleigh wind speed distribution, the wind energy density is

$$E_D = \int_0^\infty P_V p(V) dV = \int_0^\infty \frac{\pi \rho_a}{4V_m^2} V^4 \exp\left[\frac{\pi}{4}\left(\frac{V}{V_m}\right)^2\right] dV. \tag{2.77}$$

Introducing a change in variables where

$$K = \frac{\pi}{4V_m^2}, \tag{2.78}$$

the expression for E_D becomes

$$E_D = K\rho_a \int_0^\infty V^4 e^{(-KV^2)} dV. \tag{2.79}$$

Introducing a second change in variables where

$$x = KV^2, \tag{2.80}$$

so that

$$dV = \frac{dx}{2\sqrt{Kx}}, \tag{2.81}$$

yields a new expression for E_D, namely

$$E_D = \frac{\rho_a}{2K^{3/2}} \int_0^{\infty} x^{3/2} e^{-x} dx, \tag{2.82}$$

which can be reduced to a Gamma function of the form

$$E_D = \frac{\rho_a}{2K^{5/2}} \Gamma(5/2) = \frac{3}{8} \frac{\rho_a \sqrt{\pi}}{K^{1.5}}. \tag{2.83}$$

Substituting back for K in the expression one obtains

$$E_D = \frac{3}{\pi} \rho_a V_m^3. \tag{2.84}$$

The energy available for a unit rotor area over a period of time, T, is then

$$E_T = TE_D = \frac{3}{\pi} T \rho_a V_m^3. \tag{2.85}$$

To identify the most frequent wind speed, we start with the probability density function, $p(V)$, for the Rayleigh distribution that is written in terms of the constant, K, namely

$$p(V) = 2KVe^{-(KV^2)}. \tag{2.86}$$

The most frequent wind speed is then the maximum of the probability function. This is found as the condition where

$$\frac{dp(V)}{dV} = 0, \tag{2.87}$$

which yields the equation

$$2Ke^{-(KV^2)} \left(1 - 2KV^2\right) = 0, \tag{2.88}$$

which upon solving for V gives

$$V = \frac{1}{\sqrt{2K}}. \tag{2.89}$$

Checking if this condition represents a maximum, we note that

$$\frac{dp(V)}{dV} > 0 \text{ in the interval } \left[0, \frac{1}{\sqrt{2K}}\right], \tag{2.90}$$

and

$$\frac{dp(V)}{dV} < 0 \text{ in the interval } \left[\frac{1}{\sqrt{2K}}, \infty\right]. \tag{2.91}$$

Therefore, V in Equation 2.89 is the most frequent wind velocity in a Rayleigh distribution, or

$$V_{F_{\max}} = \frac{1}{\sqrt{2K}} = \sqrt{\frac{2}{\pi}}V_m. \tag{2.92}$$

The velocity contributing the maximum energy for a Rayleigh wind velocity distribution for a unit rotor area over a unit period of time is

$$E_V = P_V p(V) = K\rho_a V^4 e^{-(kV^2)}. \tag{2.93}$$

Again we seek to find the maximum which we expect to occur where

$$\frac{dE}{dV} = 0. \tag{2.94}$$

This yields the following equation

$$K\rho_a e^{-(KV^2)}\left[4V^3 + V^4(-2KV)\right] = 0, \tag{2.95}$$

which when solved gives

$$V = \sqrt{\frac{2}{K}}. \tag{2.96}$$

To prove that this is a maximum, we note that E_V is increasing in the interval

$$\left[0, \sqrt{\frac{2}{K}}\right], \tag{2.97}$$

and decreases in the interval

$$\left[\sqrt{\frac{2}{K}}, \infty\right]. \tag{2.98}$$

Therefore, the velocity that maximizes the energy for a Rayleigh wind velocity distribution for a unit rotor area over a unit period of time is

$$V_{E_{\max}} = \sqrt{\frac{2}{K}} = 2\sqrt{\frac{2}{\pi}}V_m. \tag{2.99}$$

Example 2.3

Wind velocity data (m/s) at a possible wind farm site is given in the following Table 2.4. From this, calculate the wind energy density, E_D, the monthly energy availability, E_T, the most frequent wind velocity, $V_{F_{\max}}$, and the velocity corresponding to the maximum energy, $V_{E_{max}}$, based on a Rayleigh velocity distribution.

The expressions for E_D, E_T, $V_{F_{max}}$ and V_E for a Rayleigh velocity frequency distribution were given by Equations 2.84, 2.85, 2.92, and 2.99. These involve the above monthly average wind speeds that leads to the following table.

Month	E_D (kW/m^2)	E_T (kW/m^2/month)	$V_{F_{max}}$ (m/s)	$V_{E_{max}}$ (m/s)
Jan	0.90	666.95	7.29	14.58
Feb	0.67	451.11	6.62	13.24
Mar	0.47	351.09	5.89	11.77
Apr	0.46	327.49	5.82	11.63
May	1.20	889.30	8.03	16.05
Jun	1.59	1146.72	8.83	17.66
Jul	1.76	1307.78	9.13	18.25
Aug	1.59	1184.94	8.83	17.66
Sep	1.29	931.78	8.24	16.48
Oct	0.42	313.95	5.67	11.34
Nov	0.36	258.82	5.38	10.75
Dec	0.74	551.72	6.84	13.69

We note that the wind velocity where the energy is a maximum varies from month to month. This makes it difficult to design a wind turbine that is optimum for all wind conditions at a site.

2.7 Wind Condition Measurement

The statistical analysis of the wind speed depends on accurate site measurements. The minimum information that is needed for the analysis is wind speed and direction taken at short periodic time intervals over a long enough period of time to allow for converged statistics. Wind data from nearby meteorological stations can also be quite helpful in assessing the conditions at a site. Such meteorological stations are often located at airports. However, the most precise analysis of the wind conditions at a site comes from on-site measurements.

TABLE 2.4
Monthly Average Wind Speed Data

Jan	Feb	Mar	Apr	May	Jun	Jul	Aug	Sep	Oct	Nov	Dec
9.14	8.3	7.38	7.29	10.1	11.1	11.4	11.1	10.3	7.11	6.74	8.58

The following sub-sections describe the tools that are available to perform such measurements.

2.7.1 Wind Speed Anemometers

Wind speed anemometers are transducers that deduce air velocity and provide an output (analog or digital) that is proportional to the measurement. The different types of anemometers have specific characteristics such as sensitivity and frequency response. The sensitivity determines the minimum and smallest increment in velocity that can be measured. The frequency response determines the smallest time scales of wind velocity fluctuations that can be measured.

The anemometers are usually located on tall masts or towers. The standard elevation is 10 m. However for monitoring wind conditions at an existing or proposed wind turbine site, it is useful to locate an anemometer at the hub height of the wind turbine.

2.7.1.1 Cup Anemometer

Wind speed anemometers have evolved significantly over time. One of the earliest anemometer designs is the rotating cup anemometer, invented in 1846 by John Thomas Romney Robinson. A photograph of the Robinson cup anemometer is shown in Figure 2.9.

Cup anemometers generally consist of three or four equally spaced hollow hemispherical or conical-shaped “cups”. The cups are supported off of a center shaft that rotates about its vertical axis. The cups then rotate about a horizontal plane.

Cup anemometers are drag-based devices. The concave side of the cup has a greater drag coefficient than the convex side of the cup. As a result, wind blowing towards the concave side of a cup exerts more force causing it to move. Since the cup motion is constrained to rotate, the concave side of a cup rotates out of the wind vector, where it is then replaced by the next cup in the group. This process repeats itself, causing the arrangement to rotate. Being a drag-based device, the rate of rotation is proportional to the square of the local wind speed.

Cup anemometers cannot determine wind direction. Therefore, they are often paired with a wind direction indicator which consists of a vertical tail surface that is mounted on one end of a slender body that is free to rotate in the horizontal plane that is parallel to the plane of rotation of the cup anemometer. An example is shown in the photograph in Figure 2.9. The wind direction indicator is usually connected to an angular position transducer that provides an analog or digital output that is proportional to the angular position.

Although cup anemometers are simple devices, they have a number of limitations. In particular, being mechanical devices with moving parts, their frequency response is limited by the inertia in the rotating cups. Therefore,

FIGURE 2.9
Photograph of Robinson 1846 cup anemometer.
Source: NOAA's National Weather Service (NWS) Collection, Image ID: wea00920. http://www.photolib.noaa.gov/htmls/wea00920.htm. PD-US - published in the U.S. before 1923 and public domain in the U.S.

they are not reliable to measure wind gusts. With regard to time-averaged measurements, because they are drag-based devices, their wind speed measurements depend on the density of the air, which is a function of the air temperature and humidity. Therefore for the greatest accuracy, simultaneous temperature and humidity measurements are necessary.

2.7.1.2 Propeller Anemometer

Propeller anemometers consist of a four-bladed propeller that rotates when pointed into the wind. An example is shown in Figure 2.10. The propeller anemometer is a lift-based device so that like the cup anemometer, the rate of rotation is proportional to the local wind speed squared. Unlike the cup anemometer, the output response of the propeller anemometer depends on the wind direction. The largest output (fastest rotation) occurs when the propeller anemometer is pointing directly into the wind, that is the propeller rotor disk is perpendicular to the wind direction vector. The output decreases in proportion to the cosine of the angle between the pointing angle and the wind vector angle. To account for this characteristic, propeller anemometers are generally mounted on a slender body that has a vertical tail surface. The slender body is free to rotate so that the vertical tail can keep the propeller anemometer pointed into the wind. The rotation motion can also be monitored

FIGURE 2.10
Example of a propeller anemometer that is designed to point into the wind. *Source http://www.photolib.noaa.gov/nssl/nssl0161.htm, nssl0161, National Severe Storms Laboratory (NSSL) Collection. (public domain because it contains materials that originally came from the U.S. National Oceanic and Atmospheric Administration.*

through an angular position sensor in order to record wind direction along with wind speed.

Being mechanical devices, propeller anemometers suffer from similar limitations as the cup anemometer. The inertia in the rotating propellers makes time-resolved, gust measurements unreliable. In addition, like drag-based devices, this lift-based device is sensitive to the air density, and therefore is a function of the air temperature and humidity. Thus as with cup anemometers, simultaneous temperature and humidity measurements are necessary for the greatest accuracy.

2.7.1.3 Pitot-Static Pressure Anemometers

Pitot-static pressure anemometers are another common approach for measuring wind speed. It was invented by Henri Pitot in 1732 and was modified to its modern form in 1858 by Henry Darcy. The basic Pitot probe consists of a tube pointing directly into the fluid (air with regard to wind energy) flow, and another that is perpendicular to the fluid flow direction. In the case of the former, the moving fluid is brought to rest (stagnates) as there is no outlet to allow the flow to continue. This pressure inside the tube is therefore the stagnation pressure of the moving fluid, also referred to as the total pressure, p_t. The second tube aligned perpendicular to the flow direction measures the static pressure, p_s. If the two pressures are measured at close to the same spatial location, then the difference between them is related to the local velocity of the fluid through Bernoulli's equation

$$p_t = p_s + \left(\frac{\rho V^2}{2}\right), \tag{2.100}$$

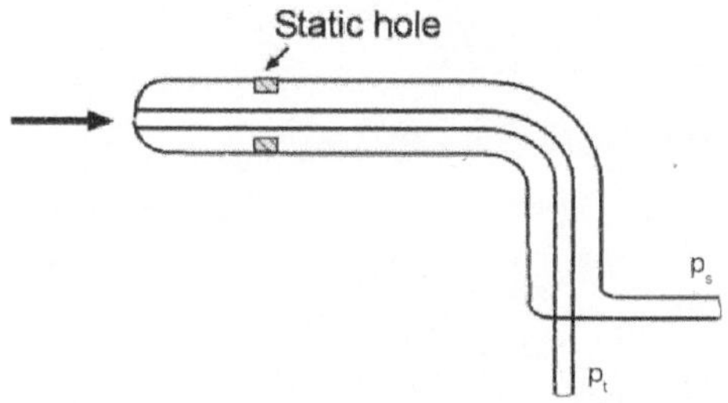

FIGURE 2.11
Schematic drawing of a Pitot-static probe anemometer.

which applies to an incompressible fluid, which is an excellent assumption in wind energy applications. Figure 2.11 shows a modern embodiment of a Pitot-static probe which is fashioned from two concentric tubes. The center tube measures the stagnation pressure, and the outer tube measures the static pressure through small holes around the perimeter of the outer tube wall.

A singular advantage of the Pitot-static probe is that it does not have any moving parts. However if it is connected through tubing to a pressure transducer to convert the pressure difference to a voltage that can be recorded, the tubing length and diameter strongly affect the frequency response of the measurement. However with moderate lengths of tubing, the frequency response can still be of the order of 10–20 Hz. which is adequate for gust measurements. A more serious problem is that Pitot probes are susceptible to fouling from dust, moisture, ice, and insects.

2.7.1.4 Sonic Anemometers

Sonic anemometers use ultrasonic sound waves to measure wind velocity. They were first developed in the 1950s. They measure wind speed based on the time of flight of sonic pulses between pairs of transducers. Measurements from pairs of transducers can be combined to provide multiple wind speed components. Figure 2.12 shows a sonic anemometer that can measure all three velocity components.

The spatial resolution of sonic anemometers is defined by the path length between the transducers, which is typically 10–20 cm. The sonic anemometer shown in Figure 2.12 has a wind speed range up to 40 m/s, with a resolution of 0.01 m/s. Being able to measure three wind speed components, it also provides the wind direction vector (in horizontal and vertical planes). Finally, since the speed of sound in air varies with temperature and is virtually constant with pressure change, sonic anemometers are also used as thermometers.

Sonic anemometers can take measurements with very fine temporal resolution, 20 Hz or better, which makes them well suited for turbulent gust measurements. The lack of moving parts makes them appropriate for long-term use, particularly in exposed automated weather stations and weather buoys

FIGURE 2.12
Photograph of a three-component sonic anemometer located on the meteorological tower of the University of Notre Dame research wind turbines.

where the accuracy and reliability of traditional cup-and-vane anemometers are adversely affected by salty air or large amounts of dust. Sonic anemometers can be affected by precipitation, where the presence of rain drops can alter the speed of sound (which is different in water compared to air).

2.7.1.5 Wind Measurement Support Equipment

Other instrumentation that is used to compile wind data for wind turbine power predictions includes independent measurements of temperature, humidity and static pressure. Each of these will provide an analog or digital output that is proportional to the respective measured quantities. These outputs are then recorded at periodic time intervals, along with the outputs from respective anemometers.

The device that records the outputs from the different transducers is referred to as a data logger. They generally consist of a dedicated digital computer with a digital-to-analog converter and a digital-to-digital interface to acquire analog and digital inputs, respectively. They often have internal memory for data storage. However for large data sets or for data archiving, they can download data to other computers for storage and post processing.

Typical data loggers can acquire data at rates from once every 10 miliseconds to 30 minutes. Shorter intervals are used to perform frequency analysis of the data time series. The longer intervals are used to compile time-averaged statistics. More typical data acquisition rates are of the order of 5 seconds to 10 minutes.

One of the primary benefits of using data loggers is the ability to automatically collect data on a 24/7-basis. Data loggers are typically deployed and left unattended to measure and record information for the duration of the monitoring period. This allows for a comprehensive picture of the wind conditions over longer uninterrupted periods.

Problems

1. The Weibull-fitted wind speed frequency distributions shown in Figure 2.7 are for different locations with wind speeds measured at an elevation of 10 m (30 ft.).

   Using this information, determine the expected number of hours per year of wind speeds between 18 and 20 mph at the Lubbock, Texas site, at the wind turbine hub-height elevation of 20 m.

2. A wind turbine is proposed for a Denver, CO site. The hub height for the wind turbine will be 100 ft. The wind turbine has a **cut-out** wind speed of 30 mph (13.4 m/s).
   (a) Using the Weibull-fitted wind speed frequency distributions shown in Figure 2.7, determine how many hours of 30 mph and higher wind speeds will occur per year.
   (b) How many hours in a 24 hours period does this correspond to?
   (c) What does that say about the wind turbine operation?
3. The object of this problem is to determine the Weibull coefficients, k and c, for the set of wind frequency data given in Table 2.2 of the text.
   (a) Table 2.2 gives columns of wind frequency (hours/month) and cumulative hours. From these, generate two additional columns of the probability of occurrence, $p(V)$, and the cumulative probability, $\mathcal{P}(V)$ (such as in Table 2.3).
   (b) Perform a graphical Weibull fit on this wind frequency data, and determine the best k and c values.
   (c) Given the best k and c values, what is the probability that the wind speeds at the site where this data was taken is between a wind turbine's cut-in and cut-out speeds of 4 and 17 m/s?
   (d) From Part c, how many hours in a 24 hours period does this correspond to?
   (e) Based on the conditions given in Part c, what would be the probability if we assume a Rayleigh wind frequency distribution?
   (f) Based on Part e, how many hours in a 24 hours period would this correspond to?
4. A wind turbine is proposed for the South Bend, IN site. The airport data obtained at an elevation of $z_{\text{ref}} = 10$ m gave a Weibull wind frequency distribution fit with coefficients of $k = 2.0$ and $c = 5.34$ m/s. The proposed wind turbine will have a hub height of 15 m.

(a) Based on the Weibull coefficient at the reference 10 m elevation, determine the new Weibull coefficients for the 15 m hub height. Assume $n = 0.23$.

(b) The rated wind speed for the wind turbine is $V_{\text{rated}} = 4$ m/s What is the probability for the wind to exceed the rated wind speed at the site?

(c) If the cut-out wind speed is $V_{\text{cut-out}} = 10$ m/s, How many hours in a 24 hours day will the wind turbine produce rated power?

5. A wind farm is proposed to be built outside of Denver Airport. The airport wind data obtained at a $z_{\text{ref}} = 10$ m, gave a Weibull wind frequency distribution fit with coefficients of $k = 1.54$ and $c = 3.78$ m/s. The wind turbines in the proposed wind farm have a rated power of $P_{\text{rated}} = 1.5$ MW, a rotor radius of $R = 35$ m, and a hub height of $H = 65$ m. The rated wind speed for the wind turbines is $V_{\text{rated}} = 8$ m/s and the cut-out wind speed is $V_{\text{cut-out}} = 20$ m/s.

   (a) Based on the Weibull coefficients at $z_{\text{ref}} = 10$ m, determine the new Weibull coefficients for the wind speed at the wind turbine hub height. Assume $n = 0.23$.

   (b) What is the probability for the wind at the hub height to be between V_{rated} and $V_{\text{cut-out}}$?

   (c) Based on this, what is the expected Annual Energy Production (AEP) [MW-h], for one of the wind turbines in the wind farm?

References

1. C. Justus, W. Hargraves, and A. Yalcin, "Nationwide assessment of potential output from wind-powered generators", *Journal of Applied Meteorology*, July 1976, 15:7.

2. J. Hennessey, Some aspects of wind power statistics", *Journal of Applied Meteorology*, February 1977, 16:2.

3

Introduction to Aerodynamics

3.1 Introduction

Horizontal axis wind turbine blades extract power from the wind using the aerodynamic forces created on the rotor blades. When the aerodynamic forces in the plane of rotation are large enough, the rotor begins to turn. The aerodynamic forces acting on the blades can be resolved into the components normal to and in the plane of rotation. The spanwise normal force distribution yields the thrust loading on the blades. Integration of the thrust loading yields the thrust load transmitted to the tower. Most wind turbine blades are securely attached to the drive shaft, and from a structural perspective, the turbine blade acts like a rotating cantilever beam. The blade thrust loading creates a bending moment that deflects the blades out of the plane of rotation toward the tower for a upwind rotor design. The component of the aerodynamic forces acting in the plane of rotation times the distance to the axis of rotation creates a torque. The torque times the angular velocity of the rotor yields the mechanical power transmitted to the wind turbine drive shaft. The mechanical power is converted into electric power by the power conversion components such as a gearbox, generator, and electric power conditioning equipment.

The efficiency of the rotor in extracting the power from the wind is a function of the aerodynamic characteristics of the airfoil sections used in the design of the rotor blades. The aerodynamic forces acting on the turbine blades are a function of the cross-sectional shape of the rotor blade. Figure 3.1 is a sketch of a rotor blade illustrating the variation in the blade cross-sectional geometry at various locations across blade. In this sketch, the blade is made up from a number of different airfoil cross-sectional shapes. The section shapes that are chosen are based on their radial position along the rotor blade, which experience different flow conditions because of the rotation. The efficiency of the rotor in extracting power from the wind is a function of the aerodynamic properties of the airfoil section shapes used in the design.

A sketch of an airfoil section at an angle of attack, α, is shown in Figure 3.2. The angle of attack is the angle that the freestream velocity makes with the chord line of the airfoil. Note that the lift force is perpendicular to the freestream velocity vector, and the drag force is parallel to the freestream velocity vector. The pitch moment acting on the airfoil is shown to be acting

DOI: 10.1201/9781003611486-3

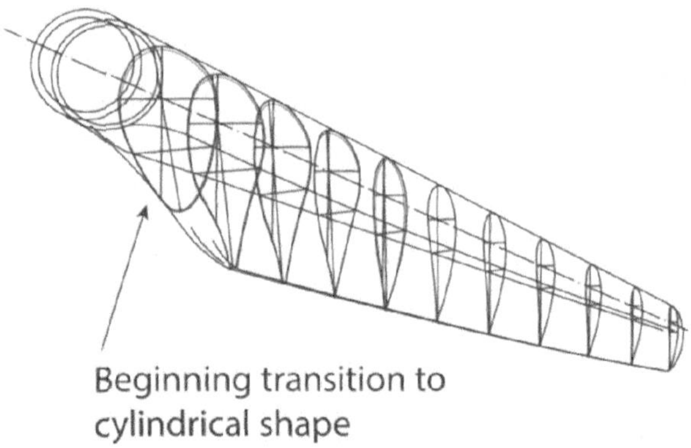

FIGURE 3.1
Sketch of a wind turbine showing the different blade section shapes across the blade span.

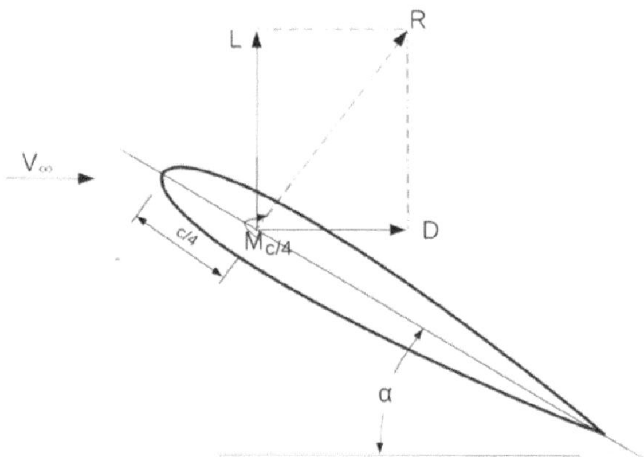

FIGURE 3.2
Aerodynamic forces and moment acting on an airfoil.

at the quarter-chord location, which is the usual case for a subsonic airfoil section.

Before discussing airfoil aerodynamics, we will digress briefly to provide a summary of the development of airfoil shapes that are used in wind turbine rotor blade design. Their early development included research by the Wright brothers who sought to understand discrepancies in the aerodynamic data by conducting their own wind tunnel experiments. They further evaluated these measurements through glider flight tests at their summer camp near Kill Devil Hill in Kitty Hawk, North Carolina. The use of wind tunnel tests followed by full-scale tests is still used today for aeronautic designs. Over several years of designing, modifying, and testing, they developed a glider that they believed could successfully achieve powered flight.

The brothers then turned their attention to modifying a gasoline motor and developing two pusher propellers that were powered by a single motor that turned the propellers by way sprockets and chain link drive train from the motor. This culminated in the first successful powered flight on December 3, 1903. On the last flight of the day, Wilbur Wright flew for 59 seconds and covered a distance of 852 ft. The Wright brothers were not successful in generating much interest in their airplane in the United States. So in 1908, they dismantled their airplane and shipped it to France where Wilbur Wright successfully flew their airplane in front of large crowds and set records in speed, altitude, distance, and time aloft.

In 1915, the United State Congress created the National Advisory Committee for Aeronautics (NACA) to provide direction for scientific and engineering solutions to problems related to flight. This was done in large part due to the rapid development of aircraft designs in Europe just before and during World War I. By 1945, the NACA had three research centers, the Langley Memorial Aeronautical Laboratory at Langley, Virginia, the Ames Aeronautical Laboratory at Moffett Field California, and the Aircraft Engine Research Laboratory at Cleveland, Ohio. The NACA Laboratories became world renown for their aeronautical research, particularly in the design and documentation of airfoil shapes. Many of these designs were used in early wind turbine rotors.

Prompted by the first successful launch of an orbital satellite by the then-Soviet Union in 1957, President Dwight D. Eisenhower signed the National Aeronautics and Space Act, which created the National Aeronautics and Space Administration (NASA). The NACA laboratories as well as several government laboratories that conducted research in rocketry and jet propulsion then became part of NASA. Research on new airfoil designs continues today at the NASA Research Centers.

3.2 Airfoil Geometry

The geometry of the airfoil section determines its aerodynamic properties. Figure 3.3 shows cross-sectional shapes for a symmetrical and a cambered airfoil. The geometry of the symmetrical airfoil is defined by its straight chord line with length, c, and its symmetric thickness distribution that is added in the normal direction to the chord line to form the upper and lower surfaces of the airfoil. The leading edge radius is the largest radius of a circle centered on the chord line and tangent to the upper and lower surfaces that form the airfoil leading edge.

With a cambered airfoil section, the mean camber line is a curved line that intersects the straight chord line at the airfoil leading and trailing edges. A

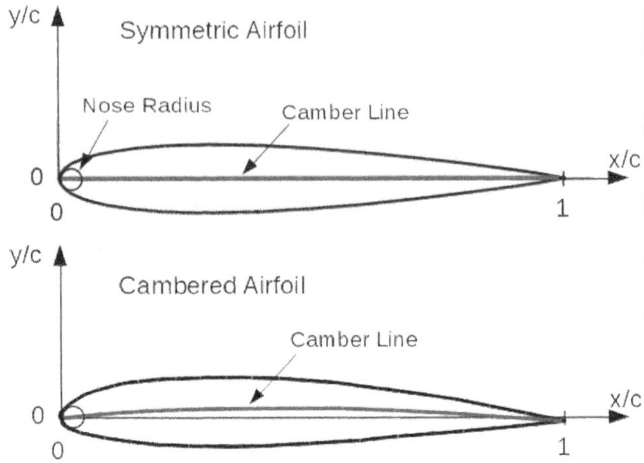

FIGURE 3.3
Geometry defining a symmetric and cambered airfoil section shapes.

uniform thickness distribution is added to the camber line to form the upper and lower airfoil surfaces. The leading edge radius in this case is on a line that is tangent to the mean camber line at the leading edge. The reasons for adding camber to an airfoil shape will be explained.

The aerodynamic forces and moment acting on an airfoil section are a function of its cross-sectional geometry, the angle of attack, and the fluid properties of the air passing over the airfoil section. The flow properties include the freestream velocity, V_∞, the air density, ρ_∞, the viscosity of air, μ_∞, and the local speed of sound, a_∞. The aerodynamic lift, drag and pitching moment for a given airfoil shape and angle of attack are a function of these flow variables, and the planform area, S, of a wing having the section shape over a span, b. In the next section, dimensional analysis will be discussed and then used to reduce the number of flow variables needed to measure the aerodynamic properties of an airfoil.

3.3 Dimensional Analysis

The Buckingham Pi Theorem is a very important theorem in dimensional analysis. It states that

> *if a physical meaningful equation has dimensional homogeneity consisting of N physical variables that are expressed in terms of K fundamental units, then the original system can be expressed in terms of $N - K$ dimensionless variables called π-products.*

By applying the Buckingham Pi Theorem to a physical equation, one can reduce the number of experimental variables to a more manageable set of dimensionless variables. The following steps provide an outline to the procedure to obtain the non-dimensional π-products.

1. Determine the number of physical variables, N, that govern the process that is being examined. The variables of the problem are $\nu_1, \nu_2, \ldots, \nu_n$.
2. List the dimensions for each of the variables from step 1. For example, the velocity, V_∞ has the dimensions of length, $\mathcal{L}$, over time, t, i.e., $(\mathcal{L}/t)$.
3. Determine the number of π-products, which is equal to the difference between the number of variables, N, and the number of fundamental dimensions, K, thus
$$\pi\text{-products} = (N - K). \tag{3.1}$$
4. Select the number of K variables from the physical variables identified in Step 1. These variables must include all of the fundamental units of the problem. In addition, no two of the selected variables can have identical dimensions, and no selected variable can be dimensionless. A dimensionless variable is a π-product.
5. Determine the π-products, $\pi_1, \pi_2, \ldots, \pi_{N-K}$ where
$$\begin{aligned} \pi_1 &= P_1^a P_2^b P_3^c P_4 \\ \pi_2 &= P_1^a P_2^b P_3^c P_5 \\ &\vdots \\ \pi_{N-K} &= P_1^a P_2^b P_3^c P_{N-K}, \end{aligned} \tag{3.2}$$
and the variables P_1, P_2, and P_3 are the repeated variables selected in Step 4.
6. Determine the exponents, a, b, and c of the π-products through dimensional analysis.

The aerodynamic characteristics of a given wind turbine rotor section having a given section shape, at a fixed angle of attack, are the lift force, l, the drag force, d, and the pitching moment about the quarter-chord location, $m_{c/4}$[1]. These characteristics are functions of fluid properties mentioned above, and the chord length, c. The lift force per unit span can be expressed in a functional form as
$$\text{Lift} \equiv l = f\left(V_\infty, \rho_\infty, \mu_\infty, a_\infty, c\right). \tag{3.3}$$

The lift force is a function five variables. To reduce the number of variables, we will apply the Buckingham Pi theorem whereby,

$$f\left(l, V_\infty, \rho_\infty, \mu_\infty, a_\infty, c\right) = 0. \tag{3.4}$$

The six physical variables in Equation 3.4 can be expressed in terms of three fundamental units: mass, m, length, l, and time, t. The equation therefore has six physical variables and three fundamental units that describe the variables. Now according to the Buckingham π theorem, we need to select three of the physical variables that include all the fundamental units. However, no selected physical variable can be dimensionless, and no two of the selected variables can have the same units. The π-products that meet the above requirements are given below. The repeated variables are ρ_∞, V_∞ and c and the non-repeating variables are l, μ_∞ and a_∞. Therefore

$$\begin{aligned}\pi_1 &= f_1\left(\rho_\infty, V_\infty, c, l\right)\\ \pi_2 &= f_2\left(\rho_\infty, V_\infty, c, \mu_\infty\right)\\ \pi_3 &= f_3\left(\rho_\infty, V_\infty, c, a_\infty\right).\end{aligned} \tag{3.5}$$

The first π-product can be expressed as

$$\pi_1 = \rho_\infty^p, V_\infty^q, c^r, l. \tag{3.6}$$

In terms of the fundamental dimensions of the π-product, π_1 must be dimensionless. Therefore, introducing the units of each of the terms,

$$[ml^{-3}]^p[lt^{-1}]^q[l]^r[mlt^{-2}] = \pi_1. \tag{3.7}$$

In order for the LHS of the equation to be dimensionless, the exponents of the mass, length and time terms must be zero. Therefore

$$\begin{aligned}\text{For the mass: } & p+1=0\\ \text{For the length: } & -3p+q+r+1=0\\ \text{For the time: } & -q-2=0.\end{aligned} \tag{3.8}$$

As a result, $p = -1$, $q = -2$, and $r = -2$. Thus the π_1-product is

$$\pi_1 = l\rho_\infty^{-1}V_\infty^{-2}c^{-1} = \frac{l}{\rho_\infty V_\infty^2 c}. \tag{3.9}$$

Now any π-product can be multiplied by a constant. Therefore, the π_1-product can be written as

$$\pi_1 = \frac{l}{\frac{1}{2}\rho_\infty V_\infty^2 c}. \tag{3.10}$$

Here we note that the quantity, $\frac{1}{2}\rho_\infty V_\infty^2$ is the dynamic pressure of the flow, often indicted as q_∞. Note that this is not to be confused with q used as one of the π exponents.

The ratio $l/\frac{1}{2}\rho_\infty V_\infty^2 c$ is called the lift coefficient and indicated as C_l, thus

$$C_l = \frac{l}{\frac{1}{2}\rho_\infty V_\infty^2 c} = \frac{l}{q_\infty c}. \tag{3.11}$$

The second π-product is a function of the repeated variables and the viscosity, μ_∞, thus

$$\pi_2 = \rho_\infty^p, V_\infty^q, c^r, \mu_\infty. \tag{3.12}$$

In terms of the fundamental dimensions of the π_2-product,

$$[ml^{-3}]^p[lt^{-1}]^q[l]^r[ml^{-1}t^{-1}] = \pi_2, \tag{3.13}$$

so that

$$\begin{aligned} \text{For the mass: } & p + 1 = 0 \\ \text{For the length: } & -3p + q + r - 1 = 0 \\ \text{For the time: } & -q - 1 = 0, \end{aligned} \tag{3.14}$$

and therefore $p = -1$, $q = -1$, and $r = -1$. As a result, the π_2-product is

$$\pi_2 = \frac{\rho_\infty^{-1} V_\infty^{-1} c^{-1}}{\mu_\infty}. \tag{3.15}$$

Now any π-product can be raised to any power. Therefore, the π_2-product can be expressed as

$$\pi_2 = \left(\frac{\rho_\infty^{-1} V_\infty^{-1} c^{-1}}{\mu_\infty}\right)^{-1} = \frac{\rho_\infty V_\infty c}{\mu_\infty}. \tag{3.16}$$

Here we note that the dimensionless quantity $\rho_\infty V_\infty c/\mu_\infty$ is the Reynolds number, often indicated as Re or Re_c where the subscript c indicates that the chord dimension is the unit of length used in defining the Reynolds number.

Finally, for the π_3-product, the fundamental dimensions yield

$$[ml^{-3}]^p[lt^{-1}]^q[l]^r[lt^{-1}] = \pi_3, \tag{3.17}$$

so that

$$\begin{aligned} \text{For the mass: } & p = 0 \\ \text{For the length: } & -3p + q + r + 1 = 0 \\ \text{For the time: } & -q - 1 = 0, \end{aligned} \tag{3.18}$$

and therefore $p = 0$, $q = -1$, and $r = 0$. As a result, the π_3-product is

$$\pi_3 = \frac{V_\infty}{a_\infty}. \tag{3.19}$$

Here we note that the dimensionless quantity V_∞/a_∞ is the Mach number, which is the ratio of the fluid velocity to the speed of sound in the fluid.

As pointed out, the π_1-product, led to the non-dimensional form which was the lift coefficient

$$C_l = \frac{l}{\frac{1}{2}\rho_\infty V_\infty^2 c}. \tag{3.20}$$

The lift on an airfoil section is a function of the angle of attack, α, and therefore so is the lift coefficient. The lift coefficient is also a function of the other two π-products, namely the Reynolds number,

$$\mathrm{Re} = \frac{\rho_\infty V_\infty c}{\mu_\infty}, \tag{3.21}$$

and the Mach number

$$M = \frac{V_\infty}{a_\infty}. \tag{3.22}$$

Based on this π-product analysis, we can state that

$$C_l = f\left(\alpha, \mathrm{Re}, M\right). \tag{3.23}$$

In a similar manner, the drag coefficient, C_d and the pitching moment coefficient about quarter-chord location, $C_{m_{c/4}}$, would also be functions of the angle of attack, Reynolds number and Mach number, namely

$$C_d = f\left(\alpha, \mathrm{Re}, M\right), \tag{3.24}$$

and

$$C_{m_{c/4}} = f\left(\alpha, \mathrm{Re}, M\right). \tag{3.25}$$

The effect of Mach number on the aerodynamic coefficients is not important until the Mach number is greater than about 0.4–0.5. With regard to a wind turbine, the velocity at any section along the rotor blade is a function of the wind speed, V_∞, and the rotational velocity of the rotor blade, Ωr, where r is a radial location along the rotor blade, and Ω is the rotation rate with units of radians/seconds. The maximum resultant velocity (at the rotor blade tip, $r = R$) is the vector sum of the two velocity components, namely

$$V_R = \sqrt{V_\infty^2 + \left(\Omega R\right)^2}. \tag{3.26}$$

For a pitched regulated wind turbine, the maximum operating wind speed of the turbine is the "cut-out" wind speed. When the cut-out wind speed is reached, the turbine blade angle of attack, α, is reduced to reduce the lift and therefore the torque on the rotor as a precautionary measure to prevent damage to the wind turbine. The cut-out wind speed for modern large wind turbines is around 25–30 m/s. Based on these maximum wind speeds and the typical rotational velocities of the rotor blades, the maximum resultant velocity is well below the Mach number where compressibility has any effect on the aerodynamic performance of the wind turbine rotor. Therefore, we can neglect the Mach number effects and the aerodynamic coefficients are only functions of the angle of attack and Reynolds number, namely

$$\begin{aligned} C_l &= f(\alpha, \mathrm{Re}) \\ C_d &= f(\alpha, \mathrm{Re}) \\ C_{m_{c/4}} &= f(\alpha, \mathrm{Re}). \end{aligned} \tag{3.27}$$

3.4 Airfoil Aerodynamics

As discussed earlier, airfoils are generally classified as having symmetrical or cambered section shapes. For a symmetrical airfoil, the lift coefficient is zero when the angle of attack is zero. The aerodynamic lift increases linearly with increasing angle of attack until at higher angles of attack, the air flow over the airfoil can no longer follow the curvature of the airfoil upper (suction) surface and the flow "separates". If the flow separation begins at the trailing edge and moves forward with increasing angle of attack, the rate of increase in the lift coefficient diminishes and then begins to decrease. This is illustrated in the lift coefficient versus angle of attack for a symmetric airfoil shown in Figure 3.4. The angle of attack where the lift coefficient reaches its maximum is referred to as the stall angle of attack, α_s. The stall exhibited in Figure 3.4 would be considered to be "very gentle". This is typical of a "thicker" airfoil section shape. The airfoil thickness is generally categorized by the ratio of its maximum thickness to its chord length, namely t/c. For "thin" airfoil sections, the air flow over the suction surface of the airfoil may separate abruptly from the leading edge, with a sharp drop in the lift coefficient. An example of this behavior is presented later in this chapter.

The aerodynamic drag on an airfoil in which Mach number effects are minimal consists of viscous drag and pressure drag. The former is due to the viscosity of the air passing over the surface of the airfoil. The latter is due to the static pressure distribution that results from the airfoil shape and angle of attack. At lower angles of attack, the viscous drag is the dominant source of aerodynamic drag on the airfoil. At higher angles of attack, pressure drag is the dominant source. As the stall angle of attack is approached, the pressure drag becomes significant. As opposed to the aerodynamic lift that diminishes in the post stall regime, the aerodynamic drag continues to increase, significantly lowering the lift-to-drag ratio, l/d, of the airfoil section.

Considering the pressure drag, the pressure on the surface of the airfoil, acting on a unit area of the surface, results in a force. The pressure force is a vector that acts normal to the local surface. Given the curved airfoil surface, the pressure force vector can be decomposed into components that are parallel to and perpendicular with the freestream velocity direction. The latter is the component lift force, and the former is the component drag force. Summing up these two forces around the surface of the airfoil gives the total lift and drag forces on the airfoil. An example of the drag coefficient as a function of angle of attack is shown in Figure 3.5. This corresponds to the lift versus angle of attack distribution that is shown in Figure 3.4.

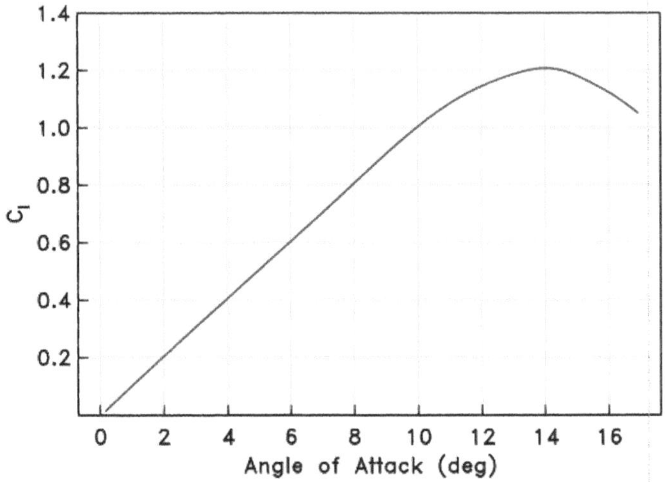

FIGURE 3.4
Sample lift coefficient versus angle of attack for a thick symmetric airfoil section.

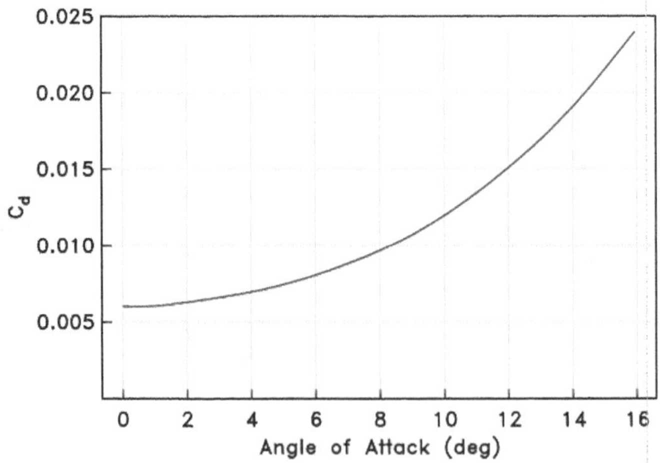

FIGURE 3.5
Drag coefficient versus angle of attack for the same airfoil section that produced the lift coefficient versus angle of attack shown in Figure 3.4.

3.5 Airfoil Geometry

This section is intended to provide an understanding of how the geometry of the airfoil influences its aerodynamic properties. This involves an examination of several of the NACA airfoil section shapes that were tested in the NACA's (now NASA) low turbulence pressure tunnels in Langley, Virginia. The selected airfoil sections are from the NACA four digit airfoil family.

In the years from the 1970s to the early 1980s, the wind turbine electric power industry used a number of airfoil designs that were developed by the NACA. Some of these airfoils were of the NACA-23XX, NACA-44XX, and NACA-63XXX series [1,2]. The NACA used a four-, five-, or six-digit numbering system to classify the cross-sectional geometry of the airfoils. With the NACA four-digit series, the first two digits indicate the camber line. The equations that describe the mean camber line are

$$y_c = \frac{m}{p^2}\left(2px - x^2\right), \tag{3.28}$$

and

$$y_c = \frac{m}{(1-p)^2}\left((1-2p) + 2px - x^2\right), \tag{3.29}$$

where m refers to the maximum ordinate of the mean camber line expressed as a percentage of the total chord, and p denotes the chordwise position of the maximum ordinate in tenths of the total chord. Equation 3.28 corresponds to the portion of the chord line that is forward of the maximum ordinate location, and Equation 3.29 corresponds to the portion that is aft of the maximum ordinate location.

The last two digits in the four-digit series correspond to the maximum airfoil thickness as a percent of the chord. The thickness distribution is given by the following equation:

$$\pm y_t = \frac{t}{0.2}\left(0.29690\sqrt{x} - 0.12600x - 0.35160x^2 + 0.28430x^3 - 0.10150x^4\right). \tag{3.30}$$

The upper and lower surface coordinates can be determined by applying the thickness distribution that is perpendicular to the mean chord line, namely

$$\begin{aligned} x_U &= x - y_t \sin\theta \\ y_U &= y_c - y_t \cos\theta \\ x_L &= x + y_t \sin\theta \\ y_L &= y_c + y_t \cos\theta, \end{aligned} \tag{3.31}$$

where subscripts U and L refer to the coordinates of the upper and lower surfaces, respectively.

The variable θ can be found by taking the derivative with respect to x of the appropriate Equations 3.28 or 3.29. If the x-location is forward or equal to the axial location of the maximum ordinate, then Equation 3.28 is used. Equation 3.29 is used if x-location is aft of the maximum ordinate. Then θ is found from

$$\theta = \arctan\left(\frac{dy_c}{dx}\right). \tag{3.32}$$

3.6 Aerodynamic Characteristic of Three NACA Airfoils

Having defined the geometry for the NACA four digit series, the aerodynamic characteristics of several airfoil sections in this series are presented. Figures 3.6 and 3.7 show the lift, drag, and pitching moment coefficients for NACA-0006 and NACA-0012 section shapes. The first two digits being zero indicate that these are symmetric airfoils (zero camber). The last two digits signify the thickness-to-chord ratio, with the 06 indicating a $t/c = 0.06$ or 6%, and the 12 indicating a $t/c = 0.12$ or 12%. Figure 3.8 shows the aerodynamic coefficients for a NACA-4412 section shape. The first two digits being 44 indicate this is a cambered airfoil. The last two digits being 12 indicate a $t/c = 0.12$ or 12%.

The aerodynamic characteristics of the NACA-0006 airfoil indicate a lift coefficient that increases linearly with angle of attack up to

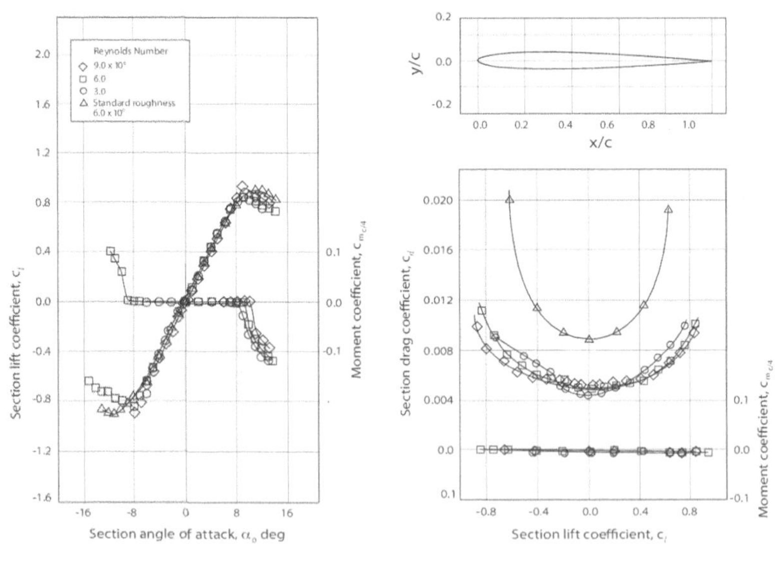

FIGURE 3.6
Aerodynamic characteristics of a NACA-0006 airfoil section.

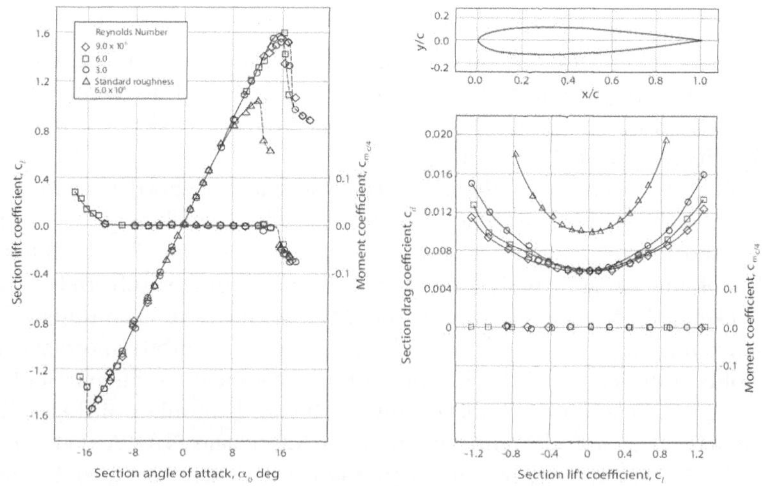

FIGURE 3.7
Aerodynamic characteristics of a NACA-0012 airfoil section.

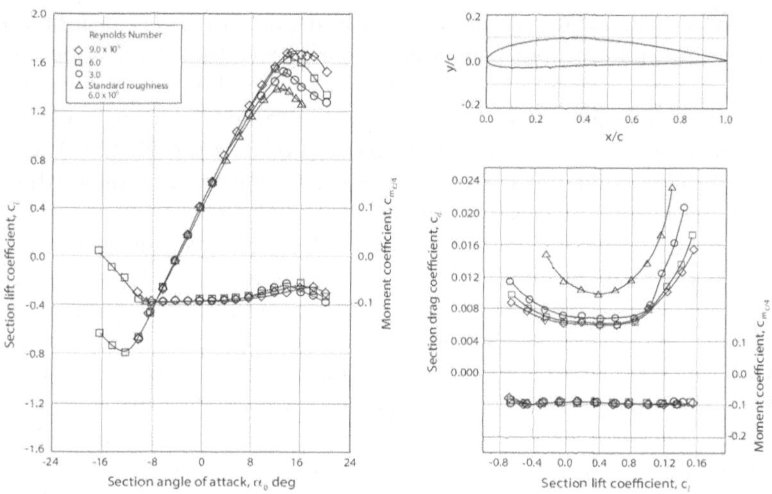

FIGURE 3.8
Aerodynamic characteristics of a NACA-4412 airfoil section.

approximately $\alpha = 8^\circ$. Above that angle of attack, the lift slope, $dC_l/d\alpha$, abruptly changes sign from positive to negative. The lift at the point of the discontinuity, $dC_l/d\alpha = 0$ is called the maximum lift coefficient, $C_{l_{\max}}$. The NACA-0006 airfoil is very thin and therefore it has a very small leading edge radius. As the angle of attack increases, the small leading edge radius causes the air flow near the leading edge to separate abruptly. The flow separation occurs at a relatively low angle of attack that results a very low $C_{l_{\max}}$. The influence of the Reynolds number on the lift coefficient is small in the linear $dC_l/d\alpha$ region. However, the aerodynamic characteristics in the post stall region is affected by the Reynolds number.

The pitching moment coefficient about the quarter-chord position, $C_{m_{c/4}}$, is constant with angle of attack up to α_s. The drag coefficient, C_d, is nearly constant, and low, at the smaller angles of attack between $0° - 4°$. This range of angles of attack at which the drag is a minimum is referred to as the "drag bucket". At higher angles of attack, C_d increases in a nonlinear fashion with increasing angle of attack. The drag coefficient exhibits more sensitivity to Reynolds number than the lift or moment coefficients. In particular, the highest drag coefficient occurs at the lowest Reynolds number of $\mathrm{Re}_c = 3 \times 10^6$.

Comparing the aerodynamic characteristics of the NACA-0006 to those of the NACA-0012 airfoil, as shown in Figure 3.7, provides insight into the effect of the thickness-to-chord ratio. In this case, the twice-larger t/c nearly doubles $C_{l_{\max}}$. The improvement in the aerodynamic lift is directly related to the larger leading radius. As a result, the range of angles of attack where the drag coefficient remains low is increased compared to the thinner airfoil.

Comparing the aerodynamic characteristics of the NACA-0012 to those of the NACA-4412 airfoil, as shown in Figure 3.8, provides insight into the effect of adding camber. The immediate difference is that the cambered airfoil produces lift at zero angle of attack. The effect of camber was to shift the angle of attack at which zero lift occurs to negative values. The angle of attack of zero lift for a cambered airfoil is denoted as α_{0L}. For the NACA-4412 this is $\alpha_{0L} = -4°$. The other consequence of adding camber is to move the center of the "drag bucket" to positive angles of attack. This is preferential since it can minimize the drag in the positive lift condition where the airfoil is designed to operate. Such as lift condition is referred to that the "Design-C_l". The camber does not affect $C_{l_{\max}}$. The effects of airfoil geometry on the aerodynamic characteristics are summarized in Table 3.1.

A very useful presentation of the aerodynamic characteristics of an airfoil is the lift-to-drag ratio, (C_l/C_d), versus angle of attack. The lift-to-drag ratio is effectively a measure of the efficiency of the airfoil. A higher lift-to-drag ratio is an important aspect of the aerodynamic performance of a wind turbine. Figure 3.9 presents a plot of the lift-to-drag ratio for the NACA-4412 airfoil for three different Reynolds numbers. The plot indicates a strong sensitivity of $(C_l/C_d)_{\max}$ on the Reynolds number. In particular, $(C_l/C_d)_{\max}$ increases with increasing Reynolds number. The increase was largest between the two lowest Reynolds numbers compared to the two largest Reynolds numbers. This indicates that Reynolds number can be important, particularly if it is too low. The Reynolds number in this instance is based on the chord dimension of the airfoil. Therefore, higher Reynolds numbers can be attained with designs that utilize airfoils with larger chord dimensions.

TABLE 3.1
Summary of Effects of Airfoil Geometry on Aerodynamic Characteristics

Reynolds number	Increasing Reynolds number delays flow separation to a higher angles of attack, increasing $C_{l_{\max}}$ and α_s.
Nose radius	Nose radius increases with increasing t/c. Increasing nose radius increases $C_{l_{\max}}$ and α_s.
Airfoil t/c	$C_{l_{\max}}$ increases with increasing t/c up to $t/c \simeq 15\%$. Further increases in t/c decrease $C_{l_{\max}}$.
Camber	Adding camber shifts the zero lift angle of attack to negative values, and shifts the drag bucket to angles of attack with positive lift, allowing those design lift conditions to have minimum drag.
Surface roughness	Surface roughness near the leading edge of an airfoil can lead to early stall that results in a lower $C_{l_{\max}}$ and increased $C_{d_{\max}}$, and as a result a lower $(C_l/C_d)_{\max}$.

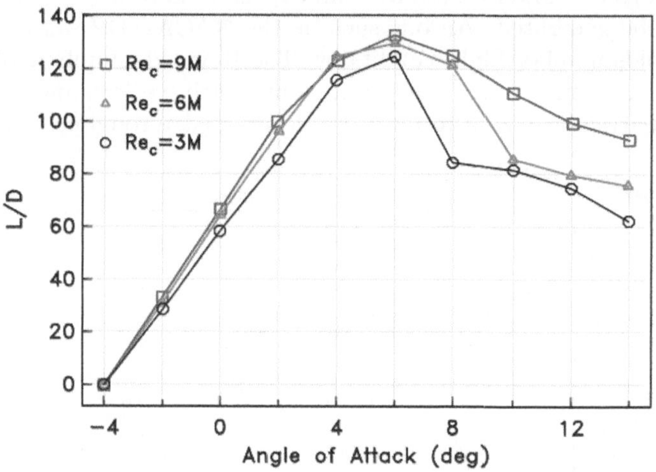

FIGURE 3.9
Effect of chord Reynolds number on the lift-to-drag ratio versus angle of attack of a NACA-4412 airfoil section.

3.7 Airfoil Sensitivity to Leading Edge Roughness

Surface roughness near the leading edge of an airfoil can significantly modify the aerodynamic characteristics. To examine the influence of surface roughness on airfoils, the NACA selected a standard form of roughness that could be applied to an airfoil model. This involved carborundum grains with a 0.011 inch diameter that were glued to the surface of a model, near the leading edge. The grains were applied from the leading edge, $x/c = 0$, down to the 8% chord location on both the upper and lower surfaces. They were sparsely spread over the selected region so that they covered from 5% to 10% of the surface area. This "standard" roughness was considered to be more severe than what would be expected under normal use of an aircraft. It did not, however, simulate roughness that could result from leading edge icing.

The effect of such standard roughness on the lift-to-drag ratio of a NACA-4412 airfoil section at a Reynolds number of $\text{Re}_c = 6 \times 10^6$ is shown in Figure 3.10. This indicates a dramatic decrease in $(C_l/C_d)_{\max}$ as a result of the roughness. On a wind turbine rotor such surface roughness could result from abrasion of the rotor leading edge, insect strikes, or ice buildup. As these results indicate, this could have a highly detrimental effect on the wind turbine performance.

The wind turbines built in the period from the 1960s to the early 1980s for the electric power industry used airfoil designs developed for airplanes such as the NACA-4412 airfoil. Unfortunately wind turbine blades using these NACA airfoils had lower efficiency than expected, lowering the electric power that could be generated. Airfoils such as the NACA-4412 were designed for high Reynolds number flight conditions. The Reynolds numbers of wind turbine rotors are much lower, and as a result, their performance significantly degraded, particularly as a result of the leading edge roughness effects.

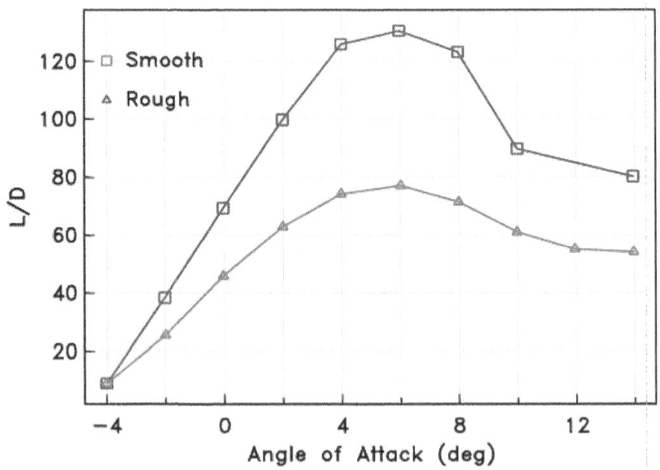

FIGURE 3.10
Effect of leading edge roughness on the lift-to-drag ratio versus angle of attack of a NACA-4412 airfoil section.

Nature provides several mechanisms that can create roughness on a wind turbine. Developers of wind farms seek areas that have a high probability that the winds will be in a range from 5 to 30 m/s at the selected site. Regions that provide such excellent wind energy resources are often located in cold, or warm-humid, or desert-like climates. Such conditions can produce operational problems that affect wind turbine efficiency. In cold climates, the air density will be higher, which would lead to more wind energy, however it can also lead to icing. Ice formation on the rotor leading edge represents surface roughness and therefore can degrade the aerodynamic performance. Ice accumulation can also cause dangerous structural loading on the blades. Even a light frost can be detrimental to the wind turbine efficiency. Heating the rotor leading edge can eliminate the ice problems for low icing conditions. In severe icing conditions, the wind turbines must be shut down to avoid serious damage.

The problem for wind turbines in warm-humid climates is surface contamination resulting from insect strikes on the leading edge. The build-up of insect

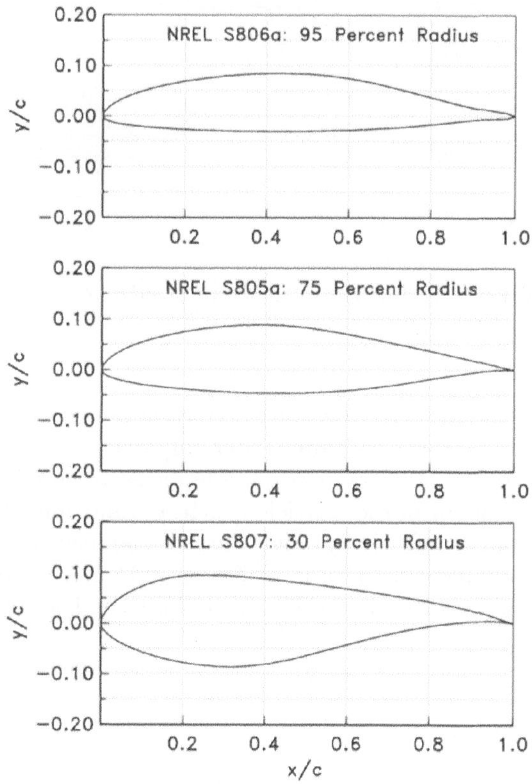

FIGURE 3.11
NREL thin-airfoil family for use in medium sized wind turbine blades.

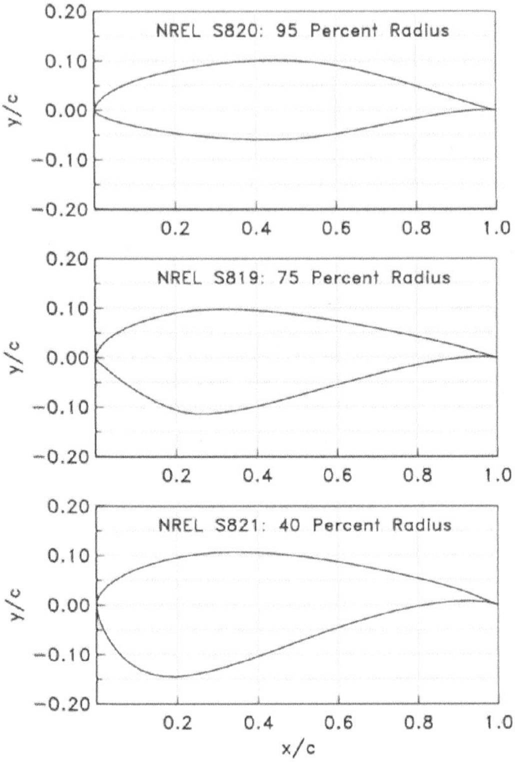

FIGURE 3.12
NREL thick-airfoil family for use in medium sized wind turbine blades.

residue acts like leading edge roughness, which subsequently lowers the wind turbine performance. Insect contamination only occurs at low wind speeds.

In desert-like regions, the source of roughness is largely do to abrasion produced by small wind-borne particles such as sand and dirt. Under these conditions, the wind is effectively sand blasting the leading edge of the rotor blades. The only solution for this is to incorporate a more resilient material for the leading edge, or a replaceable covering for the leading edge.

3.8 New Airfoil Designs for the Wind Power Industry

In the mid-1980s, research laboratories in Europe and the United States began developing new airfoil section shapes that would be less sensitive to leading edge roughness. These new designs were developed at the Delft University

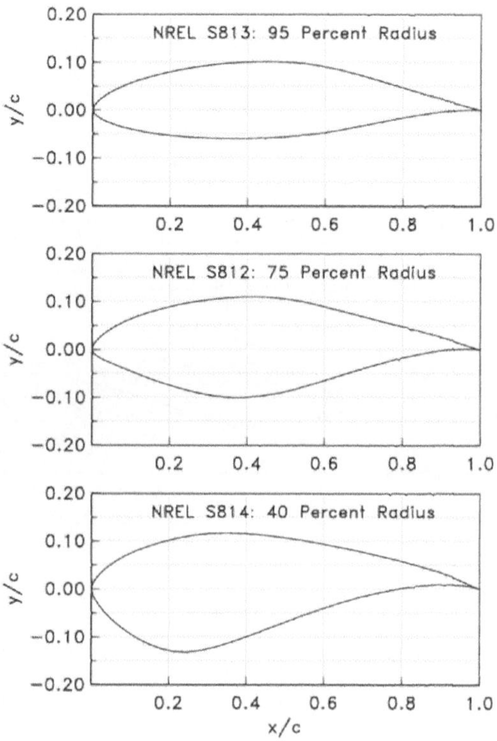

FIGURE 3.13
NREL thick-airfoil family for use in large sized wind turbine blades.

TABLE 3.2
Estimated Annual Energy Improvements from NREL Airfoil Series

Turbine Type	Roughness Insensitive $C_{l_{max}}$	Correct Reynolds No.	Low Tip $C_{l_{max}}$	Total Improvement
Stall regulated	10%–15%	3%–5%	10%–15%	23%–35%
Variable pitch	5%–15%	3%–5%	-	8%–20%
Variable RPM	5%	3%–5%	-	8%–10%

Wind Energy Research Institute, the Technical University of Denmark, the FFA in Sweden, and at the National Renewal Energy Laboratory (NREL) in the United States [3–7]. The resulting airfoil designs were suitable for stall-regulated, variable RPM and variable pitch wind turbines. Four of the NREL section shapes are shown in Figures 3.11–3.14, along with their design specifications. The expected annual improvements from the NREL airfoil designs are summarized in Table 3.2 for the different wind turbine operation.

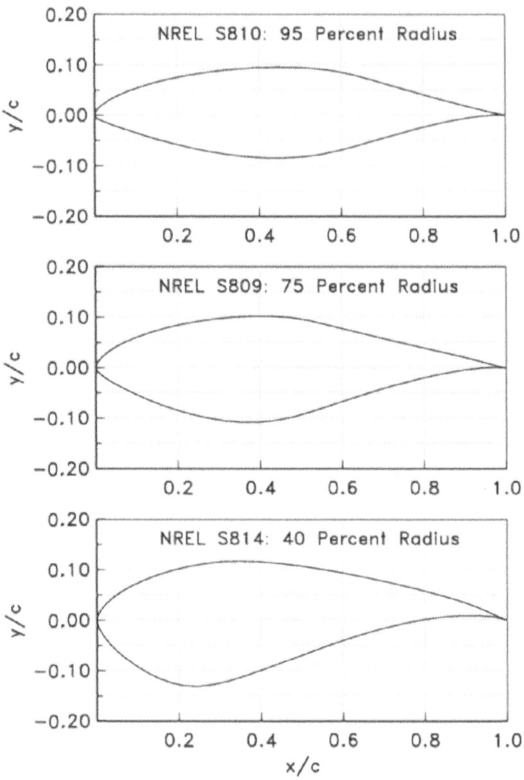

FIGURE 3.14
NREL thick-airfoil family for use in large sized wind turbine blades.

As evident in Table 3.2, the stall-regulated wind turbines achieve the largest annual energy improvement from the more roughness-tolerant airfoil designs. The annual energy improvement of the variable pitch and variable RPM wind turbines was also better, although by a lower percentage. They however demonstrate that the proper choice of the rotor airfoil section shape can have a demonstrable improvement in the performance of the wind turbine over a large range of conditions.

3.9 Summary

As demonstrated in this chapter, the aerodynamic coefficients of airfoil shapes that are relevant to wind turbines are a function of Reynolds number and the airfoil geometry. As a result, the use of aerodynamic data for Reynolds numbers that are far from the operating conditions can be very risky, particularly in the post-stall regime. The influence of leading edge roughness due to abrasion from wind-blown sand, insect strikes, and ice buildup is an important consideration in the choice of wind turbine rotor airfoil section shapes. Section shapes such as those developed at NREL can minimize the effect of such leading edge roughness on the turbine performance and therefore should be considered.

Problems

1. Consider two airfoils that have the same shape and angle of attack, α. Both will be operating in air. Airfoil 1 will be operating at sea level where the density is $\rho_1 = 1.2$ kg/m^3 and the kinematic viscosity is $\mu_1 = 1.8 \times 10^{-5}$ kg/m-s. The chord length of Airfoil 1 is $c_1 = 1.0$ m. The air velocity approaching Airfoil 1 is $V_1 = 210$ m/s. Airfoil 2 will be operating in frigidly cold conditions in which $\rho_2 = 3.0$ kg/m^3, $\mu_2 = 1.5 \times 10^{-5}$ kg/m-s and $V_2 = 140$ m/s.

   (a) Determine the chord length of Airfoil 2 that will produce the same chord Reynolds number as Airfoil 1.

   (b) What does matching the chord Reynolds number mean in terms of the lift, drag and moment coefficients between the two airfoils?

   (c) For similitude with regard to Mach number, what must be the ratio of the speeds of sound, a_1/a_2?

2. The local velocity along a spinning wind turbine scales as Ωr where Ω is the rotation speed in radians/sec, and r is the radial location along the rotor. Assuming a fixed rotation speed of $\Omega = 0.86$ s^{-1}, which is typical of a 70 m. long wind turbine rotor blade, then:

   (a) If near the root of the rotor, at $r = 10$ m, the airfoil chord length was 2 m. what is the chord length at $r = 60$ m. on the rotor to have the same chord Reynolds number?

   (b) If the air flow angle increases radially along the rotor, how must the angle of attack chamge from $r = 10$ m. to $r = 60$ m. so that the lift at the two locations is the same?

Note

1 Note that lower case characters are used to denote the lift, drag and moment. This is the convention when considering only a 2-D (infinite span) rotor. The 3-D, finite-span, quantities would be denoted by upper case characters, L, D, and $M_{c/4}$.

References

1. I. Abbott, A. Von Doenhoff, E. Albert, and L. Stivers, Jr., "Summary of airfoil data", NACA Report No. 824, 1945.
2. I. Abbott and A. Von Doenhoff, *Theory of Wing Sections*, Dover Publications, New York, NY, 1959.
3. J. Tangler, "The evolution of rotor and blade design", NREL/CP-500-28410, July 2000.
4. S. Dahl and P. Fuglsang, "Design of the wind turbine airfoil family RIS -A-XX", Ris-R-1024 (EN), Ris National Laboratory, Roskilde, Denmak, December 1998.
5. A. Timmer and R. van Rooij, "Summary of the delft university wind turbine dedicated airfoils", *ASME—Journal of Solar Energy Engineering*, 2003, 125:488–496.
6. A. Bjorck, "Coordinates and calculations for the FFA-W1-xxx, FFA-W2-xxx, and FFA -W3-xxx Series of Airfoils for Horizontal Axis Wind Turbines", Report FFA TN 1990-15, Stockholm, Sweden, 1990.
7. J. Tangler and D. Somers, "NREL airfoil families for HAWTs", NREL TP-442-7109, January 1995.

4

Aerodynamic Performance of a Wind Turbine Rotor

This chapter deals with the aerodynamic analysis of a horizontal wind turbine. It begins by considering conservation of momentum across a rotor disk that leads to a prediction for the maximum energy that can be extracted from the wind that is attributed to Betz. It then utilizes blade element momentum theory that includes sources of power loss. This chapter culminates in a sample rotor design that utilizes the blade element modeling.

4.1 Momentum Theory

Early researchers such as Rankine [1] and Froude [2] published papers for evaluating the performance of marine propellers. In their work, the marine propeller was replaced with a hypothetical "actuator disc". The basic idea of this concept can also be applied to the analysis of wind turbines. In the following sections, actuator disk and momentum theory will be used to develop some simple equations that will provide an understanding of how wind turbines extract energy from an air stream.

A useful concept in the study of a steady flowfield is that of a *streamline*. A streamline is an imaginary line in a steady (time-independent) flowfield where at every point along the line, the velocity vector is tangent to the streamline. Therefore, the velocity normal to a streamline must be zero.

Another useful concept that applies to a steady flow is a *stream-tube*. A stream-tube is a surface made up of streamlines. Therefore by definition, the stream-tube can have no flow entering or exiting through its surface. As a result, the mass flow rate is the same at any cross-section within the stream-tube. Examples of stream-tubes that relate to wind turbines are shown in Figure 4.1. The example shown at the top of Figure 4.1 shows a stream-tube that encompasses a three-bladed wind turbine. In the lower part of Figure 4.1, the wind turbine rotor is replaced by an actuator disk. The assumptions made in employing actuator disc theory are summarized in Table 4.1.

The object of the wind turbine is to extract energy from the air stream. This occurs at the actuator disk location. The energy that is extracted is

DOI: 10.1201/9781003611486-4

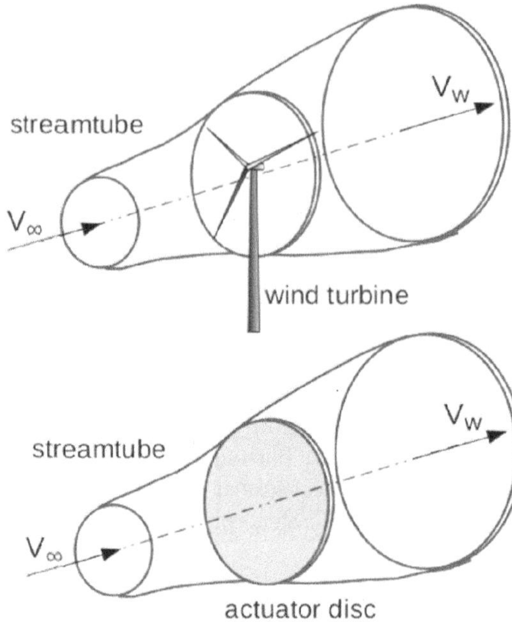

FIGURE 4.1
Flow field of a wind turbine and actuator disc.

TABLE 4.1
Properties of the Actuator Disk

1.	The flow is perfect fluid, steady, and incompressible.
2.	The actuator disc models the turbine blades, and the disc extracts energy from the flow.
3.	The actuator disc creates a pressure discontinuity across the disc.
4.	The flow is uniform through the disc and in the wake.
5.	The disc does not impart any swirl to the flow. The influence of wake rotation will be added later in this chapter.

reflected as a static pressure drop in the flowfield across the actuator disk. Following the stream-tube concept, the mass flow is constant through the actuator disk. Therefore following the Bernoulli principle which is consistent with the steady, incompressible flow assumptions, the drop in the static pressure across the actuator disk results in a reduction in the velocity in the stream-tube. Since the mass flow is constant, this requires the cross-sectional area of the stream-tube to increase the downstream of the actuator disk, which is depicted in Figure 4.1. The cross-sectional area of the stream-tube portion that is downstream of the actuator disc will continue to expand until the static pressure recovers to the free-stream static pressure. The changes in the flow properties through the stream-tube are depicted in Figures 4.2 and 4.3.

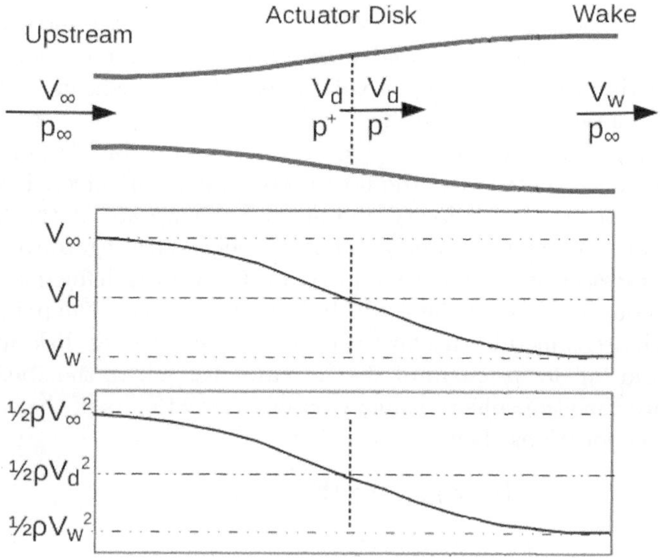

FIGURE 4.2
Variation of the velocity and dynamic pressure through the stream-tube [3].

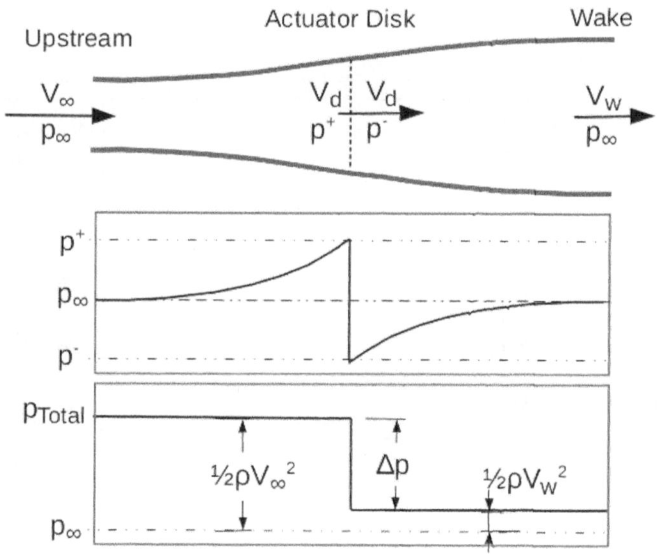

FIGURE 4.3
Variation of the static and total pressure along the steam-tube [3].

Having defined the properties of the stream-tube, it is now possible to develop expressions for thrust and power coefficients for the actuator disk. For this, a cylindrical fluid control volume is defined as shown in Figure 4.4. The control volume shown by the dashed-line box represents a 2-D section that encompasses a 2-D axial cross-section of the stream tube. At some distance upstream of the actuator disc, the flow properties are unaffected by the disc. At that location, the air velocity and static pressure are at the free-stream condition and denoted as V_∞ and p_∞, respectively. With regard to the stream-tube, the cross-section area at this upstream location is defined as A_∞.

Mass is conserved everywhere within the stream-tube. The points of interest within the stream-tube are far upstream of the actuator disk, at the actuator disk, and far downstream of the actuator disc where the static pressure again reaches the atmospheric value. Conservation of mass in the stream-tube at these three locations then requires that

$$(\rho AV)_\infty = (\rho AV)_d = (\rho AV)_w, \tag{4.1}$$

where the subscripts d and w respectively correspond to the actuator disk and far wake locations, and ρ is the air density. For wind turbine applications, it is a good assumption that the flow is incompressible, i.e., ρ is constant. Thus the mass continuity equation simplifies to

$$(AV)_\infty = (AV)_d = (AV)_w. \tag{4.2}$$

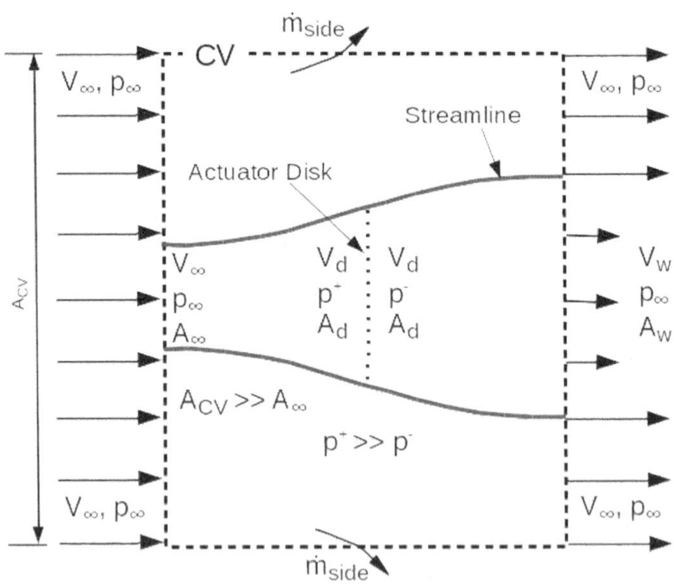

FIGURE 4.4
Cylindrical control volume surrounding the stream-tube.

As indicated in Figure 4.2, the flow in the stream-tube undergoes a pressure drop across the actuator disk. The pressure on either side of the actuator disk is denoted as p^+ and p^-. The air velocity through the actuator disk is constant and denoted as V_d. Again based on the Bernoulli principle applied from far upstream to just in front of the actuator disc, the relation between the velocity and static pressures at the two locations in the stream-tube is

$$p_\infty + \frac{1}{2}\rho V_\infty^2 = p^+ + \frac{1}{2}\rho V_d^2. \tag{4.3}$$

Similarly from the location just downstream of the actuator disc to far downstream the relation is

$$p^- + \frac{1}{2}\rho V_d^2 = p_\infty + \frac{1}{2}\rho V_w^2. \tag{4.4}$$

Considering the control volume that was shown in Figure 4.4, the difference in the momentum from the inlet to the exit is due solely to the flow exiting the stream-tube. Although the pressure at the exit of the stream tube is atmospheric, p_∞, the velocity, V_w, is lower due to the expanded cross-section area. The cross-section area at the exit of the stream tube is denoted as A_w. This is a portion of the total exit area of the control volume which is denoted as A_{CV}.

The sum of the axial (x) forces acting on the control volume is equal to the change in the axial momentum through the control volume, therefore

$$\rho V_w^2 A_w + \rho V_\infty^2 \left[A_{CV} - A_w\right] + \dot{m}_{\text{side}} V_\infty - \rho V_\infty^2 A_{CV} + \sum F_x = 0, \tag{4.5}$$

where $\dot{m}_{\text{side}}$ is the mass flow that passes through the surfaces of the control volume. Considering that the axial force acting on the control volume is equivalent to a thrust force, T, then

$$-T = \rho V_w^2 A_w + \rho V_\infty^2 \left[A_{CV} - A_w\right] + \dot{m}_{\text{side}} V_\infty - \rho V_\infty^2 A_{CV}. \tag{4.6}$$

In order that mass through the control volume be conserved, the mass flow rate through the upper and lower surfaces of the control volume must equal the difference between the mass flow rates between the inlet and the exit, or

$$\dot{m}_{\text{side}} = \rho A_{CV} V_\infty - \rho A_w V_w - \rho \left[A_\infty - A_w\right]. \tag{4.7}$$

Simplifying Equation 4.7 for $\dot{m}_{\text{side}}$ yields,

$$\dot{m}_{\text{side}} = \rho A_w \left[V_\infty - V_w\right]. \tag{4.8}$$

Substituting Equation 4.8 into Equation 4.6 and applying mass continuity in the stream-tube, namely Equation 4.2, provides an expression for the thrust given by Equation 4.9.

$$T = \rho A_w \left[V_\infty V_w - V_w^2\right] = \rho A_w V_w \left[V_\infty - V_w\right] = \rho A_d V_d \left[V_\infty - V_w\right]. \tag{4.9}$$

The thrust can also be expressed in terms of the pressure drop across the actuator disc times the area of the disc namely

$$T = \Delta p A_d. \tag{4.10}$$

Equating Equations 4.3 and 4.4, and rearranging terms yields an expression for the pressure drop across the actuator disc, namely

$$\Delta p = \frac{1}{2}\rho \left[V_\infty^2 - V_w^2\right]. \tag{4.11}$$

The thrust acting on the actuator disc can then be expressed as

$$T = \frac{1}{2}\rho A_d \left[V_\infty^2 - V_w^2\right]. \tag{4.12}$$

Equating Equations 4.12 and 4.9 results in a relationship between the velocity at the actuator disc, V_d, the free-stream velocity, V_∞, and the velocity in the wake, V_w, namely

$$V_d = \frac{1}{2}\left[V_\infty + V_w\right]. \tag{4.13}$$

A new parameter is then introduced that provides a measure of how the air velocity approaching the actuator disk has changed from its original velocity, V_∞, to V_d. This new parameter is called the *axial induction factor* which is denoted as a and defined as

$$a = \frac{V_\infty - V_d}{V_\infty}. \tag{4.14}$$

Given this, the velocity at the actuator disc, V_d, can be expressed in terms of the axial induction factor, a, namely

$$V_d = V_\infty \left[1 - a\right]. \tag{4.15}$$

Based on Equation 4.13, the wake velocity, V_w, can also be expressed in terms of the induction factor, a, namely

$$V_w = V_\infty \left[1 - 2a\right]. \tag{4.16}$$

Equations 4.15 and 4.16 illustrate that half of the axial air velocity loss occurs upstream of the actuator disc and that the other half occurs downstream in the wake region when the static pressure has returned to the upstream value, p_∞. Substituting Equations 4.15 and 4.16 into Equation 4.9 and rearranging of terms, leads to an expression for the thrust in terms of the far upstream air veocity, V_∞, the actuator disc area, A_d, and the axial induction factor, a, namely

$$T = 2\rho A_d V_\infty^2 a \left[1 - a\right]. \tag{4.17}$$

Defining a thrust coefficient as

$$C_T = T/\left[\frac{1}{2}\rho A_d V_\infty^2\right], \tag{4.18}$$

and substituting an expression for the thrust from Equation 4.17 yields an expression for the thrust coefficient that is only a function of the axial induction factor, a, namely

$$C_T = 4a\left[1 - a\right]. \tag{4.19}$$

The power extracted from the air stream by the actuator disc is equal to the produce of the thrust, T, and the wind velocity at the actuator disc, V_d, namely

$$P = TV_d. \tag{4.20}$$

Combining Equations 4.20 and 4.15, respectively, for the thrust and velocity at the actuator disc, gives the power that is extracted from the air stream in terms of the axial induction factor, namely

$$P = 2\rho A_d V_\infty^3 a\left[1 - a\right]^2. \tag{4.21}$$

The power coefficient, C_P, is defined as the ratio of the power extracted from the air stream, P, and the available power in the air stream, thus

$$C_P = P/\left[\frac{1}{2}\rho A_d V_\infty^3\right]. \tag{4.22}$$

Substituting for P from Equation 4.21 gives the power coefficient as a function of the axial induction factor, namely

$$C_P = 4a\left[1 - a\right]^2. \tag{4.23}$$

The maximum thrust and power coefficients, C_T and C_P respectively, can be determined by taking the derivative with respect to the axial induction factor, a, and then setting the resulting expressions to zero. The following is the determination of the maximum thrust coefficient.

$$\begin{aligned}\frac{dC_T}{da} &= \frac{d}{da}\left[4a(1-a)\right] \\ &= 4 - 8a \equiv 0,\end{aligned} \tag{4.24}$$

therefore

$$a = 1/2,$$

and

$$C_{T_{\max}} = 1.$$

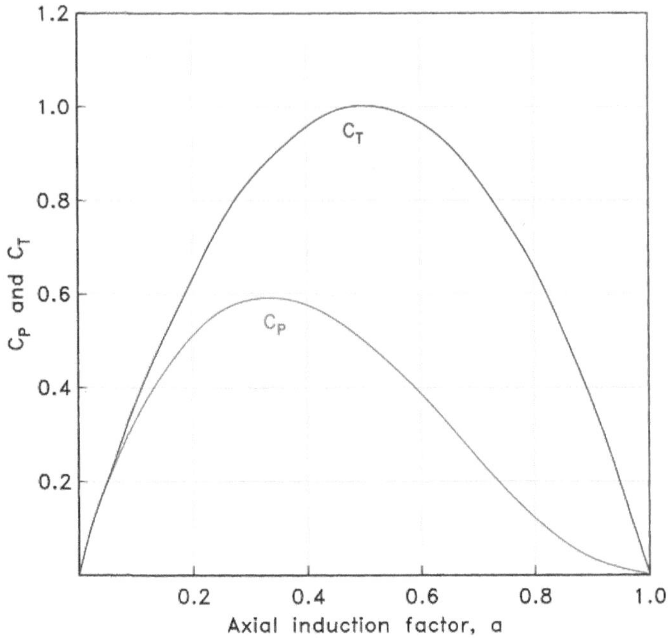

FIGURE 4.5
Variation of the rotor thrust and power coefficients, C_T and C_P, with the axial induction factor, a.

The maximum power coefficient is obtained in a similar manner.

$$\frac{dC_P}{da} = \frac{d}{da}\left[4a(1-a)^2\right] \tag{4.25}$$
$$= 1 - 4a + 3a^2 \equiv 0,$$

therefore

$$a = [1, 1/3],$$

and

$$C_{P_{\max}} = \frac{4}{3}\left[1 - \frac{1}{3}\right]^2$$
$$= \frac{16}{27},$$

or,

$$C_{P_{\max}} = 0.593.$$

Plots of C_T and C_P as functions of the axial induction factor, a are presented in Figure 4.5. These illustrate their maximums at $a = 1/2$ for $C_{T_{\max}}$, and at $a = 1/3$ for $C_{P_{\max}}$.

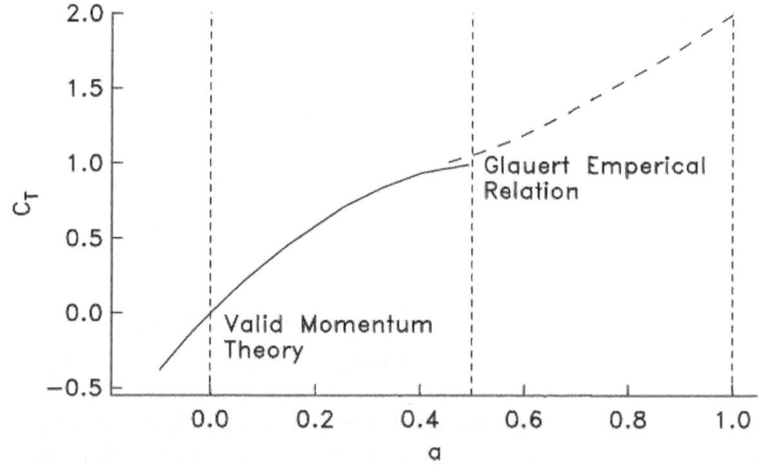

FIGURE 4.6
Thrust coefficient as a function of axial induction factor, a, indicating a valid range for momentum theory.

The maximum theoretical power coefficient, $C_{P_{\max}} = 0.593$, is often referred to as the Betz limit after Albert Betz [4], who published this finding in 1920. While Betz is given credit for identifying the theoretical maximum, several other researchers published papers citing the same conclusion. Further insight into the early work of Betz along with his contemporaries, Lanchester and Joukowsky, was provided by Van Kuik [5].

Figure 4.6 illustrates how the thrust coefficient varies as a function of the axial induction factor, a, for various rotor operation states. The operation states include, from left to right, a propeller, a wind turbine or windmill, a turbulent wake, a vortex ring, and propeller braking. When the axial induction factor is $a < 0.4$, momentum theory agrees with the experimentally obtained thrust coefficient, C_T. However, when $a > 0.4$, momentum theory breaks down. In the turbulent wake state, the flow is unsteady and characterized by large vortical structures that violate the assumptions used with actuator disc theory. The vortex ring state is of interest to helicopters during descent. The last state is the propeller brake state. This is used by propeller aircraft to apply reverse thrust to reduce landing distance during ground roll.

Example 4.1

The following figure shows a stream-tube/actuator disc model of a wind turbine. Assume that the actuator disc has a radius of 3 m. and a freestream wind speed of $V_\infty = 7$ m/s.

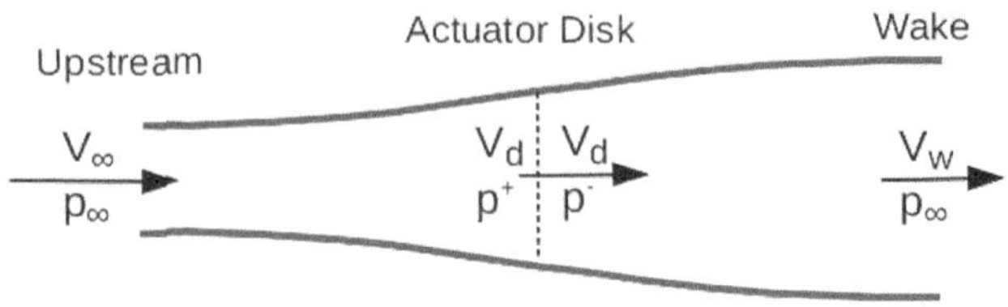

1. Estimate the maximum power that can be extracted by the idealized wind turbine.
2. Determine the velocity at the actuator disc and in the wake.
3. Determine the areas, A_∞ and A_w.

1. The power extracted by the actuator disc is given by Equation 4.22, and the maximum power coefficient, $C_{P_{max}} = 0.593$, is given by the Betz limit given in Equation 4.25.

   Therefore knowing the actuator disc radius, the disk area is

$$A_d = \pi R^2 = \pi 3^2 = 28.27\,\text{m}^2.$$

   The power extracted by the actuator disc is then

$$C_P = P/\left[\frac{1}{2}\rho A_d V_\infty^3\right],$$

   so that

$$P = 0.593(0.5)(1.22\,\text{kg/m}^3)(7\,\text{m/s})^3(28.27\,\text{m}^2) = 3.51\,\text{kW}.$$

2. The velocity at the actuator disc, V_d, and in the wake, V_w, can be calculated from Equations 4.15 and 4.16, respectively. Since the power coefficient is a maximum, then $a = 1/3$ so that

$$V_d = V_\infty\,[1 - a] = 7\,\text{m/s}\left[1 - \frac{1}{3}\right] = 4.667\,\text{m/s},$$

   and

$$V_w = V_\infty\,[1 - 2a] = 7\,\text{m/s}\left[1 - \frac{2}{3}\right] = 2.333\,\text{m/s}.$$

3. The areas A_∞ and A_w can be calculated using the continuity equation, Equation 4.2, namely,

$$(AV)_\infty = (AV)_d = (AV)_w.$$

Therefore,

$$A_\infty = \frac{A_d V_d}{V_\infty} = (28.27\,\mathrm{m}^2)(4.667\,\mathrm{m/s})/(7\,\mathrm{m/s}) = 18.85\ m^2,$$

and

$$A_w = \frac{A_d V_d}{V_w} = (28.27\,\mathrm{m}^2)(4.667\,\mathrm{m/s})/(2.333\,\mathrm{m/s}) = 56.55\,\mathrm{m}^2.$$

In this example, we see that the velocity of the wind in the wake, V_w, has been reduced to 1/3 of the ambient wind speed, V_∞, and the area of the wake, A_w is three times as large as that of the stream tube far upstream of the actuator disc, A_∞, or twice the cross-sectional area of actuator disc, A_d.

4.2 Momentum Theory with Wake Rotation

In this section we will modify the momentum analysis to allow the actuator disc to impart rotation to the flow downstream of the disc. This analysis is based upon H. Glauert's analysis [6]. It is assumed that the flow upstream of the actuator disc is not affected by the disc rotation. Immediately behind the actuator disc, a tangential flow is imparted to the downstream wake as illustrated in Figure 4.7. The tangential flow is represented by the expression $2\Omega r a'$ where a' is the angular induction factor defined as

$$a' = \frac{\omega}{2\Omega}, \tag{4.26}$$

where Ω is the angular velocity of the rotor disk, and ω is the angular velocity imparted to the wake. It is assumed that the wake rotation is much smaller the rotational velocity of the actuator disc, i.e. $\omega \ll \Omega$.

Glauert developed expressions for both the differential thrust and torque across the rotating actuator disc. The differential thrust on an annular ring of the actuator disc can be expressed as

$$dT = \Delta p(2\pi r dr) = \left[\rho\left(\Omega + \frac{\omega}{2}\right)\omega r^2\right] 2\pi r dr. \tag{4.27}$$

If the definition of the angular induction factor is substituted into Equation 4.27, and upon rearranging the expression, one obtains

$$dT = 4a'\left(1 + a'\right)\frac{1}{2}\rho\Omega^2 r^2\left(2\pi r dr\right). \tag{4.28}$$

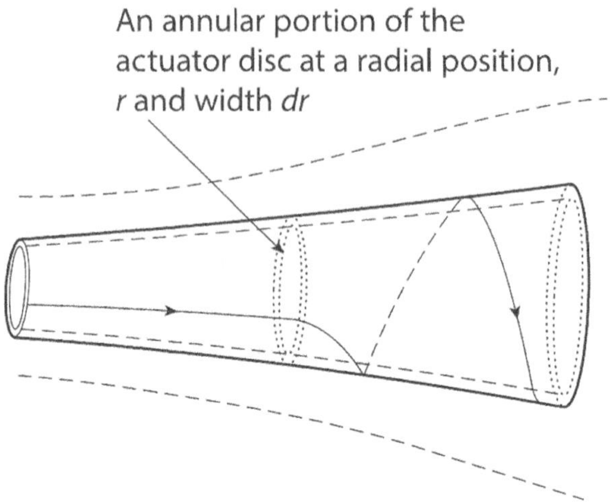

FIGURE 4.7
Schematic of the induced rotation of the flow downstream of the rotating actuator disc [7].

The thrust obtained with no wake rotation was given by Equation 4.17. This equation can be written in differential form as

$$dT = 2\rho V_\infty^2 a(1-a)(2\pi r dr). \tag{4.29}$$

Equating Equations 4.28 and 4.29, yields the following relation

$$\frac{a(1-a)}{a'(1+a')} = \lambda_r^2, \tag{4.30}$$

where λ_r is called the local speed ratio, and is defined as the ratio of the local angular velocity at a given radial position on the disc, divided by the free stream velocity, namely

$$\lambda_r = \frac{\Omega r}{V_\infty}. \tag{4.31}$$

Equation 4.30 is a useful relationship between the induction factors and λ_r. An important performance parameter for a wind turbine is the tip-speed-ratio, λ in which from Equation 4.31, $r = R$, namely the rotor disk radius. Therefore, the rotor tip-speed-ratio is

$$\lambda = \frac{\Omega R}{V_\infty}. \tag{4.32}$$

Applying conservation of angular momentum yields an equation for the differential torque acting on an angular ring at radius r of the actuator disc that is given by Equation 4.33.

$$dQ = d\dot{m}\omega r^2 = \rho V_d\,(2\pi r dr)\,\omega r^2. \tag{4.33}$$

Substituting V_d from Equation 4.15, and ω from Equation 4.26, into Equation 4.33 yields the following equation for the differential torque.

$$dQ = 2a'(1-a)\rho V_\infty \Omega r^2 (2\pi r dr). \tag{4.34}$$

The differential power, $dP = \Omega dQ$ is then

$$dP = 2a'(1-a)\rho V_\infty \Omega^2 r^2 (2\pi r dr). \tag{4.35}$$

If we equate the differential power with wake rotation given by Equation 4.35 to the differential power with no wake rotation given by Equation 4.21, another useful relationship between the axial and rotational induction factors a and a' can be developed, namely

$$\underbrace{2a'(1-a)\rho V_\infty \Omega^2 r^2 (2\pi r dr)}_{\text{with rotation}} = \underbrace{2a(1-a)^2 \rho V_\infty^3 (2\pi r dr)}_{\text{without rotation}}. \tag{4.36}$$

Simplifying Equation 4.36 yields

$$a(1-a) = a'\lambda_r^2. \tag{4.37}$$

Returning to Equation 4.35, the incremental power coefficient for an annular ring is

$$dC_P = \frac{dP}{\frac{1}{2}\rho V_\infty^3 A_d}. \tag{4.38}$$

Substituting the dP in Equation 4.38 gives the following Equation 4.39.

$$\begin{aligned} dC_P &= \frac{2a'(1-a)\rho V_\infty \Omega^2 r^2 (2\pi r dr)}{\frac{1}{2}\rho V_\infty^3 \pi R^2} \\ &= \frac{\left[8a'(1-a)\lambda_r^2 r dr\right]}{R^2}. \end{aligned} \tag{4.39}$$

Introducing the variable, μ, that is defined to be the ratio of the local radius, r, to the radius of the actuator disc, R, such that

$$\mu = \frac{r}{R}, \tag{4.40}$$

and

$$d\mu = \frac{dr}{R}, \tag{4.41}$$

then Equation 4.39 can be integrated with respect to μ to give an expression for the power coefficient, namely

$$C_P = 8\int_0^1 a'(1-a)\lambda_r^2 \mu d\mu. \tag{4.42}$$

Example 4.2

Determine the conditions on the axial induction factor that maximize the power coefficient given in Equation 4.42.

1. To determine the maximum power coefficient, we need to maximize the integrand in Equation 4.42. This can be accomplished by taking the derivative of the integrand with respect to either one of the induction factors, a or a', and setting that function equal to zero to obtain the maximum value of the selected induction factor. Therefore

$$\frac{d}{da'}\left[8a'(1-a)\lambda_r^2\mu\right]=0,$$

then

$$8\lambda_r^2\mu\left[1-a-a'\frac{da}{da'}\right]=0,$$

which yields

$$\frac{da}{da'}=\frac{1-a}{a'}.$$

2. If we differentiate Equation 4.37 with respect to d/da', then

$$\frac{da}{da'}=\frac{\lambda_r^2}{1-2a}.$$

Equating the two expressions for d/da' gives

$$\frac{1-a}{a'}=\frac{\lambda_r^2}{1-2a},$$

or rearranging terms

$$\lambda_r^2a'=(1-a)(1-2a).$$

3. Now substituting for λ_r^2a' from Equation 4.37 gives the following

$$a(1-a)=(1-a)(1-2a).$$

Solving for a we obtain

$$a=1/3,$$

which is the same for $C_{P_{\max}}$ without rotation!

4. Equation 4.37 can be rearranged so that a' is a function of the axial induction factor, a, and the local speed ratio, $\lambda_r=\Omega r/V_\infty$, namely

$$a'=\frac{a(1-a)}{\left(\frac{\Omega r}{V_\infty}\right)^2}=\frac{a(1-a)}{\lambda^2\frac{r}{R}}.$$

The above equation shows that the angular induction factor is large near the center of the disk. However if we assume a value for tip speed ratio of $\lambda = 7$ and the ideal axial induction factor, $a = 1/3$, the following values of a' show a decrease in magnitude with increasing radial position on the rotor.

r/R	a'
0.2	0.1134
0.4	0.0283
0.5	0.0181
0.6	0.0126
0.7	0.0093
0.8	0.0071
0.9	0.0056
1.0	0.0045

4.3 Blade Element Momentum (BEM) Theory

Actuator disc theory provides us with simple formulas to calculate the power extracted and thrust acting on the wind turbine rotor. The theory also provided a theoretical limit of how much power can be extracted from the air stream (wind). However to design a new wind turbine rotor, we need a different methodology that can predict the performance of the wind turbine rotor as a function of the blade design parameters such as the rotor radius, number of rotor blades, and the rotor blade geometry, including the radial variation in the airfoil shape parameters.

A sketch of the cross-section of a wind turbine blade at various radial positions is shown in Figure 4.8. This illustrates the variation in the section chord length and blade twist at selected radial locations along the blade. In this example, the airfoil sectional shape remained the same, although it often will vary along the radial span of the rotor from root to tip. The motivation behind the rotor aerodynamic design is to optimize its performance and thereby maximize the power generated by the wind turbine.

We now turn our attention towards developing the equations for the differential thrust, torque, and power developed by the aerodynamic forces generated on the turbine blades. Figure 4.9 shows an illustration of the airfoil section at some radial distance from the axis of rotation of a wind turbine rotor.

The angle of attack of the airfoil section is the angle between the airfoil chord line and the resultant velocity the airfoil section experiences. Once the turbine begins to rotate, the resultant velocity, V_R, is made up of the vector

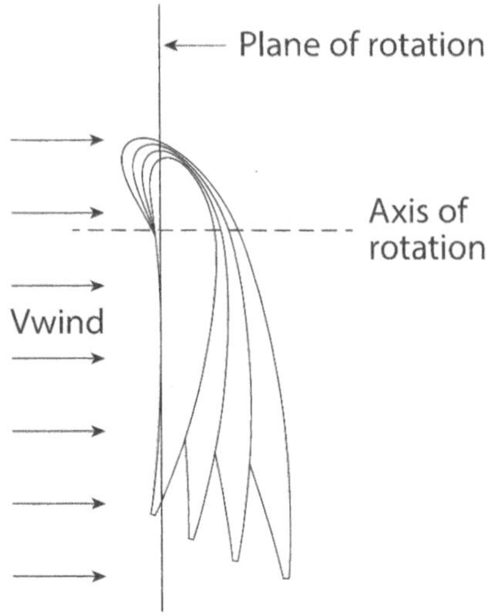

FIGURE 4.8
Example of the variation in chord and geometric twist along the radial distance of a wind turbine rotor blade.

sum of the wind speed and the rotational speed of the blade section, thus

$$V_R = \sqrt{[V_\infty(1-a)]^2 + [\Omega r(1+a')]^2}, \tag{4.43}$$

where again, Ω is the angular rotation rate of the rotor.

Both the wind speed and rotation velocities are modified by the axial and angular induction factors previously developed by applying actuator disk theory. The angle that the resultant velocity makes with respect to the plane of rotation is denoted as ϕ. Based on the geometry indicated in Figure 4.9,

$$\tan\phi = V_\infty \frac{1-a}{\Omega r(1+a')}, \tag{4.44}$$

so that

$$\phi = \tan^{-1}\left[\frac{V_\infty(1-a)}{\Omega r(1+a')}\right]. \tag{4.45}$$

As illustrated in Figure 4.8, the turbine blade must have a built-in twist distribution from the hub to the tip, so that each blade section will be at an angle of attack that is near the angle required to produce the maximum lift to drag ratio, l/d. In addition, the blade can be mounted into the hub at some desired angle that will be referred to as θ_{cp}. For a fixed pitch blade, θ_{cp} is a constant, and is usually measured as the pitch angle that the tip section of the rotor makes with the plane of rotation. In a pitch controlled wind turbine,

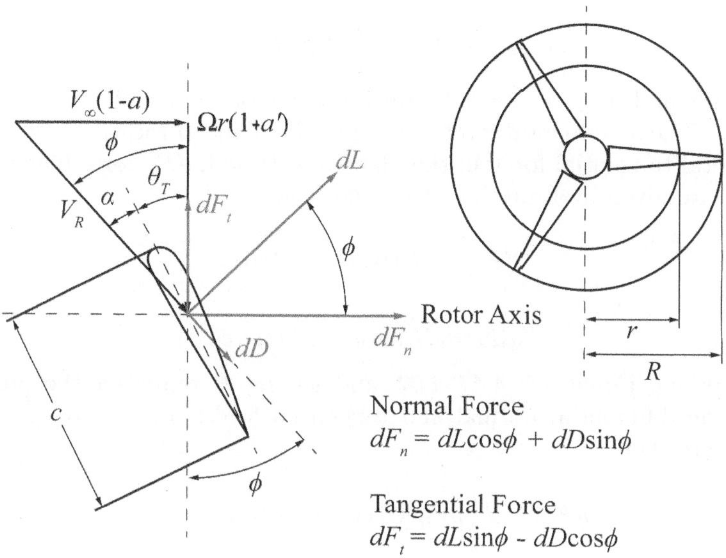

FIGURE 4.9
Illustration of the aerodynamic forces acting on a wind turbine blade section at a distance r from the axis of rotation.

θ_{cp}, is varied to control the power output of the wind turbine. This occurs between the rated and cut-out wind speeds.

The local angle of attack, α, at any radial location is the sum of the local resultant velocity vector angle, $\phi(r)$, minus the local twist angle, $\theta_T(r)$ and the pitch angle, θ_{cp}, namely

$$\alpha(r) = \phi(r) - [\theta_T(r) + \theta_{cp}]\,. \tag{4.46}$$

If the turbine blade is divided into a finite number of segments from the blade root to the blade tip, we can estimate the thrust and torque produced by each of the blade segments. The thrust force acting on a blade section acts normal to the plane of rotation of the blade. The torque on a blade section is equal to the net aerodynamic force in the plane of rotation times its distance to the axis of rotation. The normal force and tangential force on each blade segment can be expressed in terms of the lift and drag forces. The differential lift and drag forces that act on a segment of the rotor can be expressed as given in Equations 4.47 and 4.48. In these, C_l and C_d are the respective lift and drag coefficients for the particular rotor section shape, and c is the respective section chord dimension.

$$dL = C_l \frac{1}{2} \rho V_R^2 c dr, \tag{4.47}$$

$$dD = C_d \frac{1}{2} \rho V_R^2 c dr. \tag{4.48}$$

The lift and drag coefficients are functions of the airfoil section angle of attack, α. The incremental force normal to the plane of rotation, dF_n, and the incremental tangential force in the plane of rotation, dF_t, for a blade element segment are given by Equations 4.49 and 4.50.

$$dF_n = dL \cos\phi + dD \sin\phi, \tag{4.49}$$

$$dF_t = dL \sin\phi - dD \cos\phi. \tag{4.50}$$

Combining Equations 4.47–4.50, and letting B represent the number of blades, the differential normal and tangential forces for at any given radial position are then

$$dF_n = B \frac{1}{2} \rho V_R^2 \left[C_l \cos\phi + C_d \sin\phi \right] c dr, \tag{4.51}$$

and

$$dF_t = B \frac{1}{2} \rho V_R^2 \left[C_l \sin\phi - C_d \cos\phi \right] c dr. \tag{4.52}$$

To simplify these equations, we will define the normal and tangential force coefficients to be the expressions contained within the brackets in Equations 4.51 and 4.52 as C_n and C_t, respectively, so that

$$C_n = C_l \cos\phi + C_d \sin\phi, \tag{4.53}$$

and

$$C_t = C_l \sin\phi - C_d \cos\phi, \tag{4.54}$$

and therefore

$$dF_n = B \frac{1}{2} \rho V_R^2 C_n c dr, \tag{4.55}$$

and

$$dF_t = B \frac{1}{2} \rho V_R^2 C_t c dr. \tag{4.56}$$

The differential torque, $dQ = r dF_t$, and the differential power, $dP = \Omega dQ$, are then respectively

$$dQ = r dF_t = B \frac{1}{2} \rho V_R^2 C_t c r dr, \tag{4.57}$$

and

$$dP = \Omega dQ = B \Omega \frac{1}{2} \rho V_R^2 C_t c r dr. \tag{4.58}$$

The differential thrust, torque, and power are each functions of the blade section aerodynamics coefficients, which are functions of the axial and rotational induction factors, a and a'. Therefore we wish to incorporate these induction factors into the formulations for the differential thrust, torque and power. The thrust determined by momentum theory with no wake rotation that was developed in Section 4.2, can be expressed in differential form, namely

$$dT = 2\rho V_\infty^2 a(1-a)2\pi r dr. \tag{4.59}$$

The differential thrust, dT, is equivalent to the differential normal force, dF_n. Therefore equating Equation 4.17 in differential form from momentum theory, and Equation 4.55 above, we obtain

$$\underbrace{2\rho V_\infty^2 a(1-a)2\pi r dr}_{\text{Momentum Theory}} = \underbrace{B\frac{1}{2}\rho V_R^2 C_n c dr}_{\text{BEM Theory}}. \tag{4.60}$$

As illustrated in Figure 4.9, the relative velocity, V_R, can be expressed as

$$V_R = \frac{V_\infty(1-a)}{\sin\phi}. \tag{4.61}$$

Substituting Equation 4.61 into Equation 4.60 and rearranging terms then yields

$$\frac{a}{1-a} = \frac{BC_n c}{8\pi r\sin^2\phi}. \tag{4.62}$$

Now defining a new parameter, σ_r where

$$\sigma_r = \frac{Bc}{2\pi r}, \tag{4.63}$$

and rearranging Equation 4.62 leads to a useful relation for the axial induction factor, namely

$$a = \frac{1}{\frac{4\sin^2\phi}{\sigma_r C_n} + 1}. \tag{4.64}$$

In a similar manner, we can equate the differential torque equation with wake rotation from momentum theory given in the left-hand-side of Equation 4.36 with that based on BEM theory given in Equation 4.34, namely

$$\underbrace{2a'(1-a)\rho V_\infty \Omega r^2 2\pi r dr}_{\text{Momentum Theory}} = \underbrace{B\frac{1}{2}\rho V_R^2 C_t c r dr}_{\text{BEM Theory}}, \tag{4.65}$$

and from this, develop a useful relation for the angular induction factor, namely

$$a' = \frac{1}{\frac{4\sin\phi\cos\phi}{\sigma_r C_t} - 1}. \tag{4.66}$$

4.4 Prandtl's Tip Loss Factor

Before discussing how we can solve the BEM equations to predict the performance of the wind turbine rotor, we need to include a correction factor to account for aerodynamic losses near each rotor blade tip. In developing the momentum theory, the rotor was modeled as an actuator disc that represents an infinite number of blades. The loading on the actuator disc is assumed to be uniform across the disc. The blade element momentum (BEM) technique divides the rotor into radial segments and assumes that the aerodynamics associated with each segment is independent of the other segments. This is a reasonable assumption for the inboard portion of the rotor blade, however because of rotation and the finite radius of the rotor, significant interference occurs on the outboard radial portion of the blades. One example is *tip loss* which is caused by air flow that passes around the blade tip from the high pressure side to the low pressure side of the blade. A manifestation of this is the formation of a "tip vortex" such as illustrated in Figure 4.10, that originates from each of the rotor blade tips and convects in the downstream direction. A photograph [8] of such rotor tip vortices eminating from a two-bladed wind turbine in a wind tunnel is shown in Figure 4.11. The vortices were made visible by introducing smoke at a location that was upstream of the rotor, at a height where the smoke streak would intersect the rotor blade tips.

The effect of the blade tip vortices is to lower the lift and therby the generated torque, at the outboard portion of the blade. Ludwig Prandtl [9] developed an equation to estimate the blade tip losses. A detailed development of Prandtl's analysis was given by Glauert [6]. To account for the effect of the

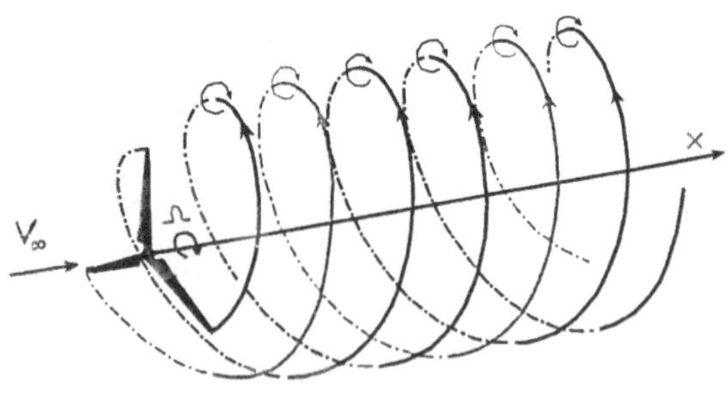

FIGURE 4.10
Illustration of rotor tip vortices from a three-bladed wind turbine rotor.

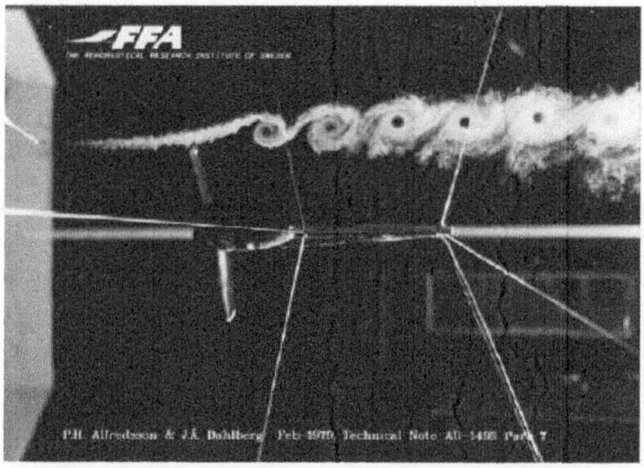

FIGURE 4.11
Photograph of the cross-section of the tip vortices from a two-bladed wind turbine that was visualized in a wind tunnel experiment [8].

tip vortices, Prandtl introduced a tip loss factor, F, given as

$$F = \frac{2}{\pi}\cos^{-1}\left(e^{-f}\right), \tag{4.67}$$

where

$$f = \frac{B}{2}\frac{R-r}{r\sin\phi}. \tag{4.68}$$

Here again, B is the number of rotor blades, r is the local radius on the rotor, R is the rotor radius, and ϕ is the local angle that the resultant velocity vector makes with the rotor disk plane of rotation at the local radius.

The tip loss factor is introduced into the differential thrust (Equation 4.59) and torque (Equation 4.34) equations such that

$$dT = 2F\rho V_\infty^2 a(1-a)2\pi r dr, \tag{4.69}$$

and

$$dQ = 2Fa'(1-a)\rho V_\infty \Omega r^2(2\pi r dr). \tag{4.70}$$

The differential torque then relates to the differential power as

$$dP = \Omega dQ. \tag{4.71}$$

As noted, the tip loss factor F is a function of the number of blades, the local radius, and the angle, ϕ, that the resultant wind velocity, V_R, makes with the airfoil section chord line. Generally for the inboard section of the rotor, $r/R \leq 0.6$, $F \simeq 1$. However on the outboard section of the rotor blade, $r/R > 0.6$, the tip loss factor has a pronounced effect. This is demonstrated in Figure 4.12 which is a plot of the tip loss factor, F, at varius radii along a wind turbine blade.

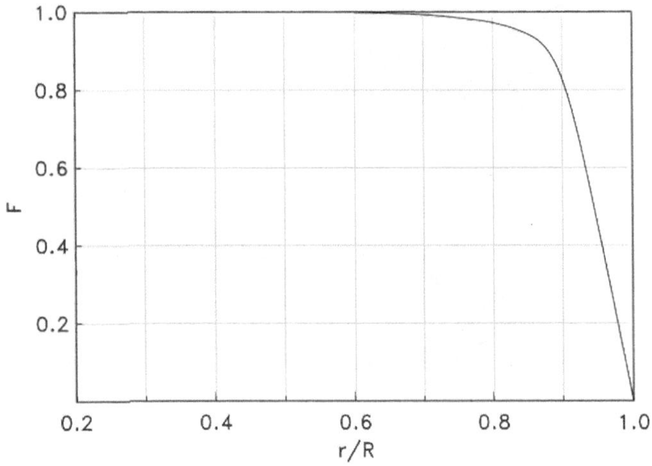

FIGURE 4.12
Prandtl tip loss factor along the span of a wind turbine rotor.

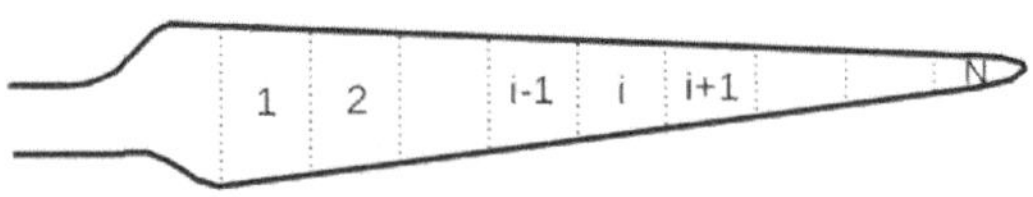

FIGURE 4.13
Example of a wind turbine blade divided into ten sections for BEM analysis.

Equating the differential momentum equation for thrust and torque including the Prandtl tip loss factor, with the corresponding differential thrust and torque equations from blade element theory yields equations for the axial and angular induction factors in terms of F, namely

$$a = \frac{1}{\frac{4F\sin^2\phi}{\sigma_r C_n} + 1}, \tag{4.72}$$

and

$$a' = \frac{1}{\frac{4F\sin\phi\cos\phi}{\sigma_r C_t} - 1}. \tag{4.73}$$

It is important to note that if Equations 4.72 and 4.73 are used for a and a' to include the Prandtl tip loss factor, then the equations for the thrust, torque, and power given in Equations 4.69–4.71 would use $F = 1$. Otherwise the effect of the tip loss would be double counted.

4.5 Solution of the BEM Equations

Now that we have established the relationship between the induction factors of momentum theory including the tip loss factor with the aerodynamic and geometric characteristics of the turbine blades the thrust, torque and power generated by the wind turbine rotor can be estimated. For a given tip speed ratio, λ, and a wind speed, V_∞, the axial and angular induction factors, a and a', respectively can be calculated. Then by dividing the turbine blade into a finite number of radial segments as shown in Figure 4.13, an iterative approach can be used to determine the axial and rotational induction factors at a given radial segment on the blade. Once the induction factors are known for that radial segment, the segment differential thrust, torque and power can be determined. This process is continued for each of the radial segments across the blade. The differential components of thrust, torque and power can then be numerically integrated (added) to obtain the *total* thrust, torque and power generated by the rotor blade. A flow chart that illustrates this approach is shown in Figure 4.14. The corresponding equations used at each step in the process is listed in Table 4.2.

TABLE 4.2
Equations Used in the Solution of the BEM Equations in Analysis of a Wind Turbine Rotor

Step 1	Divide the blade into n, spanwise segments and input the geometric blade information for each segment.
Step 2	Start at the most inboard segment.
Step 3	Set the axial and tangential induction factors, a and a' to zero.
Step 4	Compute the angles ϕ and α using Equations 4.45 and 4.46.
Step 5	Knowing the angle of attack, α, the lift and drag coefficients, C_l and C_d, can be computed from polynomial expressions that are a fit to the lift and drag coefficient data for the airfoil section shape at the given spanwise segment of the rotor.
Step 6	Calculate the normal and tangential force coefficients, C_n and C_t, from Equations 4.53 and 4.54.
Step 7	Calculate a and a' from Equations 4.64 and 4.66.
Step 8	Compare the new values of a and a' with the previous values. Does the difference meet the convergence criteria? If "No" go to Step 9 using the new values of a and a'. If "Yes" go to Step 10.
Step 9	Use the values of a and a' from Step 7 and go to Step 4.
Step 10	Calculate the differential thrust, dT, torque, dQ, and power, dP, for the blade segment using Equations 4.69–4.71. If this is the last (most outboard) blade segments go to Step 11. Otherwise move to the next blade segment and repeat the process starting at Step 3.
Step 11	Calculate the total thrust T, torque, Q, and power, P as the sum of the differential power from each of the spanwise segments.

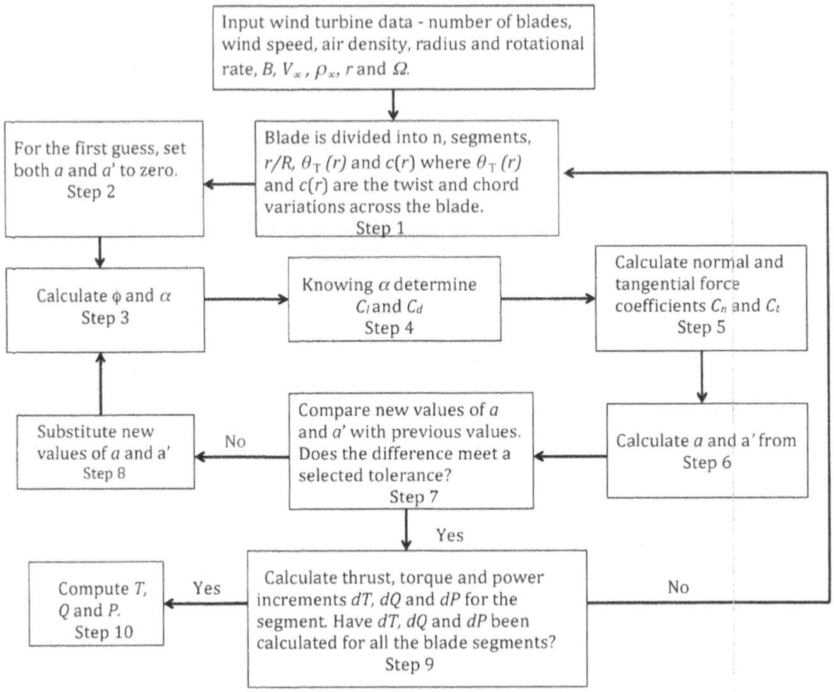

FIGURE 4.14
Flow chart for the iterative procedure used in solving the BEM equations.

4.5.1 Example BEM Equation Solution

The turbine selected for this example is one of the research wind turbines used by the Department of Aerospace and Mechanical Engineering at the University of Notre Dame in Notre Dame, Indiana. A photograph of the wind turbines, along with the companion instrumented meteorological tower, is shown Figure 4.15. These are three-bladed wind turbine that employ variable pitch control to maintain rated power.

The geometric and aerodynamic characteristics of the Notre Dame wind turbines are given in Tables 4.3 and 4.4. The spanwise chord and twist distributions listed in Table 4.4 are plotted in Figures 4.16 and 4.17. The wind turbines are designed to generate a rated electric power of 25 kW. The combined efficiency of the power train components, bearings, gearbox, generator, etc. was assumed to be $\eta = 0.9$. That is 90% of the power extracted by the rotor is converted to electrical power. A MATLAB® code, listed in Appendix, was developed based on the BEM theory outlined in the flow chart in Figure 4.14. A description of the steps for the solutions and relevant equations were listed in Table 4.2.

TABLE 4.3
Characteristics of the University of Notre Dame Wind Turbines

Re_c	0.5×10^6
$C_l(\alpha)$	$0.327 + 0.1059\alpha - 0.0013\alpha^2$
$C_d(\alpha)$	$0.006458 - 0.000272\alpha + 0.000219\alpha^2 - 0.0000003\alpha^3$
α	$-2° \leq \alpha 12°$
B	3
λ	7
R	4.953 m.
$V_{\text{cut-in}}$	3.0 m/s
V_{rated}	11.6 m/s
$V_{\text{cut-out}}$	37.0 m/s

Figure 4.18 shows the angles ϕ and θ_T as a function of the non-dimensional radial location, r/R. The difference between these angles corresponds is the aerodynamic angle of attack along the span of the rotor for the given tip-speed-ratio. The angle of attack across the span of the rotor blade varies changes by only a few degrees, and is very near the angle of attack of the maximum C_l/C_d.

The rotor spanwise variation of the induction factors a and a' are shown in Figure 4.19. The axial induction factor, a, increases slightly with increasing non-dimensional distance from the axis of rotation until approximately $r/R = 0.85$. The tangential induction factor, a', is approximately 0.05 at

FIGURE 4.15
Photograph of the University of Notre Dame Research Wind Turbines and Meteorological tower.

TABLE 4.4
Rotor Geometry of the University of Notre Dame Wind Turbines

r/R	**Chord (mm)**	**Blade Twist (°)**
0.2414	467.62	14.39
0.2835	421.45	11.89
0.3257	382.21	9.92
0.3678	349.07	8.34
0.4100	323.59	7.05
0.4521	303.19	5.98
0.4943	287.05	5.08
0.5364	274.53	4.31
0.5785	259.42	3.64
0.6207	249.51	3.07
0.6628	239.74	2.56
0.7050	230.16	2.11
0.7471	220.04	1.71
0.7893	211.77	1.34
0.8314	204.56	1.03
0.8736	200.88	0.73
0.9157	196.84	0.47
0.9579	192.37	0.22
1.0000	188.02	0

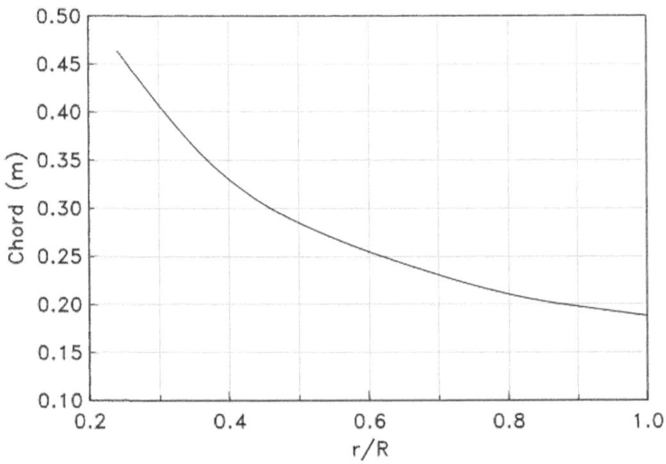

FIGURE 4.16
Blade chord distribution for the University of Notre Dame Research Wind Turbines.

the most inboard location, and decreases monotonically as the radial location approaches the blade tip.

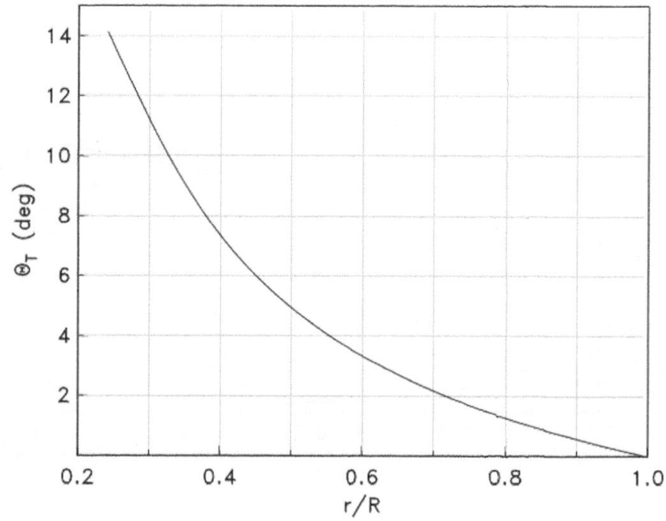

FIGURE 4.17
Blade twist distribution for the University of Notre Dame Research Wind Turbines.

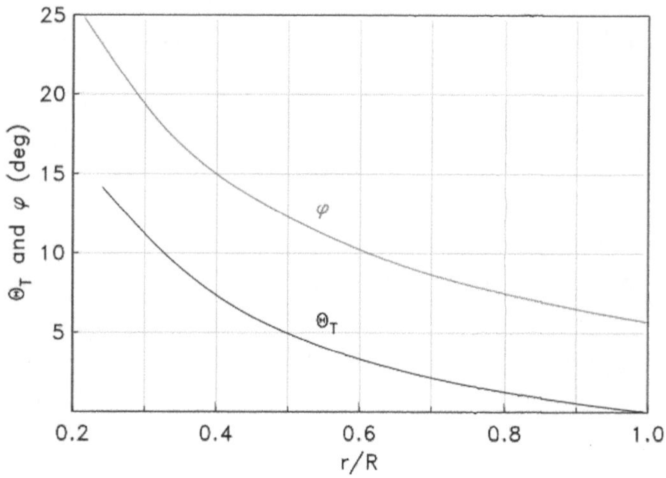

FIGURE 4.18
Spanwise distribution of the rotor blade angles ϕ and θ_T for the University of Notre Dame Research Wind Turbines.

The spanwise distribution of the lift-to-drag ratio for the rotor blade is shown in Figure 4.20. This shows the l/d-ratio to increase along the span of

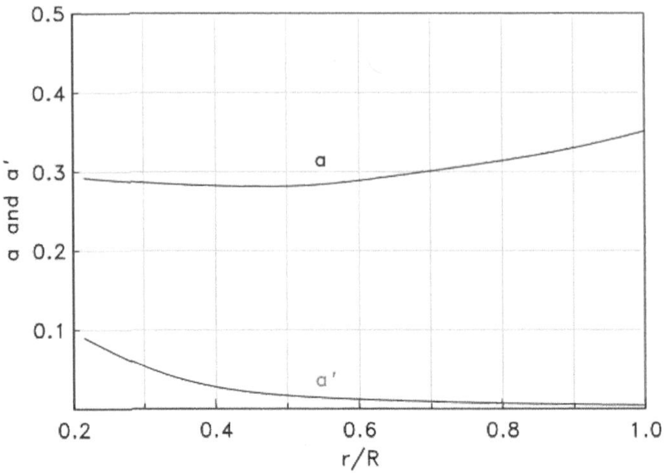

FIGURE 4.19
Spanwise distribution of the induction factors, a and a' for the University of Notre Dame Research Wind Turbines.

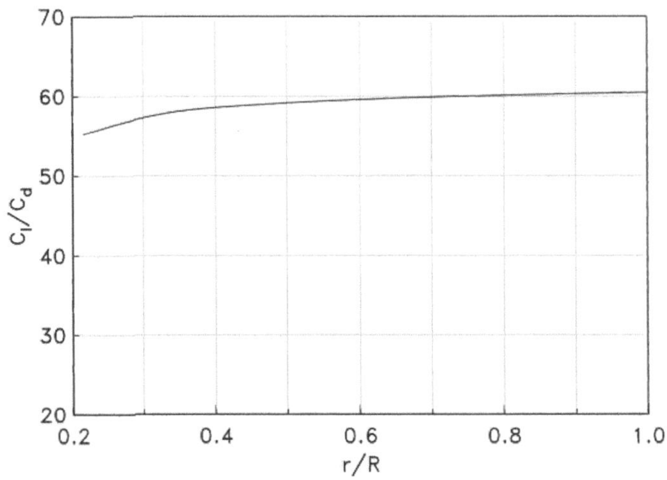

FIGURE 4.20
Spanwise distribution of the lift-to-drag ratio for the University of Notre Dame Research Wind Turbines.

the rotor, with a maximum at the rotor tip. This l/d does not account for the tip loss.

The spanwise distribution of the Prandtl tip loss factor, F, is shown in Figure 4.21. The tip loss factor has no effect on the blade loading for $r/R < 0.7$.

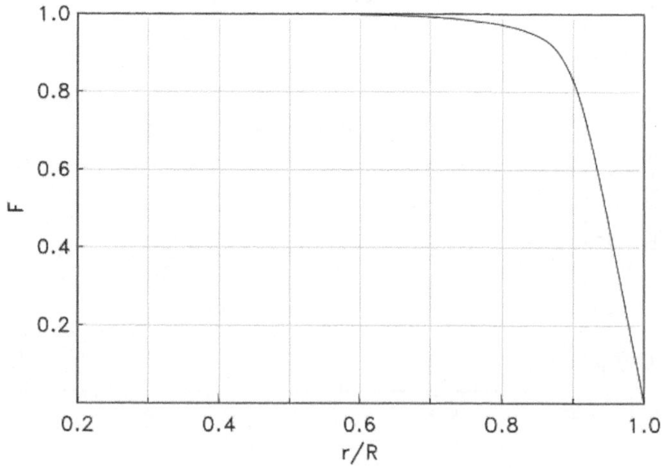

FIGURE 4.21
Spanwise distribution of the Prandtl loss coefficient for the University of Notre Dame Research Wind Turbines.

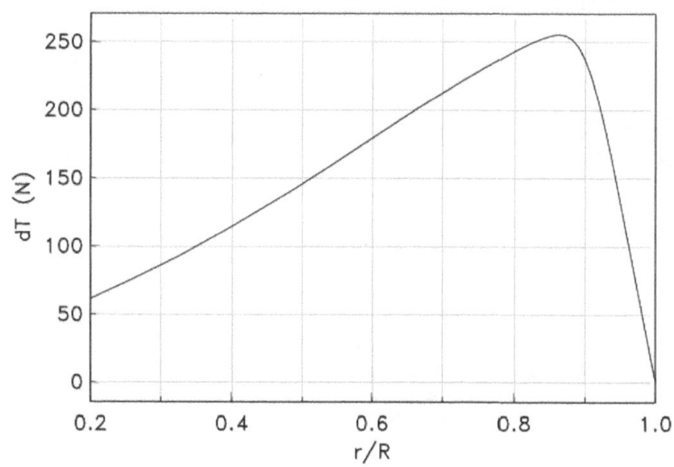

FIGURE 4.22
Spanwise distribution of the differential thrust for the University of Notre Dame Research Wind Turbines.

However it decreases rapidly further outboard, reaching a value of 0.65 close to the rotor tip. This will have the effect of lowering the torque produced by the rotor near the tip.

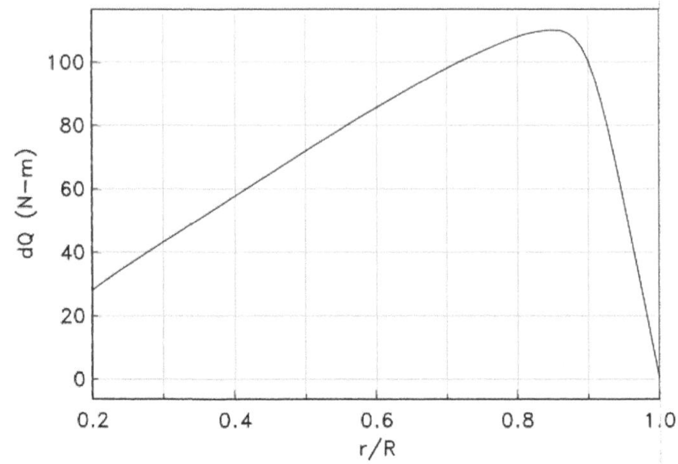

FIGURE 4.23
Spanwise distribution of the differential torque for the University of Notre Dame Research Wind Turbines.

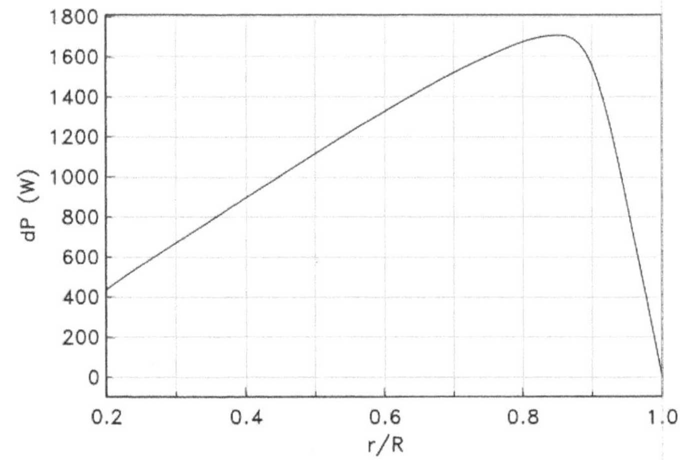

FIGURE 4.24
Spanwise distribution of the differential power for the University of Notre Dame Research Wind Turbines.

The spanwise distribution of the differential thrust and torque for the rotor blade is shown in Figures 4.22 and 4.23. The respective areas under the two curves yield the total thrust force produced by the rotor, and the torque transmitted to rotor rotation axis. The thrust coefficient and thrust force for

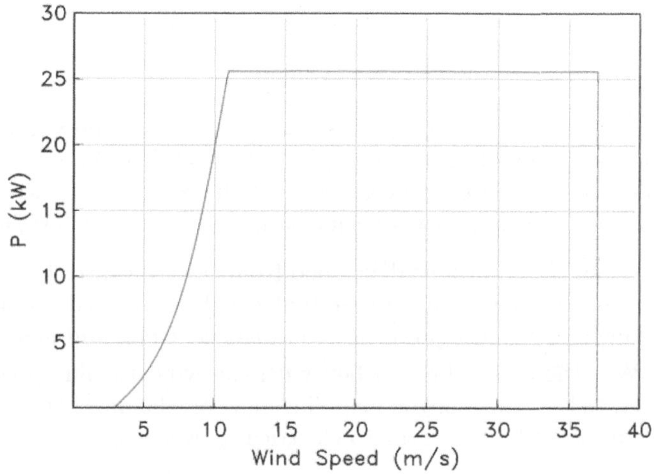

FIGURE 4.25
Power curve for the University of Notre Dame Research Wind Turbines.

the *three* blades was found to be $C_T = 0.70$ and $T = 3,984$ N. The torque delivered to the rotor shaft was found to be $Q = 1,827$ N-m.

Finally, the spanwise distribution of the differential power for the rotor blade is presented in Figure 4.24. Integrating the area under the differential power curve yields the total power generated by the wind turbine. The power coefficient and power generated for the three blades was found to be $C_P = 0.45$ and $P = 28,416$ N-m/s. The conversion of the mechanical power generated by the wind turbine into electric power involves the efficiency of the bearings, gear-box, and generator. These were stated to be a combined efficiency of $\eta = 0.90$. Therefore the *electric* power deliver to the power grid is equal to the mechanical power times the efficiency of the power train components, namely $(\eta)(P) = (0.9)(28.4) = 25.6$ kW.

This example corresponds to the conditions at the rated wind speed, and optimum tip-speed-ratio. The power for the started-up wind-speed region can be determined for example, if we assume the tip-speed-ratio remains constant and optimum. The can similarly be computed at other wind speeds in order to build up the power versus wind speed that is shown in Figure 4.25. The power generated for $V_{\text{rated}} \leq V_\infty \leq V_{\text{cut-out}}$ is maintained to be constant by reducing the pitch of the rotor. This is performed by the wind turbine power control system. When $V_{\text{cut-out}}$ is reached, the control system reduces the blade pitch to the point where no lift (torque) is generated, and applies breaking load to the generator to stop the rotor from rotating. The computed power characteristics of the University of Notre Dame wind turbines agrees well with the experimentally measured characteristics presented by Cooney [10].

Problems

1. The owner of a factory has purchased a three-bladed wind turbine having a radius of $R = 6.62\,\text{m}$. He was told that the wind turbine would generate 50 kW of electricity at the sites average wind speed of 11.2 m/s. Assume the density of air to be $\rho = 1.225\,\text{kg/m}^3$.

   (a) Using the equations developed from actuator disc theory determine the maximum power that can be produced by the wind turbine. Neglect power losses in the gear-box and generator.

   (b) Will the owner be satisfied with the wind turbine power generated if the conversion efficiency of the mechanical power produced by the rotor to electrical power is $\eta = 0.85$?

2. A new wind turbine company claims that their new three bladed wind turbine with a rotor radius of $R = 9.0\,\text{m}$ can produce 100 kW of power at the rated wind speed of 12 m/s. Use a simple analysis to determine whether this design can actually produce 100 kw.

   Assume the following:

   (a) the density of air is $\rho = 1.225\,\text{kg/m}^3$, and

   (b) the conversion efficiency of the mechanical power produced by the rotor to electrical power is $\eta = 0.85$.

3. The object of this problem is to determine the power generated by a designated segment of a 3-bladed wind turbine rotor with a blade radius of 4.953 m. The tip-speed-ratio for the wind turbine is $\lambda = 7$, and the rated wind speed is $V_{\text{rated}} = 11.62$ m/s.

   The segment of interest is at $r/R = 0.5$ on the rotor and has the following characteristics:

   - NACA 4415 airfoil
   - $C_L = 0.368 + 0.0942\alpha$
   - $C_D = 0.00994 + 0.000259\alpha + 0.0001055\alpha^2$
   -

r/R	r(m)	c(m)	θ_T (deg)
0.5	2.447	0.259	7.4

   - The blade pitch is fixed at an angle of $\theta_{cp} = -2°$.

   Follow the steps in the flow chart in Figure 4.14, which is also stated in the Table 4.2.

   (a) Obtain the total power (kW) generated by the designated segment at the specified r/R location of the rotor.

(b) What is the total thrust (N) generated by that segment at the specified r/R location of the rotor?

4. Provide a brief discussion of why blade twist is important for a rotor blade?

5. In selecting the airfoil section for a new wind turbine blade list all of the characteristics would you like the airfoil to have?

6. A three bladed wind turbine has the following geometric and aerodynamic characteristics.

   Number of blades, $B = 3$

   Tip speed ratio, $\lambda = 7$

   Blade radius, $R = 4.953$ m.

   Rated wind speed, $V_\infty = 11.62$ m/s

   Rotor section shape, NACA 4415 airfoil

   $C_l = 0.368 + 0.0942\alpha$

   $C_d = 0.00994 + 0.000259\alpha + 0.0001055\alpha^2$

   The angle of attack, α, has units of degrees

   Rotor $\theta_{cp} = -2°$

r/R	$r(m)$	c(m)	θ_T
0.1	0.495	0.411	45
0.2	0.991	0.455	25.6
0.3	1.486	0.384	15.7
0.4	1.981	0.311	10.4
0.5	2.477	0.259	7.4
0.6	2.972	0.223	4.5
0.7	3.467	0.186	2.7
0.8	3.962	0.167	1.4
0.9	4.458	0.137	0.4
1.0	4.953	0.107	0.00

Given this information, apply the BEM approach to the rotor divided into the 9 spanwise segments listed in the previous table. For this,

(a) Calculate and plot a and a' as a function of the radial position on the rotor. Include the Prandtl tip loss factor. Comment on how it compares to the Betz optimum.

(b) Calculate and plot the lift and drag coefficients as a function of the radial position on the rotor.

(c) Calculate and plot the differential normal and tangential force coefficients, dF_n and dF_t as a function of the radial position on the rotor.

(d) Calculate and plot the differential power, dP, as a function of the radial position on the rotor.

(e) Calculate total power generated by the wind turbine, $\sum_{i=1}^{9}(dP_i)$

References

1. W. J. M. Rankine, "On the mechanical principles of the action of propellers", *Transactions of the Institution of Naval Architects*, 1865, 6:13–39.
2. R. E. Froude, "On the part played in propulsion by difference in pressure", *Transactions of the Institution of Naval Architects*, 1889, 30:390–423.
3. D. M. Eggleston and F. S. Stoddard, *Wind Turbine Engineering Design*, Van Nostran and Reinhold, New York, 1987.
4. A. Betz, "Das Maximum der theoretisch m'oglichen Ausn'utzung des Windes durch Windmotoren", *Zeitschrift 'fur das gesamte Turbinenwesen*, 1920, 26:307–309.
5. G. A. M. Van Kuik, "The Lanchester-Betz-Joukowsky limit", *Wind Energy*, 2007, 10:289–291.
6. H. Glauert, "Airplane propellers," In: W. F. Durand, Ed., *Aerodynamic Theory*, Vol. 4, p. 332, Springer, New York, 1934.
7. E. H. Lysen, "Introduction to wind energy", Steering Committee: Wind Energy Developing Countries, Amersfoort, NL, 1983.
8. P.-H. Alfredson and J.-A. Dahlberg, "A preliminary wind tunnel study of windmill wake dispersion in various flow conditions", Technical Note AI-1499, Part 7, FFA, Stockholm, Sweden, September 1979.
9. L. Prandtl, "Application of modern hydrodynamics to aeronautics", Translated by Staff of the National Advisory Committee for Aeronautics, NACA Report No. 116, 1923.
10. J. Cooney Jr.,"Enhanced wind turbine energy capture through active flow control", Ph.D. Dissertation, University of Notre Dame, Notre Dame, IN, May 2015.

5

Horizontal Wind Turbine Rotor Design

5.1 Designing a New Wind Turbine

In Chapter 4, the blade element momentum (BEM) theory was introduced as a means of assessing the performance of a new wind turbine rotor. The BEM method requires information on the blade radius, the variation of the rotor blade chord and blade twist as a function of the blade radius, as well as the airfoil section shapes used for the rotor, and their corresponding aerodynamic characteristics. To design a new rotor, one needs to know the amount of power the new rotor is designed to produce for a prescribed wind condition at a proposed site.

Any new wind turbine design begins by identifying the user requirements. These generally reduce to producing a prescribed annual amount of electric power at a given site. Based on the statistical wind conditions at the proposed site, the number of wind turbines with a given rated power and rated wind speed are determined to meet the annual power requirement. The final decision on proceeding with a new design is based upon an economic analysis to determine if the cost per kilowatt hour of the electricity generated by the wind turbines is competitive, and the owners can make a profit. The focus of this chapter is on the steps involved in developing a new horizontal wind turbine design. The material draws on experimental and analytical research on the aerodynamics of wind turbines[1-7].

5.2 Initial Blade Sizing

A new horizontal wind turbine design can begin once the rated power, and the range of operational wind speeds have been identified. For this, the power extracted from the wind by the rotor, P_{rotor}, at the rated wind speed, V_{rated} is

$$P_{\text{rotor}} = \frac{C_p \rho V_{\text{rated}}^3 A_{\text{rotor}}}{2}. \tag{5.1}$$

DOI: 10.1201/9781003611486-5

TABLE 5.1
Power Train Efficiencies for Modern Wind Turbines at Rated Power Conditions

Gearbox	η_{GB}	0.94–0.98
Generator	η_G	0.95–0.97
Converter	η_{Conv}	0.96–0.98

The mechanical power delivered by the rotor to the electric turbine shaft is subsequently

$$P_{\text{rated}} = \eta P_{\text{rotor}}, \tag{5.2}$$

where η is the efficiency of the mechanical power train driving the generator. The mechanical efficiency is always less than 1, so that the power extracted by the wind turbine rotor must be greater than the rated electric power in order to compensate for the power losses during the electric conversion.

The power train efficiency can be expressed as

$$\eta = \eta_{GB}\eta_G\eta_{\text{Conv}}, \tag{5.3}$$

where, η_{GB} is the efficiency of the gearbox in transmitting the mechanical power to the generator, η_G is the efficiency of the generator in converting the mechanical power into electric power, and η_{Conv} is the efficiency in converting the electrical power from the generator to that required by the electrical grid. Table 5.1 provides a summary of typical efficiency values for these power train components. The values in this table are reasonable estimates for modern wind turbine systems operating at normal power levels. However, these efficiencies are lower when the wind turbine is operating at lower power conditions.

The power coefficient for a horizontal wind turbine that was previously developed from momentum/actuator disc theory in terms of the axial induction factor, a, as

$$C_p = 4a(1-a)^2. \tag{5.4}$$

The maximum power coefficient was shown by Betz to be 0.593. This occurs when the axial induction factor is $a = 1/3$. However, Figure 5.1 shows how the rotor power coefficient, C_p, can vary with the rotor tip speed ratio, $\lambda = \Omega R/V_{\text{rated}}$, for modern wind turbines. The power coefficient is a maximum at a tip speed ratio of 7, but in this case, $C_{p_{\max}} = 0.47$. This value of C_p is about 20% lower than the Betz limit! The reason is that the Betz limit is a theoretical maximum limit that *does not account for rotor tip losses and other losses from aerodynamic drag.*

Combining Equations 5.1 and 5.2, the rated power that accounts for the losses on the drive train is

$$P_{\text{rated}} = \frac{C_p \eta \rho V_{\text{rated}}^3 A_{\text{rotor}}}{2}. \tag{5.5}$$

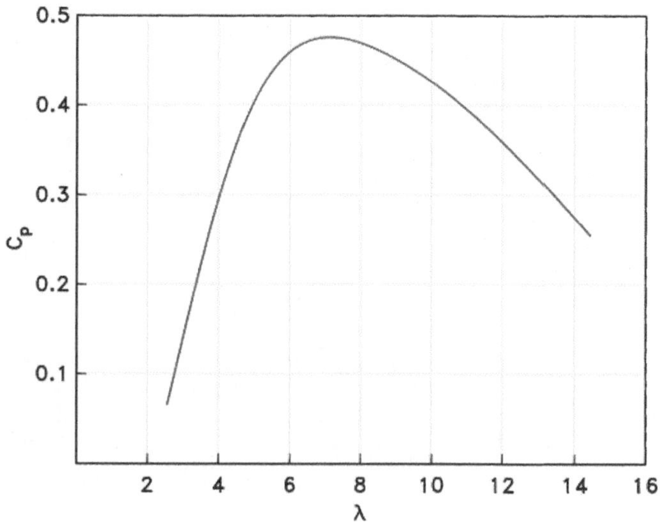

FIGURE 5.1
Power coefficient as a function of the rotor tip speed ratio.

Based on this, and given that the area of the rotor is $A_{\text{rotor}} = \pi R^2$, where R is the radius of the rotor, the required rotor radius of a wind turbine to produce the rated power is

$$R_{\text{rotor}} = \left[\frac{2P_{\text{rated}}}{C_p \eta \rho V^3_{\text{rated}} \pi}\right]^{1/2}. \tag{5.6}$$

We observe that the required blade radius to produce the rated power is a function of the wind rated wind velocity, the rotor power coefficient, and the efficiency in converting the mechanical power delivered by the rotor into electric power suitable for the grid.

The blade radius predicted using the Betz maximum power coefficient of $C_p = 0.593$ will yield a rotor blade radius that is smaller than the blade radius on modern wind turbines. Therefore, in sizing a new rotor a $C_p = 0.45$ would be more appropriate. This is illustrated in Figure 5.2 which shows curves of the rated power for a three-bladed rotor as a function of the blade radius for the Betz $C_p = 0.593$ and for the more realistic $C_p = 0.45$, against the values (data points) for several modern wind turbines. Therefore, a better historical estimate of rotor radius of a new wind turbine would account for the less than Betz efficiency, namely C_p in Equation 5.7 would be 0.45 so that

$$R_{\text{rotor}} = \left[\frac{2P_{\text{rated}}}{0.45 \eta \rho V^3_{\text{rated}} \pi}\right]^{1/2}. \tag{5.7}$$

The next step in the rotor design is to determine the amount of blade twist, $\theta_T(r)$ and the variation in the rotor chord, $c(r)$, which are both functions of

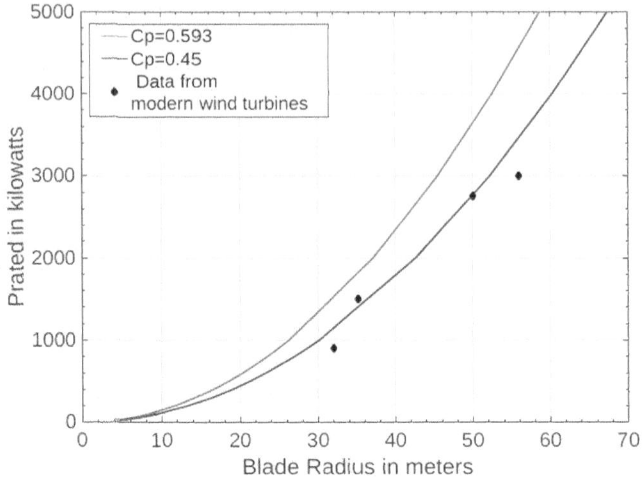

FIGURE 5.2
Comparison of theoretical wind turbine rated power versus rotor diameter for two C_p values and that of modern wind turbines.

the radial distance along the rotor from the axis of rotation. If one assumes that there is no wake rotation, namely $a' = 0$, that there is zero aerodynamic drag, namely $C_d = 0$, and that there are no rotor tip losses, namely $F = 1$, the momentum and blade element equations from are

$$\begin{aligned} dT &= \rho V_\infty^2 4a(1-a)\pi r dr, && (5.8)\\ dF_n &= \frac{1}{2} B\rho V_R^2 C_l \cos(\phi) c(r) dr. && (5.9) \end{aligned}$$

Equating these two equations, and substituting for V_R as

$$V_R = \frac{V_\infty(1-a)}{\sin(\phi)}, \quad (5.10)$$

one obtains

$$\rho V_\infty^2 4a(1-a)\pi r dr = \frac{1}{2} B\rho V_\infty^2 \left(\frac{(1-a)^2}{\sin^2(\phi)}\right) C_l \cos(\phi) c(r) dr. \quad (5.11)$$

After canceling and rearranging terms, one obtains

$$\frac{BC_l c(r)}{4\pi r} = \frac{2a(\sin^2(\phi)}{(1-a)\cos(\phi)} = \frac{2a}{(1-a)} \tan(\phi)\sin(\phi). \quad (5.12)$$

In the previous chapter, it was shown that

$$\tan(\phi) = \frac{V_\infty(1-a)}{\Omega r(1+a')}. \quad (5.13)$$

TABLE 5.2
Summary of Equations for Estimating the Blade Chord and Twist Angle as a Function of the Local Rotor Radius

Blade radius, R	$R_{\text{rotor}} = \left[\frac{2P_{\text{rated}}}{0.45\eta\rho V_{\text{rated}}^3\pi}\right]^{1/2}$
The resultant velocity angle with respect to the plane of rotation, $\phi(r)$	$\phi(r) = \tan^{-1}\left(\frac{1-a}{\lambda_r}\right)$ where $\lambda_r = \frac{\Omega r}{V_\infty}$
Blade twist angle, $\theta_T(r)$	$\theta_T(r) = \phi(r) - \alpha(r)$ where $\alpha(r)$ can be taken as the angle of attack where $(C_l/C_d)_{\max}$ occurs for the airfoil section or sections
Local blade chord, $c(r)$	$c(r) = \frac{8\pi ra\sin(\phi)}{BC_l\lambda_r}$ where C_l can be taken where $(C_l/C_d)_{\max}$ occurs for the airfoil section or sections

In the present analysis, we assumed that $a' = 0$, therefore

$$\tan(\phi) = \frac{V_\infty(1-a)}{\Omega r} = \frac{1-a}{\lambda_r}, \tag{5.14}$$

or

$$\phi = \tan^{-1}\left(\frac{1-a}{\lambda_r}\right). \tag{5.15}$$

Substituting Equation 5.14 into Equation 5.12 and canceling like terms leads to the following:

$$\frac{BC_lc(r)}{4\pi r} = \frac{2a\sin(\phi)}{\lambda_r}. \tag{5.16}$$

Solving Equation 5.16 for the chord distribution results in the following:

$$c(r) = 8\pi ra\frac{\sin(\phi)}{BC_l\lambda_r}. \tag{5.17}$$

The equations that are needed to determine the chord, $c(r)$, and the blade twist angle, $\theta_T(r)$, are summarized in Table 5.2. The blade twist angle is a function of the angle $\phi(r)$, which is the angle that the resultant velocity makes with the plane of rotation and the local angle of attack of the rotor. The angle of attack should be the angle where the maximum lift-to-drag ratio is a maximum, that is at $(C_l/C_d)_{\max}$.

If we assume the Betz optimum, $a = 1/3$, the equations for the optimum design for the chord length and blade twist are given by the following

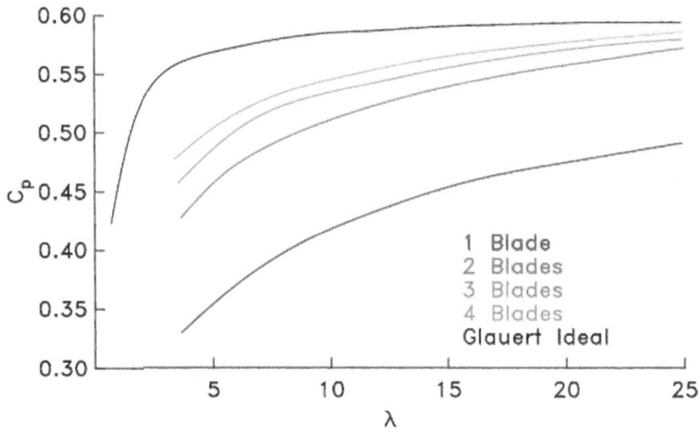

FIGURE 5.3
Power coefficient as a function of tip-speed-ratio for different numbers of rotor blades based on inviscid simulation.

equations.

$$\phi(r) = \tan^{-1}\left(\frac{2}{3\lambda_r}\right), \tag{5.18}$$

$$\theta_T(r) = \phi(r) - \alpha(r), \tag{5.19}$$

$$c(r) = \frac{8\pi r \sin(\phi)}{3BC_l\lambda_r}. \tag{5.20}$$

These simple equations allow a designer to develop an initial estimate of the blade twist, chord, and radius. The only remaining variable that is needed is the number of rotor blades. For reference, J. L. Tangler [1] provided an excellent review of the evolution of rotor blade design from years 1980 to 2000. Figure 5.3 shows how the power coefficient, C_p, varies as a function of the tip-speed-ratio for rotors with different numbers of blades. This is a modified version of a figure presented by Rohrback et al. [2]. This indicates that the power coefficient increases with increasing blade number, but the change becomes smaller as the number of blades increase, with the change in C_p between rotors with three and four blades being minimal. Note that Figure 5.3 is based on an inviscid analysis and as a result leaves out viscous losses that result in a drop in C_p at higher tip-speed-ratios, such as presented in Figure 5.1.

The reasons why large commercial wind turbines use three bladed rotors involves a number of factors. Tangler [1] pointed out that the choice of the number of blades is influenced by rotor noise, dynamic loading, and aesthetic considerations. The noise level created by a three bladed rotor is lower than that of a two bladed rotor for the same output power. The dynamic loading increases with an increasing number of rotor blades. As a result, the slightly higher power coefficient of a four-blade rotor over that of a three-blade rotor is offset by the increase in the dynamic loading. Combining all of these factors, the standard for modern wind turbines has settled on three bladed rotors.

5.2.1 Example Rotor Design

A new design for a variable rotational speed three-blade horizontal wind turbine is proposed that provides a rated power of 100 kW at a wind speed of 12 m/s, and a tip speed ratio of $\lambda = 7$. The cut-in and cut-out wind speeds are taken to be 5 and 25 m/s, respectively.

The design of the three-blade rotor begins by using the Betz optimum blade shape equations. The maximum (Betz) power coefficient occurs with an axial induction factor of $a = 1/3$.

For $a = 1/3$,

$$C_p = 4a(1-a)^2 = 0.593, \tag{5.21}$$

which is the Betz limit on the power coefficient.

The rotor blade section shape has the following characteristics:

$$C_l = 0.9 \,\text{at}\, (C_l/C_d)_{\text{max}},$$

and

$$\alpha = 6^\circ \,\text{at}\, (C_l/C_d)_{\text{max}}.$$

The first step in the design is to determine the radius of the turbine blades using Equation 5.6. Rather than the Betz optimum $C_p = 0.593$, a more realistic value of $C_p = 0.45$ is assumed. The efficiency of the electrical power conversion equipment is assumed to be $\eta = 0.9$. By substituting these values into Equation 5.6, one obtains the following equation for the rotor radius,

$$R = \left[\frac{1 \times 10^5}{(0.45)(0.9)(0.5)(1.225)(12^3)(\pi)}\right]^{0.5} = 8.62\,\text{m}. \tag{5.22}$$

Note that if the Betz C_p were used, the radius would have been 14.7% smaller or $R_{\text{min}} = 7.51$ m.

With the turbine radius determined, the next step is to estimate the blade twist and chord distributions. These can be found using Equations 5.18, 5.19, and 5.20. The radial distributions for the relative wind vector angle, $\phi(r)$, found from Equation 5.18, and the relative blade twist angle, $\theta_T(r)$, found from Equation 5.19, are shown in Figure 5.4. As previously noted, it is desirable to have the effective angle of attack of each radial segment of the blade be that where (C_l/C_d) is a maximum. We note that this results in the blade tip segment θ_T is slightly negative.

The radial distribution of the chord length, $c(r)$, found from Equation 5.20, is presented in Figure 5.5. The first thing one notices is the large growth in the chord length near the root ($r/R \leq 0.2$) portion of the blade. The manufacture of this blade design would be very costly for three reasons. First, the mold for

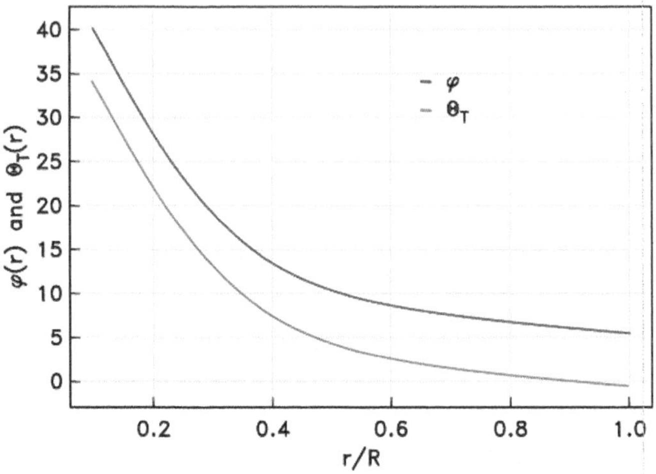

FIGURE 5.4
Relative wind angle, $\phi(r)$, and blade twist angle, $\theta_T(r)$ along the rotor radius for a Betz optimum design.

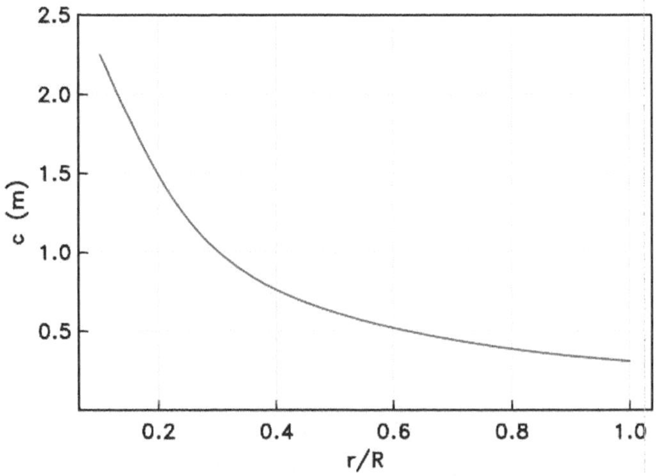

FIGURE 5.5
Radial distribution of the local rotor chord length of a rotor for a Betz optimum design.

fabricating the blades would be more expensive because of the complicated shape of the blade near the inboard portion. Second, the large increase in the chord from ($0.1 \leq r/R \leq 0.4$) would add considerable weight to the blade where there is very little contribution to the wind turbine power generation.

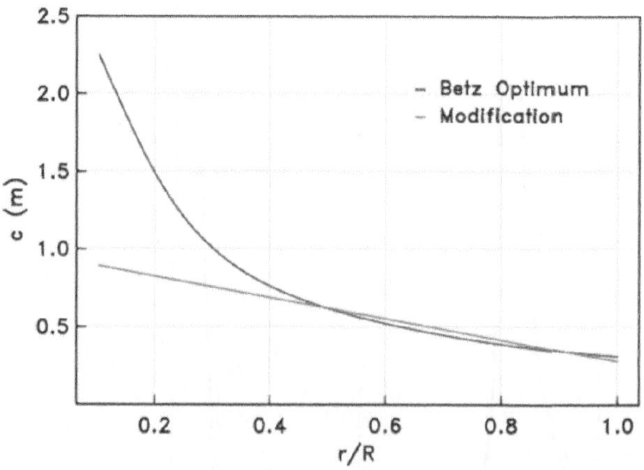

FIGURE 5.6
Example of a modification to the Betz optimum chord distribution to reduce the weight of the rotor.

Finally, the size of the inboard rotor section would add considerable weight to the rotor blade. The weight of the blade affects the cost of every major component that makes up a wind turbine. For example, an increase in blade weight requires a stronger drive shaft, gearbox, tower, and foundation that ultimately adds to the purchase cost of a new wind turbine. In a later chapter, methods for predicting the cost of a new design will be presented. Empirical models for predicting the cost of many of the wind turbine components are mostly based on the weight of the rotor. Therefore to be competitive in the wind turbine market, a new wind turbine design must have a competitive cost.

An approach to reduce the weight of the rotor involves tapering the blade chord length in the inboard radii of the rotor. An example is shown in Figure 5.6 in which two points at $r/R = 0.5$ and $r/R = 0.9$ are fitted with a straight line. The straight line is extrapolated to $r/R = 0.1$ and $r/R = 1.0$ to form a linear chord distribution. The difference between the tapered blade chord and the Betz Optimum chord distribution is very small, especially considering that the oputboard 50% of the rotor is primarily responsible for producing the wind turbine power.

The following then summarizes the steps for generating a realistic radial blade chord distribution from the Betz Optimum design.

1. Determine the blade radius based on a realistic power coefficient such as from Figure 5.1.
2. Apply the Betz Optimum Method to determine the blade twist angle and chord as a function of the rotor radial location using Equations 5.18–5.20. Assume the Betz maximum $C_p = 0.593$, which is based on an axial induction factor of $a = 1/3$.

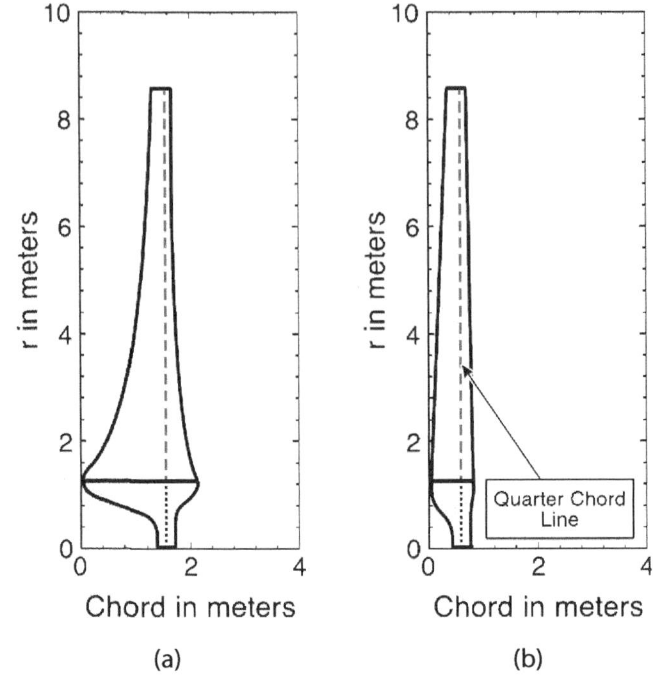

FIGURE 5.7
Transformation from Betz optimum chord distribution to a realistic tapered distribution that is a close approximation to the optimum for $r/R \geq 0.5$. (a) Belz design and (b) tapered design.

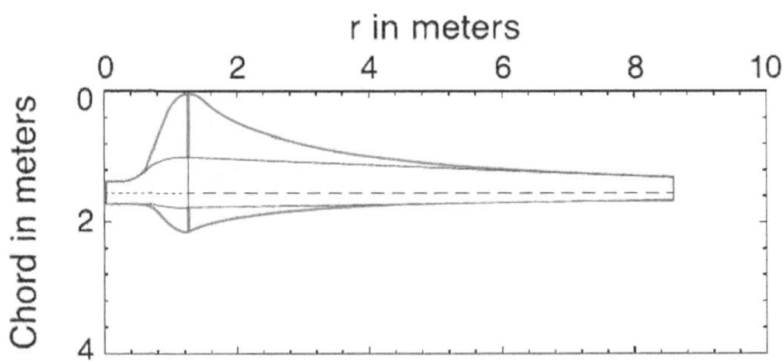

FIGURE 5.8
Overlay of Betz optimum chord distribution and a realistic tapered distribution from Figure 5.7.

3. Create a tapered blade from the chord distribution predicted by Betz Optimum Design Method by drawing a straight line from $r/R = 0.15$ to $r/R = 1.0$ that intersects the Betz Optimum chord distribution at radial positions of $r/R = 0.5$ and $r/R = 0.9$.

An example of the rotor chord distribution starting from the optimum Betz and ending in the distribution following the preceeding steps is shown in Figure 5.7. The radial chord distribution was designed so that the line passing through the quarter-chord locations were perpendicular to the blade-root hub. Figure 5.8 shows the two chord distributions superimposed on the same plot. This clearly shows that the chord distribution of the outboard 50% of the blade in nearly identical to the Betz optimum distribution.

Allowing for this new chord distribution, it is then possible to determine the performance of this design using BEM equations. The design could be modified until the required rated power is met.

Problems

1. An established wind turbine manufacturer has an extensive data base on the wind turbine blades that they have designed. This data is generally called proprietary data and is considered the intellectual property of the company and is not available to the public. However, the wind turbine companies do publish some basic information on their machines in advertising brochures. Data on a variety of wind turbines is contained in the Appendix. Using this data, develop a curve similar to that in Figure 5.2 showing the rated power versus the blade radius or length.
2. Determine the blade radius, twist and chord distributions for a two-bladed rotor that will produce a rated power 150 kW.
3. Using the modified Betz Method discussed in this chapter, design a three bladed rotor that can produce a rated power output or 250 kW. Determine the rotor radius, blade twist, and chord distribution.
4. Using the data from Problem 3, develop a tapered blade starting from the Betz optimum design.

References

1. J. L. Tangler, "The evolution of rotor and blade design", NREL/CP-500-28410, July, 2000.
2. C. Rohrback, H. Wainauski, and R. Worbel, "Experimental and analytical research on the aerodynamics of wind driven turbines", Hamilton Standard, COO-2615-T2, 1977.
3. P. J. Schubel and R. J. Crossley, , "Wind turbine blade design", *Energies*, 2012, 5:3425–3449.
4. R. Gash, and J. Twele, *Wind Power Plants-Fundamentals, Design, Construction and Operation*, James and James Science Publishers Ltd, London, 2002.
5. J. F. Manwell, J. G. McGowan, and A. L. Rogers, *Wind Energy Explained: Theory, Design and Application*, John Wiley & Sons, Hoboken, NJ, 2009.
6. T. Burton, D. Sharpe, N. Jenkins, and E. Bossanyi, *Wind Energy: Handbook*, John Wiley & Sons, Hoboken, NJ, 2001.
7. M. O. L. Hansen, *Aerodynamics of Wind Turbines*, 2nd edition, Earthscan, London, 2008.

6

Wind Turbine Control

The control system on a wind turbine is designed to (1) seek the highest efficiency of operation that maximizes the coefficient of power, C_P, and (2) ensure safe operation under all wind conditions. Wind turbine control systems are typically divided into three functional elements:

1. the control of groups of wind turbines in a wind farm,
2. the supervising control of each individual wind turbine, and
3. separate dedicated dynamic controllers for different wind turbine sub-systems.

A flowchart of these wind turbine functional control elements are shown in Figure 6.1.

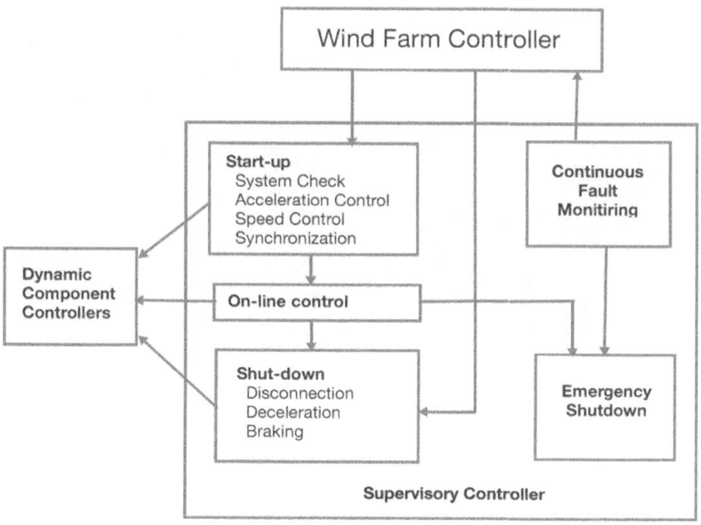

FIGURE 6.1
Schematic of the wind turbine functional control elements.

DOI: 10.1201/9781003611486-6

The wind farm controller's function is "power management". It can initiate and shut down turbine operation as well as coordinate the operation of numerous wind turbines in response to environmental and operating conditions.

The wind turbine supervisory controller manages the individual turbine operation including power production, low-wind shutdown, high-wind shutdown, high load limits, and orderly start-up and shut-down. In addition, it provides control input to the dynamic controllers for such things as r.p.m. control to maintain an optimum tip-speed-ratio, blade pitch control, and power level control.

The wind turbine dynamic controllers make continuous high-speed changes in the operating conditions such as blade pitch, yaw, and power management. As mentioned, these receive input from the supervisory controller.

Figure 6.2 shows a cut-away view of a modern wind turbine that illustrates various components that make up the monitoring and control systems. Figure 6.3 shows a schematic of the closed-loop wind turbine control system that makes up the supervisory and dynamic control components. The control system is designed to maintain a desired rotor frequency, f_d. This is controlled through pitch control (if it exists) and torque control, which occurs as a result of the power load torque or "braking torque" generated by the power converter. The aerodynamic torque is a function of the blade pitch, rotor tip-speed ratio, λ, as well as the wind speed, and any off-design conditions such as yaw error, wind shear, etc.

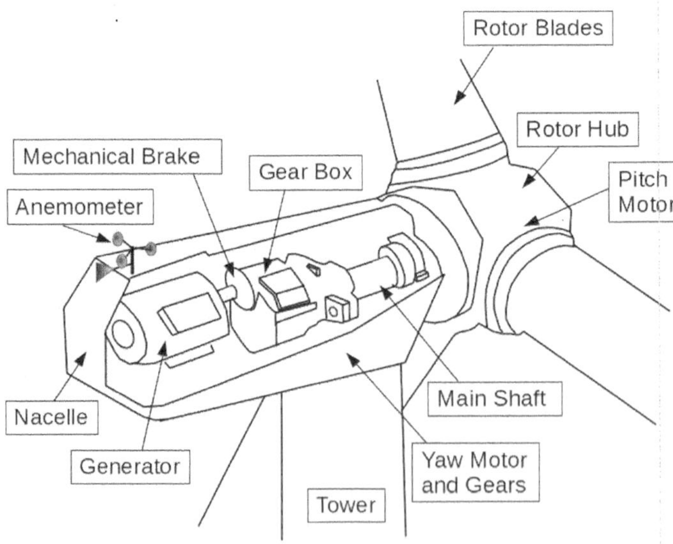

FIGURE 6.2
Section view of typical components of a wind turbine that are involved in its monitoring and control.

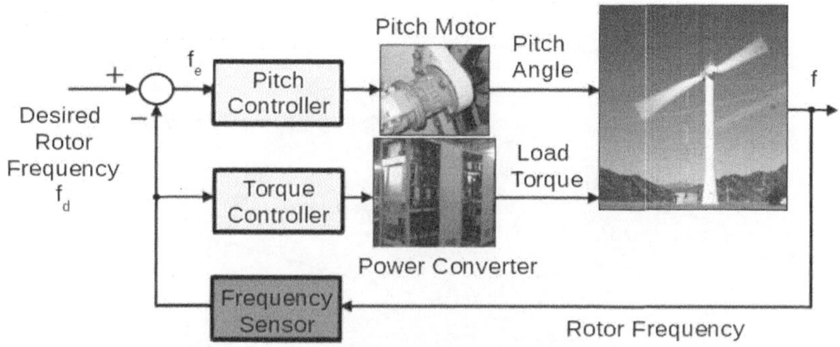

FIGURE 6.3
Schematic of a wind turbine closed-loop control system.

An example [1] of the relation between the tip-speed ratio and rotor pitch angle on the coefficient of power for a sample 600 kW two-bladed horizontal wind turbine is shown in Figure 6.4. This indicates that an optimum power condition occurs with a tip-speed ratio of approximately 7. The power coefficient is observed to drop off rather steeply from the optimum condition.

The exact optimum tip-speed ratio will depend on the individual wind turbine design. It generally ranges from about six to eight for wind turbines covering a large range of rated powers. The sensitivity of the coefficient of power to the tip-speed ratio is what motivates the closed-loop control focusing on the rotation frequency as is shown in Figure 6.3. As pointed out, this control comes from balancing the aerodynamic torque and the electrical (braking) torque. The following sections discuss how this can be accomplished.

6.1 Aerodynamic Torque Control

As discussed, one of the approaches to control the rotor tip-speed ratio is through control of the rotor aerodynamic torque, which ultimately comes by controlling the rotor aerodynamic lift. For lift control, there are two approaches that have been commonly used: (1) stall-regulated rotor designs and (2) pitch regulated rotor designs.

Stall-regulated rotors are ones that are designed with section shapes and mean angles of attack to cause the rotor to stall at higher wind speeds, beginning at rated power wind speeds. When the rotor stalls it loses lift and increases drag, which causes a reduction of aerodynamic rotor torque.

Pitch-regulated rotors reduce the aerodynamic torque by reducing the pitch and thereby the local angle of attack of the rotor sections. The lower angles of attack reduces the section lift coefficient and thus the aerodynamic torque on

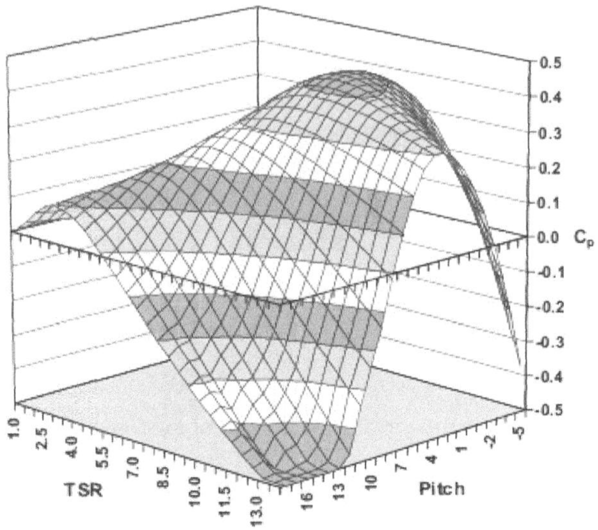

FIGURE 6.4
Example of the relation between the rotor tip-speed ratio and rotor pitch angle on the coefficient of power for a 600 kW two-bladed horizontal wind turbine [1].

the rotor. The pitch control is initiated when the wind velocity is sufficient to generate the turbine rated power level. It continues to reduce the pitch to seek to maintain an optimum tip-speed ratio while maintaining a constant rated power until the cut-out wind speed is encountered.

6.1.1 Electrical Torque Control

The approaches to electrical torque control can involve different designs of electric power generators used in the wind turbines. The most common of these are synchronous generators.

A synchronous machine is an alternating current (AC) rotating machine whose speed, ω, under steady-state condition is proportional to the frequency of the current in its armature. The magnetic field created by the armature currents rotates at the same speed as that created by the field current on the rotor, which is rotating at the synchronous speed, and results in a steady torque.

Synchronous machines are commonly used as generators especially for large power systems, such as turbine generators and hydroelectric generators in the grid power supply. The reactive power generated by a synchronous machine can be adjusted by controlling the magnitude of the rotor field current, unloaded synchronous machines are also often installed in power systems solely for power factor correction, or for control of reactive kV-A flow.

Figure 6.5 shows a schematic drawing of a four-pole synchronous machine along with the sinusoidal waveform of the induced electromotive force (emf), which has units of volts, that is produced by the rotation of the center rotor. Defining θ_m as the angular position of the mechanical rotor, and θ as the phase angle of the generated sinusoidal emf, for the 4-pole machine, one revolution of the rotor, namely $\theta_m = 2\pi$, results in an emf phase angle of $\theta = 4\pi$. Therefore, the relation between the mechanical phase angle, θ_m, and the emf phase angle, θ, is

$$\theta = 2\theta_m. \tag{6.1}$$

For a general case of a synchronous machine with P poles, the relationship between the electrical and mechanical phase angles is then

$$\theta = \frac{P}{2}\theta_m. \tag{6.2}$$

Taking a time derivative of both sides of Equation 6.2, to put it in terms of angular velocity, ω, then

$$\omega = \frac{P}{2}\omega_m. \tag{6.3}$$

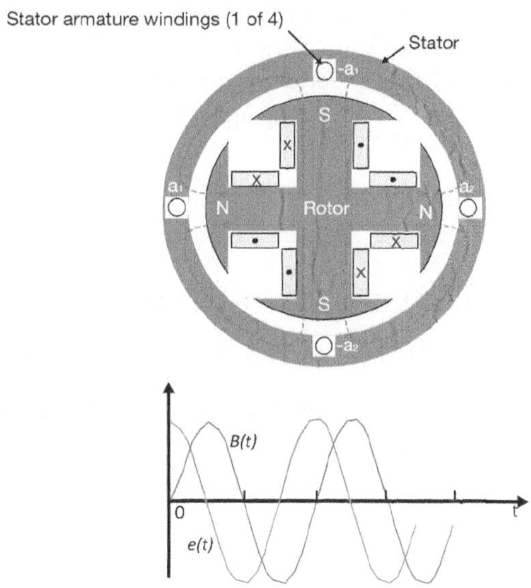

FIGURE 6.5
Schematic drawing of a four-pole synchronous machine along with the sinusoidal waveform of the induced electromotive force (emf) which has units of volts, that is produced by the rotation of the rotor.

Converting Equation 6.3 into physical frequency, f, with units of Hertz,

$$f = \frac{P}{2}\frac{n}{60}, \tag{6.4}$$

or

$$n = \frac{120f}{P}, \tag{6.5}$$

where $\omega = 2\pi f$ and $\omega_m = 2\pi n/60$, with n being the rotor speed in revolutions/minute.

Most wind turbine generators have four poles. Therefore, based on Equation 6.5 to produce the 60 Hz. frequency that is the U.S. power standard, the rotor would need to spin at 1800 r.p.m! The typical rotor r.p.m. of a multi-megawatt wind turbine is about 0.8 r.p.m. Therefore a gear box would be designed so that at the optimum tip-speed ratio, the generator rotor would spin at the necessary r.p.m. to produce 60 Hz AC. This approach is quite restrictive and leads to an alternate approach in which the AC power is generated at any frequency then converted to DC power, after which it is converted back to AC power with the U.S. standard 60 Hz AC frequency.

There is still an advantage to increasing the r.p.m. of the AC generator in that the power generated is linear with the angular velocity of the rotor. This can be accomplished either by increasing the diameter of the rotor or by increasing the r.p.m. The former is suitable for lower power levels and therefore eliminates the gear box. However at multi-megawatt power levels, the necessary diameter of the generator to produce those power levels makes that approach prohibitive. Thus, a step-up gear box is used.

6.2 Wind Turbine Operation Strategy

There are generally four strategic objectives to wind turbine operation:

1. to maximize energy production while keeping operation within speed and load constraints,
2. to prevent extreme loads and to minimize fatigue damage that can occur as a result of repeated bending caused by weight on the rotors and unsteady aerodynamics loads,
3. to provide acceptable power quality at the point of connection to the power grid, and
4. to provide safe operation.

The control approach depends on the wind turbine design such that

- For ($V_{\text{cut-in}} < V_\infty < V_{\text{rated}}$) the object is to maximize power production.

- For ($V_{\text{rated}} < V_\infty < V_{\text{cut-out}}$) the object is to limit power to the rated value.

The two approaches to accomplish this are: (1) fixed speed designs and (2) variable speed designs. These two approaches are discussed in the following two sections.

6.2.1 Fixed Speed Designs

Fixed speed designs fall under two categories: (1) stall regulated and (2) active pitch regulated.

6.2.1.1 Stall Regulated Fixed Speed Control

In stall regulated designs, the rotor blades are at a fixed pitch angle. They are designed to stall at higher wind speeds to passively regulate the generated power.

Stall regulated wind turbines are designed to operate near the optimum tip-speed ratio at lower speeds, below V_{rated}. As the wind speed increases, the effective angle of attack of the rotor sections, α, increases. To illustrate this, the effective angle of attack of any spanwise section of the rotor is

$$\alpha = \phi - \theta_T - \theta_{cp}, \tag{6.6}$$

where θ_T is the local twist angle, θ_{cp} is the rotor constant pitch angle, and ϕ is the aerodynamic angle of attack which again is given as

$$\phi = \tan^{-1}\left[\frac{1-a}{1+a'}\frac{V_\infty}{\Omega r}\right] = \tan^{-1}\left[\frac{1-a}{(1+a')\lambda_r}\right], \tag{6.7}$$

where r is the local radius on the rotor, Ω is the rotor rotation rate, and a and a' are the axial and tangential induction factors, respectively.

For fixed global pitch angle, and a fixed rotor twist angle at some radial location on the rotor, the effective angle of attack, α, is only a function of ϕ. For a constant tip-speed ratio, $\lambda = \lambda_{\text{optimum}}$, and near optimum power coefficient where $a \simeq 1/3$ and $a' \simeq 0$, then

$$\alpha \simeq \phi \sim \tan^{-1}(V_\infty)\,. \tag{6.8}$$

Therefore there is a direct link between the effective angle of attack and the free-stream wind speed. When the effective angle of attack exceeds the rotor section shape stall angle of attack, α_{stall}, the rotor section lift will equilibriate or decrease, and the rotor section drag will increase. The result will be a decrease in the aerodynamic torque and generated power. This is the fundamental mechanism of passive stall regulated fixed speed control.

6.2.1.2 Active Pitch Regulated Fixed Speed Control

In active pitch regulated wind turbines, the blade pitch is changed to provide power smoothing in high wind conditions. Below the rated wind speed, V_{rated}, the blade pitch is kept fixed. This is the chosen approach to limit the pitch mechanism wear, although there would be a power coefficient benefit if the rotor pitch were varied between $V_{\text{cut-in}}$ and V_{rated}.

At the rated wind speed, the blade pitch is dynamically varied to seek to hold a constant power level. Above the cut-out wind speed, the blade is pitched to a position that minimized the rotor aerodynamic torque. This minimizes the rotor rotation and potential damage during high wind speeds.

6.2.2 Variable Speed Designs

Variable speed designs also fall under the categories of (1) stall regulated and (2) active pitch regulated. These differ from the fixed speed designs in that electrical torque control is also utilized.

6.2.2.1 Stall Regulated Variable Speed Control

In stall regulated wind turbines, variable speed control comes by regulating the generator torque. At low speeds, below V_{rated}, variable speed control is used to maintain the optimum tip-speed ratio, thereby seeking to maximize the coefficient of power. As the wind speed increases to the rated velocity, the rotor r.p.m. is decreased and the rotor blades are allowed to stall. This is illustrated in Figure 6.6 which shows a power curve for a stall regulated wind turbine with variable speed control. The solid curve corresponds to the r.p.m. schedule that is read on the right vertical axis. The dashed curve corresponds to the power being generated, and is read on the left vertical axis. Also indicated is the point at which the blade is designed to stall to hold a constant power level. To accomplish this, the rotor r.p.m. is gradually decreased.

6.2.2.2 Active Pitch Regulated Variable Speed Control

With active pitch regulated wind turbines, at lower wind speeds between $V_{\text{cut-in}}$ and V_{rated}, the rotor pitch remains fixed, similar to fixed speed designs. However variable speed control is performed to seek to maintain an optimum tip-speed ratio through the addition of electrical torque control.

From V_{rated} to $V_{\text{cut-out}}$, the generator torque is used to maintain constant power. Pitch control is used to regulate the rotor r.p.m., seeking to maintain the optimum tip-speed ratio.

6.2.3 Variable Speed Adaptive Torque Control

The control strategy discussed in this section is based on an NREL Report [1]. It seeks to maximize energy capture in Region 2 of the power curve, namely where $V_{\text{cut-in}} \leq V_\infty \leq V_{\text{rated}}$. In Region 2, the control of a variable speed wind turbine is often accomplished by setting the control torque (i.e.,

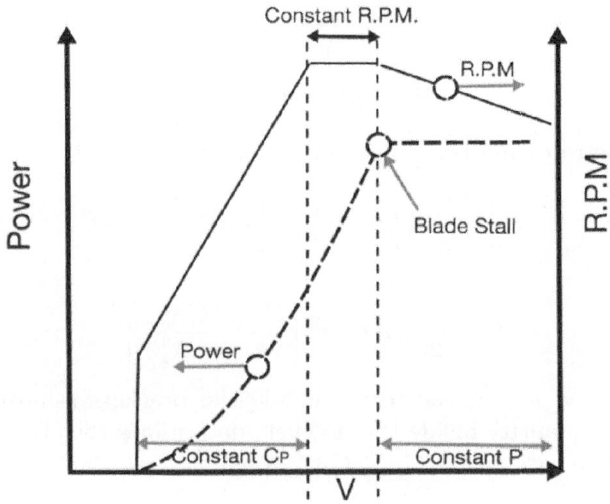

FIGURE 6.6
Power curve for a stall regulated wind turbine with variable speed design.

generator torque) equal to a gain times the rotor speed squared, namely

$$Q_C = k\Omega^2, \tag{6.9}$$

where Ω is the rotor speed, and k is a Gain Factor given by

$$k = \frac{1}{2}\rho A R^3 \frac{C_{P_{\max}}}{\lambda_*^3}, \tag{6.10}$$

A =rotor sweep area
R =rotor radius

$$C_P = \frac{P}{\frac{1}{2}\rho A V_\infty^3}, \tag{6.11}$$

$C_{P_{\max}}$ = maximum power coefficient
$P = \Omega Q_{\text{aero}}$

$$Q_{\text{aero}} = \frac{1}{2}\rho A R C_Q V_\infty^2, \tag{6.12}$$

$$\lambda = \frac{\Omega R}{V_\infty}, \tag{6.13}$$

$\lambda_* = \lambda$ at $C_{P_{\max}} = \lambda_{\text{opt}}$.

The coefficient C_Q, is defined as the rotor torque coefficient where

$$C_Q = f(\lambda, \theta_{cp}) = \frac{C_P(\lambda, \theta_{cp})}{\lambda}, \tag{6.14}$$

and θ_{cp} is the rotor constant pitch angle as before.

The angular acceleration of the rotor is

$$\dot{\Omega} = \frac{1}{J}\left(Q_{\text{aero}} - Q_C\right), \tag{6.15}$$

where J is the rotor inertia. Substituting for Q_{aero} and Q_C,

$$\dot{\Omega} = \frac{1}{J}\left[\frac{1}{2}\rho ARC_Q V_\infty^2 - \frac{1}{2}\rho AR^3 \frac{C_{P_{\max}}}{\lambda_*^3}\Omega^2\right], \tag{6.16}$$

or

$$\dot{\Omega} = \frac{1}{2J}\rho AR^3\Omega^2\left[\frac{C_P}{\lambda^3} - \frac{C_{P_{\max}}}{\lambda_*^3}\right]. \tag{6.17}$$

In the expression for $\dot{\Omega}$, the term outside the brackets is positive definite. Therefore, the quantity inside the brackets determines the sign of $\dot{\Omega}$.

Consider the case where $C_P \leq C_{P_{\max}}$ then

1. if $\lambda > \lambda_{\text{opt}}$ then $\dot{\Omega} < 0$ and the rotor will decelerate toward $\lambda = \lambda_{\text{opt}}$
2. if $\lambda < \lambda_{\text{opt}}$ then $\dot{\Omega} > 0$ and the rotor will accelerate toward $\lambda = \lambda_{\text{opt}}$.

Generally, therefore $C_P = (C_{P_{\max}}/\lambda_{\text{opt}})\lambda^3$ is a control trajectory. This control trajectory is plotted as the dotted curve in Figure 6.7 which shows the power coefficient versus tip speed ratio for the sample turbine performance with a fixed pitch angle of $\theta_{cp} = -1°$ that was shown in Figure 6.4. This illustrates that the control trajectory properly seeks out the optimum tip-speed ratio that maximizes C_P.

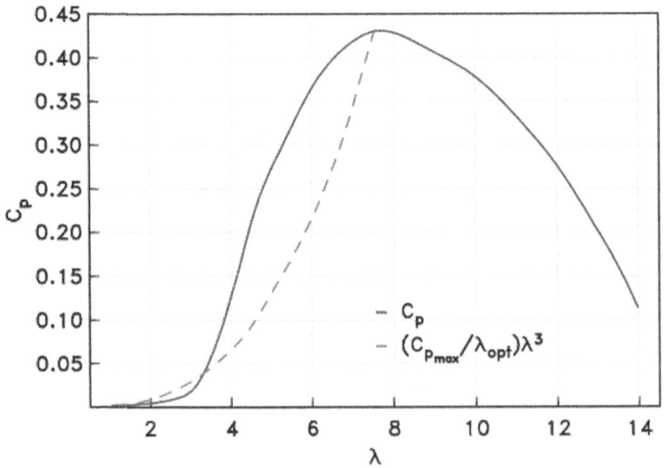

FIGURE 6.7
Example of control trajectory to seek the optimum tip-speed ratio for the wind turbine performance shown in Figure 6.4 with $\theta_{cp} = -1°$ [1].

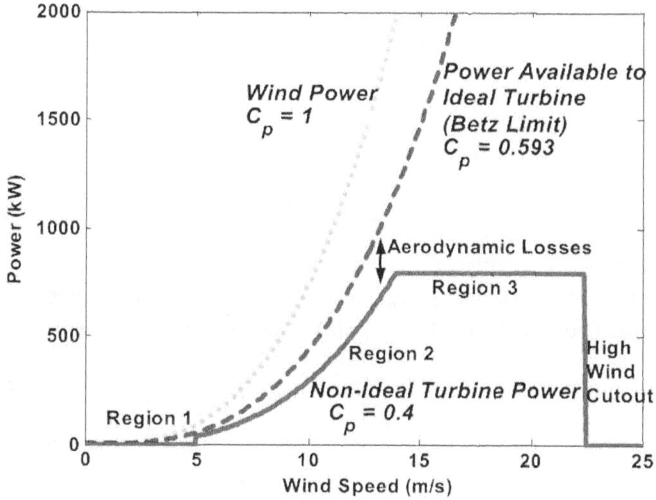

FIGURE 6.8
Generic power curve for a wind turbine illustrating optimum (Betz) and actual performance in Region II.

6.3 Axial Induction Control

Standard control of wind turbines has focused on changing the pitch of the rotor and control of the rotor r.p.m. in order to maintain an optimum tip speed ratio. The standard practice for rotor pitch control is to have a fixed pitch angle for Region II wind speeds ($V_{\text{cut-in}} \leq V_\infty \leq V_{\text{rated}}$), then to change the pitch to maintain a constant rated power for Region III wind speeds ($V_{\text{rated}} \leq V_\infty \leq V_{\text{cut-out}}$). The fixed pitch in Region II wind speeds is intended to maximize the average efficiency over the wind speeds from cut-in to rated. However, as will be apparent, for a rigid rotor with fixed twist and constant pitch, the optimum (Betz) efficiency is only approached *at best* at a single wind speed. As a result, present generation wind turbines generally fall well short of optimum performance.

Figure 6.8 shows a generic power curve for wind turbine operation at different wind speeds. In the Region II wind speed range, the discrepancy between the actual power (solid red) and the ideal power (dashed blue) is the result of aerodynamic losses. To understand the roots of the aerodynamic inefficiency of modern wind turbines, the factors governing the aerodynamic performance are examined. This involves the tip speed ratio, coefficient of power, and the axial induction factor.

Recall that the rotor blade tip speed ratio, λ is

$$\lambda = \frac{\Omega R}{V_\infty}. \tag{6.18}$$

The power generated from the wind is

$$P_{aero} = Q\Omega, \tag{6.19}$$

where Q is the total torque generated by the rotor.

The coefficient of power, C_p, is the ratio of the aerodynamic power extracted from the wind and the available aerodynamic power or,

$$C_p = P_{\text{aero}}/P_{\text{available}}. \tag{6.20}$$

The local axial and tangential induction factors are defined as

$$a = 1 - \frac{V_x}{V_\infty}, \tag{6.21}$$

and

$$a' = \frac{V_y}{\Omega r} - 1, \tag{6.22}$$

where V_x and V_y are the respective axial and tangential velocities in the rotor plane, and r is the radial position along the rotor, measured from the rotor hub.

The local flow angle at a given radial location on the rotor is then

$$\phi_r = \tan^{-1}\left(\frac{V_y}{V_x}\right) = \tan^{-1}\left(\frac{V_\infty(1-a)}{\Omega r(1+a')}\right) = \tan^{-1}\left(\frac{(1-a)}{(1+a')\lambda_r}\right), \tag{6.23}$$

where λ_r is the local tip speed ratio at the radial position, r.

The local effective rotor angle of attack at any radial location is then

$$\alpha_r = \phi_r - \theta_{T_r} - \theta_{cp}, \tag{6.24}$$

where ϕ_r is again the local flow angle, θ_{T_r} is the local rotor twist angle, and θ_{cp} is the global rotor pitch angle which is constant over the rotor radius.

The local lift and drag coefficients, $C_l(r)$ and $C_d(r)$, at a radial location on the rotor are then

$$C_l(r) = C_y \cos(\phi_r) - C_x \sin(\phi_r), \tag{6.25}$$

and

$$C_d(r) = C_y \sin(\phi_r) + C_x \cos(\phi_r), \tag{6.26}$$

where C_x and C_y are the force coefficients in the tangential and normal directions of the rotor section at the effective angle of attack, α_r. Note that C_x and C_y respectively are the drag and lift coefficients for the local (r) 2-D rotor section shape at the effective angle of attack, α_r.

The differential torque produced by radial segment of the rotor at radius, r, is

$$dQ = 4\pi\rho V_\infty(\Omega r)a'(1-a)r^2dr - \frac{1}{2}\rho V_R^2 BcC_d\cos(\phi_r)rdr. \tag{6.27}$$

where again V_R is the resultant velocity component which is a combination of the free-stream wind speed and the rotation speed of the rotor, and B is the number of rotor blades.

In order to simplify the calculation, the second term in Equation 6.27 is dropped. This is equivalent to neglecting the drag on the rotor, which is a good assumption as long as the rotor is not stalled (i.e., the local angle of attack is in the linear lift vs. angle of attack region). This gives the following form for the differential torque

$$dQ = 4\pi\rho V_\infty(\Omega r)a'(1-a)r^2dr. \tag{6.28}$$

Substituting for a' in terms of a gives

$$dQ = 4\pi\rho V_\infty^2\frac{a(1-a)^2r^2}{\lambda}dr. \tag{6.29}$$

Assuming constant wind conditions (ρ and V_∞) and a fixed tip speed ratio, λ, then

$$dQ = C_1a(1-a)^2r^2dr, \tag{6.30}$$

where C_1 lumps all of the constants into one.

For analysis purposes, the axial induction factor, ideal or otherwise, is assumed to be constant along the entire rotor span. Then the total torque is proportional to the axial induction factor namely,

$$Q \propto a(1-a)^2. \tag{6.31}$$

In terms of the aerodynamic power,

$$P_{\text{aero}} = Q\Omega, \tag{6.32}$$

and therefore

$$P_{\text{aero}} \propto a(1-a)^2. \tag{6.33}$$

Figure 6.9 shows a plot of the right-hand side of Equation 6.33, $a(1-a)^2 \equiv A$ versus a. This illustrates that the maximum occurs at $a = 1/3$, which agrees with the rotor disk analysis that predicted the Betz power limit at $a = 1/3$.

To help in quantifying the possible gains in power if the optimum $a = 1/3$ were achieved, the ratio of the "ideal" $A = a(1-a)^2$ where $a = 1/3$, designated A_I, to the "non-ideal" A_{NI} are plotted in Figure 6.10. This is represented as a percent improvement in the power coefficient in Figure 6.11. This indicates that there is a greater penalty in having a lower than optimal axial induction factor than having a higher than optimal value. As an example based on

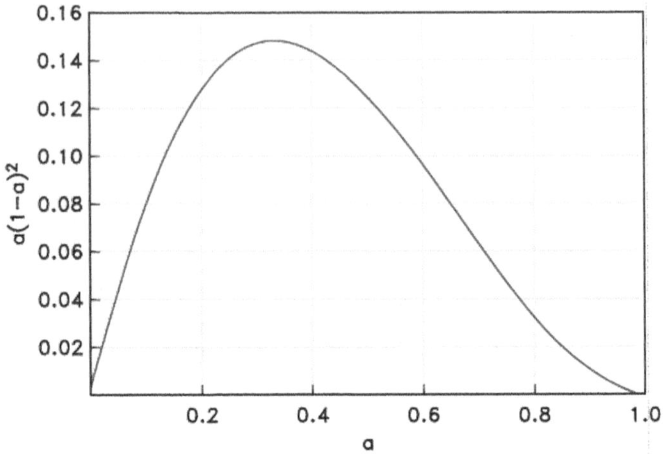

FIGURE 6.9
Plot of $A = a(1-a)^2$ versus a showing that the maximum occurs at $a = 1/3$ [2,3].

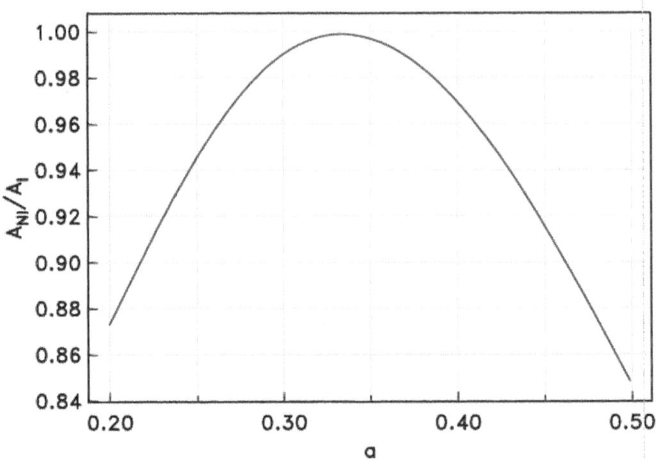

FIGURE 6.10
Plot of ratio of the "not ideal" (NI) to the "ideal" (I) values of $a(1-a)^2$ versus a [2].

Figure 6.9, an a that is 30% lower than optimal will result in a 10% loss in power. However, a value of a can be 40% higher before reaching the same 10% loss in power.

Figures 6.12–6.15 examine the effect of an imperfect axial induction factor on a current generation multi-megawatt wind turbine. This is performed for

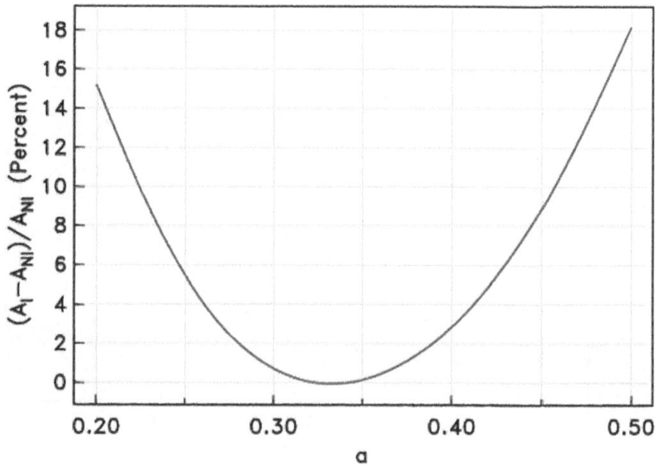

FIGURE 6.11
Plot of percent improvement obtained by optimizing the axial induction factor [2,3].

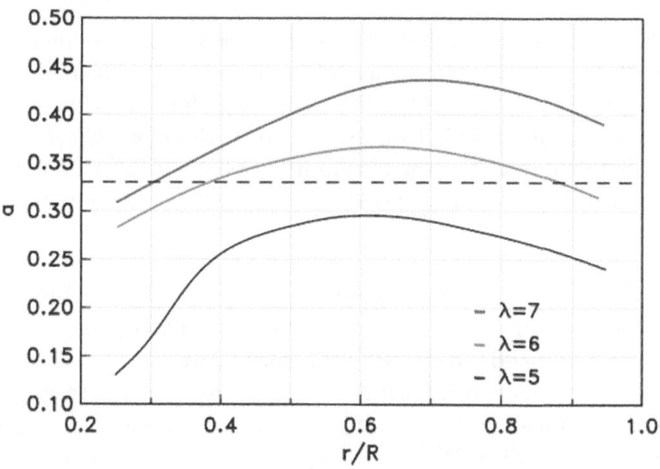

FIGURE 6.12
Plot of the rotor radial distribution of the axial induction factor for three tip speed ratios of an existing current-generation wind turbine [2,3].

three tip speed ratios of $\lambda = 5$, 6 and 7, which bracket the optimum tip speed ratio for the wind turbine.

Figure 6.12 shows the radial distribution of the axial induction factor for the three tip speed ratios. This illustrates that axial induction factor varies significantly along the rotor span, and seldom is the ideal 1/3 value.

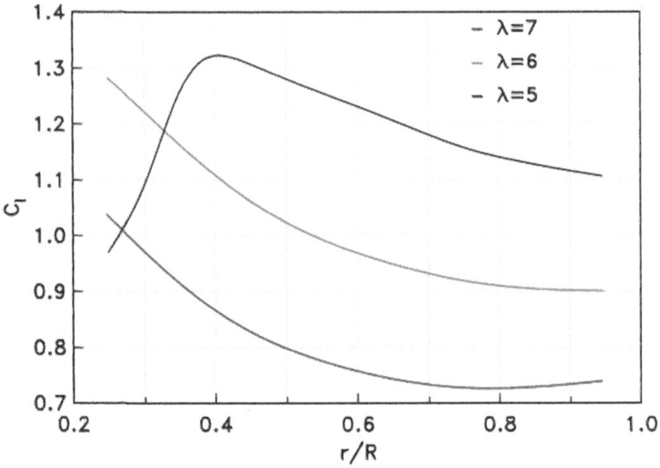

FIGURE 6.13
Plot of the rotor radial distribution of the lift coefficient for three tip speed ratios of an existing current-generation wind turbine [2].

Figure 6.13 shows the radial distribution of the lift coefficient that corresponds to the axial induction factor as shown in Figure 6.12. Figure 6.14 shows the radial distribution of the lift coefficient that would produce the ideal axial induction factor of 1/3. Finally, Figure 6.15 shows the radial distribution of the difference between the actual lift coefficient distributions at a given tip speed ratio in Figure 6.13, and the "ideal" lift coefficient distributions in Figure 6.14.

The change in the lift distribution that is shown in Figure 6.15 is required to achieve the Betz limit for this current generation wind turbine. If this were to occur, it would result in increases in the coefficient of power of 4.1%, 0.03% or 2.9% for the tip speed ratios of 5–7, respectively.

To put this in perspective, a wind farm rated at 100 MW (approximately 65 1.5 MW wind turbines) and operating with a reasonable 35% capacity factor would produce about 307 GW-h of energy in a given year. If the cost of energy was $0.04 per kW-h, each GW-h is worth about $40,000, meaning that a 1% loss of energy on this wind farm is equivalent to a loss of $123,000 per year. A 4% improvement in the power would result in approximately $500 K profit for the wind farm.

Finally the analysis assumes ideal conditions, that is a uniform wind distribution, from a single wind direction, without gusts, and free of wakes of other wind turbines. Under non-ideal conditions such as these, the improvement in the coefficient of power from lift control aimed at optimizing the axial induction factor, could double.

The next topic is how the rotor lift could be controlled in a responsive manner. This goes back to basic aerodynamic lift control for airfoils.

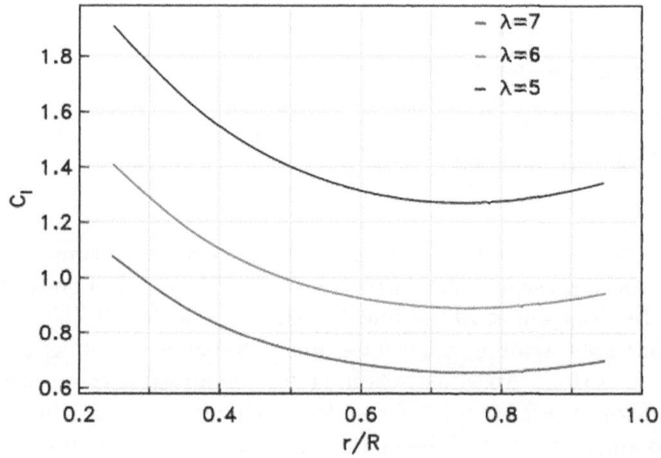

FIGURE 6.14
Plot of the rotor radial distribution of the lift coefficient for which the axial induction factor is the ideal 1/3 for three tip speed ratios of an existing current-generation wind turbine [2,3].

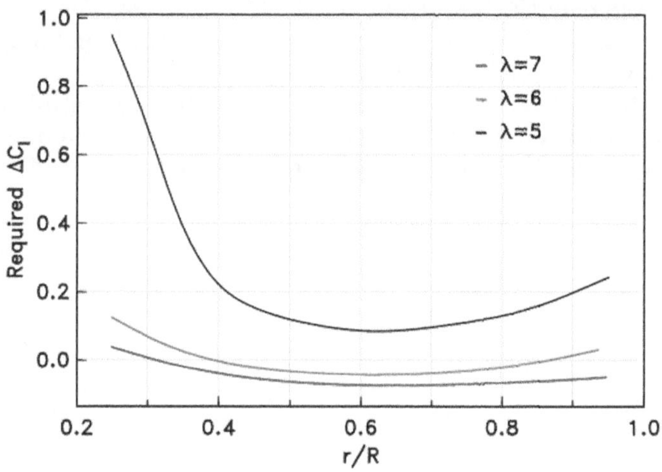

FIGURE 6.15
Plot of the rotor radial distribution of the change needed in the lift coefficient to achieve the ideal 1/3 axial induction factor for three tip speed ratios of an existing current-generation wind turbine [2,3].

6.4 Lift Control

Lift control techniques that have been developed for general airfoils can be applied to wind turbine rotors. These include

1. plane trailing edge flaps
2. split trailing edge flaps
3. Gurney flaps
4. trailing edge blowing
5. plasma actuators.

Figure 6.16 provides a comparison of the lift control performance of many of the lift control approaches. A comprehensive comparison of these approaches is provided by Johnson et al. [4] and Barlas and van Kuuik [5].

Plane and split trailing edge flaps have the same effect as changing the camber of an airfoil. An illustration [4] of an airfoil section with positive camber is shown in Figure 6.17. An airfoil with zero camber will produce zero lift at a zero angle of attack. The angle of attack where zero lift occurs is called the "zero lift angle of attack" and denoted as α_{0L}. An airfoil with positive camber will move α_{0L} to a negative angle of attack so that at a zero angle of attack, lift is produced. Importantly, the minimum drag will occur at α_{0L}.

Plane and split trailing edge flaps produce the same effect as adding camber to a wing section. A downward deflection of a trailing edge flap is equivalent to adding positive camber. A plane flap pivots the whole trailing edge. A split flap pivots only the bottom half of the trailing edge. The top half of the trailing edge remains fixed. The lift versus angle of attach and drag polar are shown for a plane flap in Figure 6.18.

As indicated by Figure 6.15, spanwise lift distribution control on the rotor is needed to realize an optimum axial inflow coefficient of 1/3 over the entire rotor. One approach is the use of spanwise segmented trailing edge flaps such

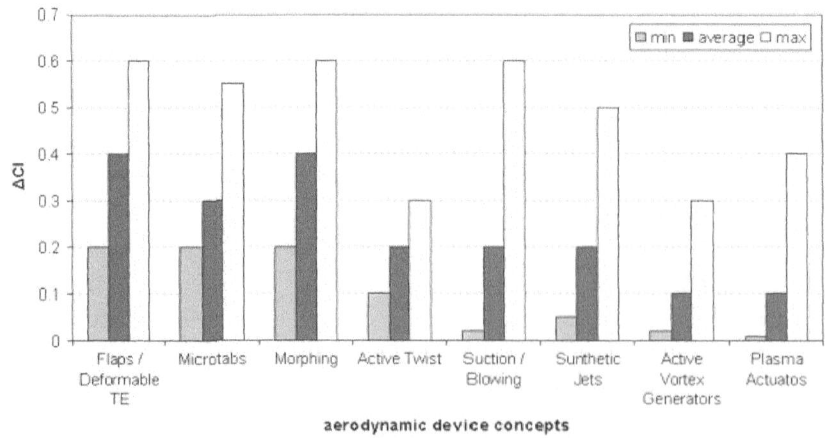

FIGURE 6.16
Comparison of the performance of different active lift control approaches [5].

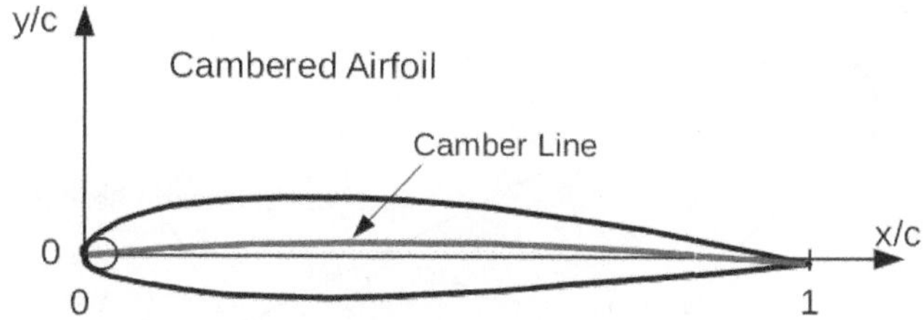

FIGURE 6.17
Airfoil section illustrating positive camber.

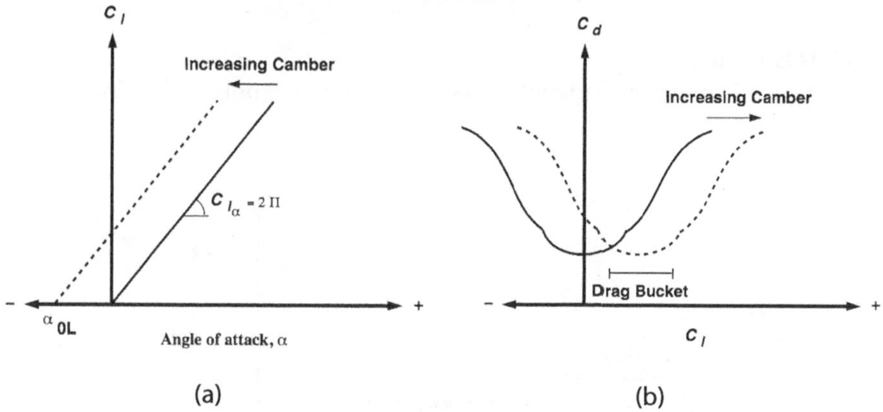

FIGURE 6.18
Lift as a function of angle of attack (a) and drag polar (b) for a zero camber airfoil (solid curve) and with a plane trailing edge flap with downward deflection (dashed curve) [6].

as shown in Figure 6.19. Another approach given by Williams et al. [7,8] involved a spanwise variation in the trailing edge geometry obtained through a multi-objective design optimization approach.

A variation on a split flap is a Gurney flap. This consists of a vertical fence that sits on the surface of an airfoil near the trailing edge. A schematic showing a Gurney flap at the very trailing edge of an airfoil section is shown in Figure 6.20. The Gurney flap causes a flow separation to occur upstream and downstream of the flap which changes the pressure distribution at the trailing edge, and subsequently the lift force on the airfoil. A Gurney flap on the lower surface (pressure side) of an airfoil will increase lift. This is the example shown in Figure 6.20. A Gurney flap on the upper surface (suction side) will produce negative lift. The general rule of thumb for Gurney flaps is that their height should range between 1% and 1.5% of the airfoil chord length, and that their

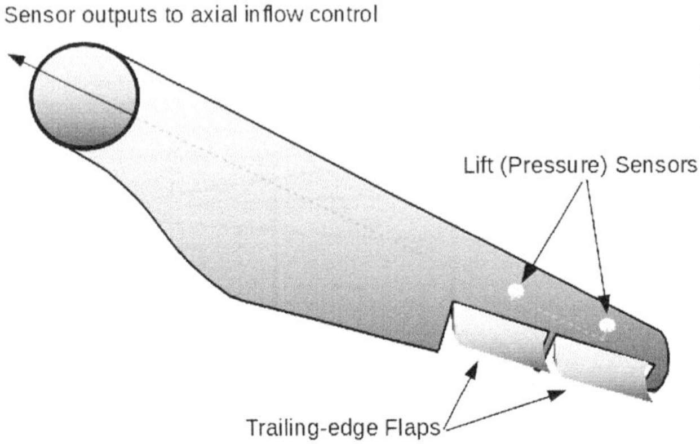

FIGURE 6.19
Illustration of spanwise segmented flaps as it might apply to a wind turbine rotor.

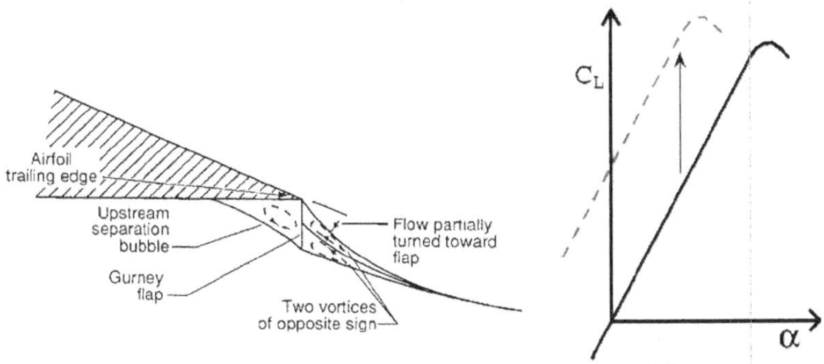

FIGURE 6.20
Illustration of a Gurney flap for lift control [9].

position should be from 0% to 10% of the chord length from the trailing edge of the airfoil [4]. The largest effect occurs when the Gurney flap is placed at the exact trailing edge. An illustration of multiple spanwise Gurney flaps for spanwise varying lift control [10] is shown in Figure 6.21. In this arrangement, the Gurney flap segments would be extended or retracted to provide spanwise lift control.

In contrast to trailing-edge flaps, plasma flow control does not involve any moving parts. It consists of two electrodes that are separated by a thin dielectric material layer. The electrodes are usually staggered. The arrangement is usually applied to an aerodynamic surface such as a wing, or in this case a wind turbine rotor blade. The electrodes are powered by an AC voltage.

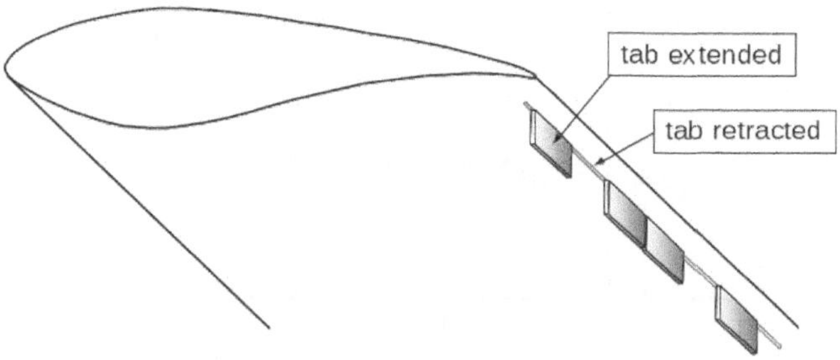

FIGURE 6.21
Illustration of multiple spanwise Gurney flaps for spanwise varying lift control.

When the voltage is large enough, the air over the surface of the dielectric layer ionizes, causing it to be conductive. The conductive air in the presence of the electric field results in a body force vector field. In the staggered electrode arrangement, this induces a flow field that is similar to a tangential wall jet. When this is located near the trailing edge, it produces an effect that is similar to a trailing edge flap to control lift. When placed near the leading edge it produces an effect like a slotted leading-edge to control leading-edge flow separation. A comprehensive review of plasma flow control is given by Corke et al. [11].

A variation on the Gurney flap concept shown in Figure 6.21 that uses plasma actuators is shown in Figure 6.22. In this case the upstream side of the Gurney flaps are rounded to allow the placement of a plasma actuator. The Gurney flaps always remain in the retracted position. The plasma actuator controls the degree of flow separation that occurs downstream of the Gurney flaps, which then controls the amount of added lift produced by the Gurney flap. As illustrated at the top of the figure, when the plasma actuator is not operating (Off) the Gurney flap produces extra lift. When the plasma actuator is operating (On) the effect of the Gurney flap is removed, and the lift decreases. Williams [8] performed a parametric design optimization on plasma Gurney flaps to determine the optimum size and placement to maximize the change in lift for lift control on the Notre Dame wind turbine rotors.

Problems

1. Consider the the power coefficient versus tip speed ratio for the sample turbine performance with a fixed pitch angle of $\beta = -1°$ that was shown in Figure 6.7. This was based on a 600 kW wind turbine with a 43 m. diameter rotor.

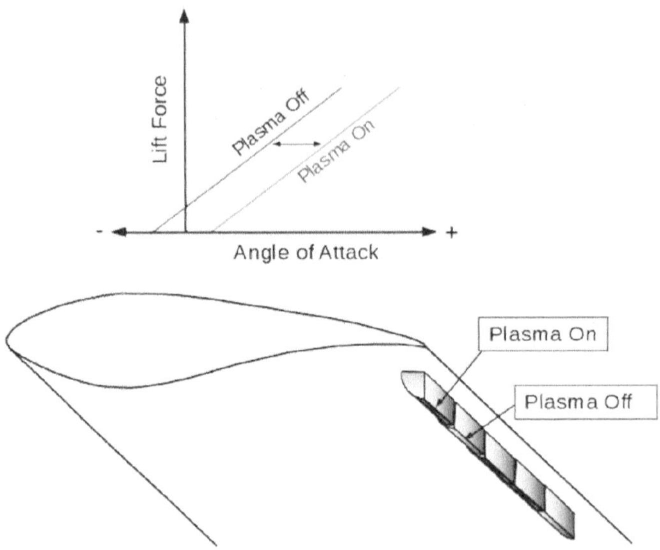

FIGURE 6.22
Illustration of multiple spanwise plasma Gurney flaps for spanwise varying lift control.

(a) Utilizing the relationships given in Section 6.2.3, starting with

$$P = Q_{\text{aero}}\omega, \tag{6.34}$$

show that

$$P = \frac{1}{2}\rho A V^3 C_p, \tag{6.35}$$

(b) Based on the power coefficient versus tip-speed ratio for the wind turbine, plot the normalized power generated, P/P_* as a function of the ratio λ/λ_* where P_* is the power at the maximum C_p, and λ_* is the optimum tip-speed ratio.

Do this for wind speeds of 7 and 10 m/s and for a range of tip speed ratios from 4 to 14.

(c) What is the implication of the need for wind turbine optimum tip-speed ratio control?

(d) If the cost of electricity was \$0.08/kW-h, plot the power as a function of tip speed ratio as dollars revenue. What is the impact of optimum tip-speed ratio control in terms of generated revenue?

2. Based on the power curve shown in Figure 6.8, compare the power generated in wind speed Region 2 (between $V_{\text{cut-in}}$ and V_{rated}) for the Ideal (Betz) curve versus the Non-ideal curve. Do this by

summing the area under the respective curves for the complete Region 2.

(a) What is the percentage of power of the ideal power that is lost by the non-ideal operation?

(b) If the cost of electricity was \$0.08/kW-h, what is the amount of lost revenue that results from the non-ideal operation?

3. For the case study resulting in Figure 6.15, a nominal increase in the rotor blade lift coefficient of $\Delta C_l = 0.2$ is sufficient to produce an ideal axial induction factor for Betz efficiency.

(a) Based on Figure 6.16, which of the lift control approaches would satisfy this requirement? Base this selection on their average operation.

(b) For more optimal tip-speed ratios for this wind turbine of $6 \leq \lambda \leq 7$, the necessary increase in the lift coefficient for ideal operation is $\Delta C_l \simeq 0.1$. In this case, which of the active lift control approaches satisfies this requirement under average operation?

(c) Based on the previous $\Delta C_l \simeq 0.1$ requirements, which of the active lift control approaches under minimum operation, would be suitable?

References

1. K. Johnson, "Adaptive torque control of variable speed wind turbines", NREL/TP-500-36265, August, 2004.
2. J. Cooney, "Increasing power generation in horizontal axis wind turbines using optimized flow control", Ph.D. Thesis, University of Notre Dame, 2014.
3. J. Cooney, T. Williams, and T. Corke, "Improve power coefficient of horizontal axis wind turbines using optimized lift control", *32nd ASME Wind Energy Symposium*, Paper AIAA-2014-1218, AIAA SciTech, January, 2014, DOI: 10.2514/6.2014-1218.
4. S. Johnson, C. P. van Dam, and D. Berg, "Active load control techniques for wind turbines", Sandia Report SAND2008-4809, 2008.
5. T. Barlas and G. A. M. van Kuuik, "Review of state of the art in smart rotor control research for wind turbines", *Progress in Aerospace Sciences*, 2010, 46:1–27.

6. T. C. Corke, *Design of Aircraft*, Prentice-Hall Publishers, New York, 2002.
7. T. Williams, A. Jemcov, and T. Corke, "Airfoil shape optimization for dielectric barrier discharge plasma compliant flows", *AIAA Journal*, 2015, 53(10):3125–3129.
8. T. Williams, "Compliant flow designs for optimum lift control of wind turbine rotors", Ph.D. Thesis, University of Notre Dame, 2013.
9. Liebeck, "Design of subsonic airfoils for high lift", J. Aircraft, Also NASA TM-4071 (1978).
10. S. Johnson, J. Baker, C. P. van Dam, and D. Berg, "An overview of active load control techniques for wind turbines with an emphasis on microtabs", *Wind Energy*, 2013, 13:239–253. DOI:10.1002/we.356.
11. T. Corke, L. Enloe, and S. Wilkinson, "Dielectric barrier discharge plasma actuators for flow control", *Annual Review of Fluid Mechanics*, 2010, 42:505–529.

7

Structural Design

This chapter deals with the structural design of the rotor and tower for a horizontal axis wind turbine. This naturally follows from the aerodynamic design from which the aerodynamic loads are derived. As often happens in the design of aerodynamic systems, there needs to be a compromise between the aerodynamic optimum and the structural optimum. The latter seeks to optimize strength, weight and cost. Catastrophic failures of wind turbine structures are rare, but not impossible. Failures include delamination of the composite rotor structure, structural cracks in the rotor, the loss of a rotor, and buckling collapse of the tower.

Conditions leading to such structural failures include extreme winds, an inadequate control system, or cyclic-load fatigue that leads to cracks in the structure, particularly the rotor blades. Fatigue is a very important issue since wind turbines are designed to operate for a minimum of 20 years over which the rotor will rotate on the order of 10^9 revolutions! Some of the loads repeat with every revolution of the rotor, which results in a cyclic straining of the structure that could lead to strain hardening and brittle fracture.

There are four primary sources of loads that are relevant to horizontal axis wind turbines. These are

1. aerodynamic loads,
2. gravitational loads,
3. dynamic loads, and
4. control loads.

7.1 Aerodynamic Loads

Aerodynamic loads includes the lift, drag, and pitch moment on the rotor which can be determined by the BEM method as presented in Chapter 4. The resulting force vectors that act at a given radial location on the rotor are shown in the left part of Figure 7.1. When the forces are integrated along the rotor span, they result in spanwise distributions such as illustrated in the right part of Figure 7.1. As will be discussed in a later section, structurally, the rotor is a cantilever beam with a fixed attachment at the rotor hub. As

DOI: 10.1201/9781003611486-7

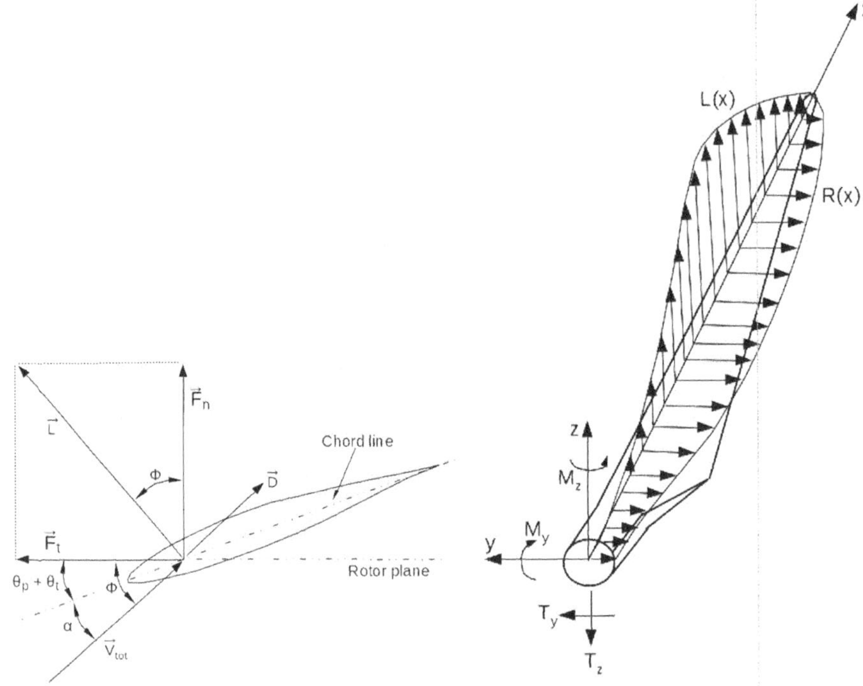

FIGURE 7.1
Force vectors based on BEM analysis (left) and illustration of 3-D lift and drag force distribution resulting in maximum shear forces and bending moments at the rotor root.

a result, the rotor root location experiences the largest bending moment and shear forces. The material stresses associated with these loads determine the structural design, which will be discussed in further detail.

The forces that act on the rotor can be transmitted through the rotor shaft to the gear box and tower. Structural failure of the gear box continues to be an important issue with horizontal axis wind turbines.

7.2 Gravitational Loads

Gravitational loads are primarily associated with the weight of the rotor blades. This is a cyclic loading whose magnitude on a radial element is

$$dF_g = \vec{g} dm \cos(\psi), \tag{7.1}$$

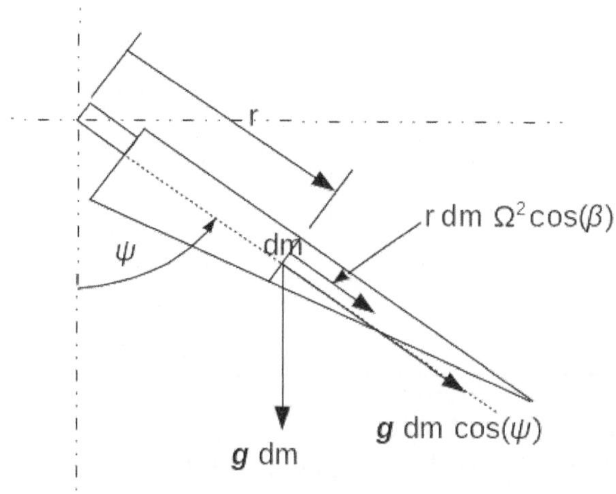

FIGURE 7.2
Illustration of gravitational and centrifugal loads acting on a spinning wind turbine rotor.

where dm is the mass of a radial element of the rotor at some radius, and ψ is the azimuthal angle of the rotor with $\psi = 0$ corresponding to the bottom dead center of the rotation cycle. This is illustrated in Figure 7.2. This loading alternately produces cyclic extension, compression and bending of the rotor with each rotation. The cyclic gravitational loading on the rotor is converted into a cyclic torque variation on the rotor shaft that is then transmitted to the gear box.

The gravitational loading generally acts through the rotor plane axis, except if the rotor bends out of plane, which is referred to as "flapping". The out of plane or flapping angle is defined as β. Figure 7.3 illustrates types of flapping motions. The left illustration, β_0, shows a rotor plane that is aligned with the wind direction, although the rotors are angled in the upwind direction, which is referred to as "coning". In this case, the loading on the blades is steady with respect to the rotor rotation angle, ψ.

In the middle illustration, β_{1c}, the axis of the rotor is aligned with the wind direction, but the coned rotor plane is canted upward. As a result, the rotor location that is tilted upwind (bottom portion) will have a larger effective angle of attack compared to the rotor that is tilted downwind. This will produce a cyclic loading with a magnitude that varies as $\cos(\psi)$, where again $\psi = 0$ corresponds to the bottom of the rotation cycle.

In the right illustration, β_{1s}, the axis of the coned rotor is yawed with respect to the wind direction. This also produces a cyclic loading whereby the rotor that tilts upwind (right portion) will have an effectively larger angle of attack compared to the rotor that tilts downwind. This will produce a cyclic loading with a magnitude that varies in this case, as $\sin(\psi)$.

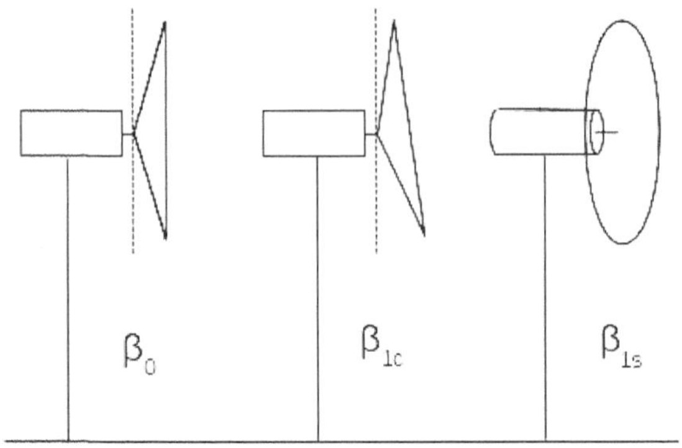

FIGURE 7.3
Illustration of types of coned or "flapping" rotor conditions of the horizontal axis wind turbine.

The cyclic loading produced in these last two cases is transmitted through the rotor to the main rotor shaft and gear box. In addition, it can result in forced vibration of the rotor that can lead to structural fatigue.

It is reasonable to sum the effects of the three coned rotor conditions illustrated in Figure 7.3 to obtain an effective flapping angle, β given as

$$\beta = \beta_0 + \beta_{1c}\cos(\psi) + \beta_{1s}\sin(\psi). \tag{7.2}$$

In this case β_0 represents the collective or coned response, and β_{1c} and β_{1s} are the coefficients representing the respective cosine and sine cyclic responses.

7.3 Dynamic Loading

Dynamic loading is the result of changes in the motion of rotor. An example is the centrifugal force generated by the rotation of the rotor. This is also illustrated in Figure 7.2 where the centrifugal force acting on a radial element of the rotor at some radius is

$$dF_c = rdm\Omega^2\cos(\beta), \tag{7.3}$$

where again β is the effective flapping angle given by Equation 7.2. The centrifugal force can be considered as a point load that acts on the center of mass of the rotor blade, and is directed perpendicular to the axis of rotation.

For a non-zero flapping angle β, the centrifugal force acting on the rotor will produce a bending moment at the rotor root location. The moment produced by the centrifugal force acting on a differential element at radius r

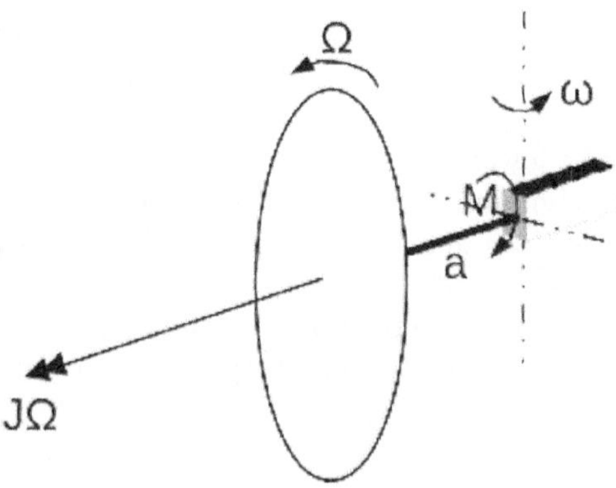

FIGURE 7.4
Illustration of the gyroscopic restoring moment produced by the yawed motion of the rotor.

is

$$dM_c = r\sin(\beta)\left[rdm\Omega^2\cos(\beta)\right]. \tag{7.4}$$

Another prominent example of dynamic loading are gyroscopic loads that are produced by yaw or flapping motions of the spinning rotor. Figure 7.4 illustrates the gyroscopic forces and moments that would act on a main rotor shaft of a horizontal wind turbine. The rotor is shown mounted on the main rotor shaft. The shaft is supported by a bearing block, which is considered to be a rigid. The unsupported length of the shaft is L.

Assuming that the rotor has a polar moment of inertia of J, and spins at a rate Ω, it will have an angular momentum of $J\Omega$. This is indicated by the double-arrow in Figure 7.4. Based on the theory of gyroscopes, if a body with angular momentum of $J\Omega$ is rotated about an axis that is perpendicular to the rotor Ω plane, it will generate a moment equal to the cross product, $\omega \times J\Omega$, where ω is the yawing rate. This yaw motion corresponds to the illustration shown in the right portion of Figure 7.3. The generated bending moment acts on the bearing block as indicated in Figure 7.4. A rotor pitching motion such as illustrated in the middle portion of Figure 7.3 would produce a bending moment that is 90° opposed to the one shown in Figure 7.4. These bending moments put stress on the rotor shaft and bearing block that could lead to structural failure unless compensated for in the design.

7.4 Control Loads

As was previously discussed in Chapter 6, wind turbines employ a control system that is designed to seek the highest efficiency of operation and ensure safe operation under all wind conditions. The wind turbine dynamic controllers make continuous high-speed changes in the operating conditions such as blade pitch, yaw and power management. Pitch-regulated rotors reduce the aerodynamic torque by reducing the pitch and thereby the local angle of attack of the rotor sections. The lower angles of attack reduce the section lift coefficient and thereby the aerodynamic torque on the rotor. They also employ electric torque control to seek to maintain an optimum tip-speed-ratio. Finally when the wind speed exceeds the cut-out value, the rotor is braked to a stop. These control operations can produce intermittent loads on the rotor, shaft and gear box that also need to be accounted for in the structural design.

7.5 Rotor Response to Loads

The horizontal axis wind turbine rotor is designed to be stiff and lightweight. To accomplish this, it is generally fabricated with a thin fiberglass-epoxy skin that is bonded to a central box-beam spar. The spar is designed to add stiffness to the rotor to resist bending and twist. An example of the construction is shown in Figure 7.5.

As suggested by the schematic of the 3-D rotor blade that is shown in the right part of Figure 7.1, the blade can be modeled as a cantilever beam. As such, classical beam theory can be applied whereby based on the loads and beam stiffness at different spanwise locations, the stresses and deflections can be computed. To accomplish this, the rotor blade is divided into small spanwise segments (similar to the BEM approach). This is illustrated in Figure 7.6 in which a segment of width dx is specified. The external loading of the rotor segment, pdx is known from the BEM analysis. This results in the shear forces, T and $T + dt$, and bending moments, M and $M + dM$ that act on the element.

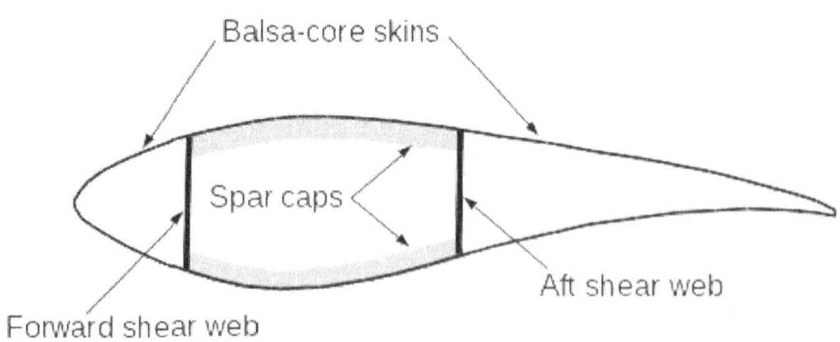

FIGURE 7.5
Section view of a HAWT rotor illustrating the internal structure.

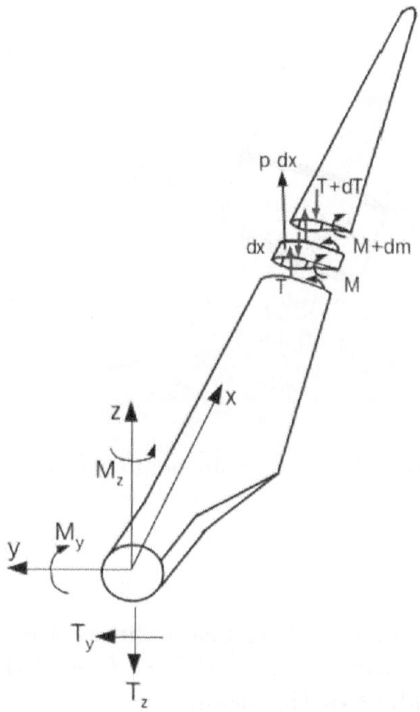

FIGURE 7.6
Illustration of shear force and bending moment on a small spanwise element of the loaded rotor.

A balance of forces and moments gives the following equations.

$$\frac{dT_z}{dx} = -p_z(x) + m(x)\frac{d^2u_z(x)}{dt^2} \tag{7.5}$$

$$\frac{dT_y}{dx} = -p_y(x) + m(x)\frac{d^2u_y(x)}{dt^2} \tag{7.6}$$

$$\frac{dM_y}{dx} = T_z \tag{7.7}$$

$$\frac{dM_z}{dx} = -T_y \tag{7.8}$$

The time derivative terms represent the inertia in the blade motion, where $m(x)$ is the mass of the blade element. If the blade is steadily deflected, the inertial terms are zero.

In order to determine the bending deflections of the rotor blade, it is necessary to determine the principle bending axes. In simple cross-section shapes (box beams and I-beams), this is straight forward. For an airfoil-shaped rotor, it is sometimes more complicated. Figure 7.7 illustrates possible principle axes

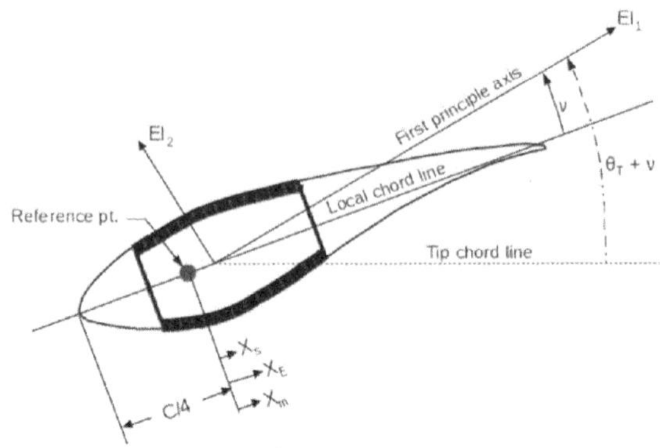

FIGURE 7.7
Spanwise element of rotor blade used in beam analysis to determine principle bending axis.

for a rotor blade section. Based on beam theory, the point of bending elasticity is defined as that where a normal force (out of the plane in Figure 7.7) does not produce bending of the beam. The shear center is defined as the point where an in-plane force will not rotate the beam section. If a beam is bent about one of the principle axes, it is only bent about that axis. With the first principle bending axis located, the bending stiffness about that axis is defined as EI_1 where E is the Young's modulus of elasticity of the material, and I is the bending moment of inertial (moment of area) of the cross-section. The bending stiffness about the second principle axis is defined as EI_2. The quarter-chord location is taken as a reference location against which other distances are defined. The quarter-chord location is generally the center of lift for subsonic airfoils and the point about which the pitching moment acts. The distance X_E is defined to be the distance of the point of elasticity from the reference point. Similarly, X_m is the distance of the center of mass from the reference point, and X_s is the distance of the shear center from the reference point. The twist angle of the airfoil section relative to the tip location is defined as before as θ_T. The angle ν is the angle between the chord line and the first principle axis. Finally, $\theta_T + \nu$ is the angle between the tip chord line and the first principle axis.

The transformation of the bending moments due to the loads in Figure 7.6 to those along the principle axes is then

$$M_1 = M_y \cos(\theta_T + \nu) - M_z \sin(\theta_T + \nu), \tag{7.9}$$

and

$$M_2 = M_y \sin(\theta_T + \nu) - M_z \cos(\theta_T + \nu). \tag{7.10}$$

If the airfoil section is symmetric (no camber), the first principle axis lies along the chord line, that is $\nu = 0$. Also for normally twisted blades, $\theta_T \leq 0$, although $(\theta_T + \nu)$ is considered to be positive.

From beam theory, the curvatures about the principle axes are

$$\kappa_1 = \frac{M_1}{EI_1}, \tag{7.11}$$

and

$$\kappa_2 = \frac{M_2}{EI_2}. \tag{7.12}$$

These curvatures are transformed back to the y and z axes by

$$\kappa_z = -\kappa_1 \sin(\theta_T + \nu) + \kappa_2 \cos(\theta_T + \nu), \tag{7.13}$$

and

$$\kappa_y = \kappa_1 \cos(\theta_T + \nu) + \kappa_2 \sin(\theta_T + \nu). \tag{7.14}$$

The angular deformations are then calculated as

$$\frac{d\theta_y}{dx} = \kappa_y, \tag{7.15}$$

and

$$\frac{d\theta_z}{dx} = \kappa_z. \tag{7.16}$$

Based on the angular deformations, the deflections, u_z and u_y are found by integrating

$$\frac{du_z}{dx} = -\theta_y, \tag{7.17}$$

and

$$\frac{du_y}{dx} = -\theta_z. \tag{7.18}$$

If the number of spanwise elements along the rotor blade are large enough, we can assume a linear variation in the loads between elements. This makes integrating the previous relations trivial, replacing integrals with summations using differential calculus.

As an example, if the rotor blade is divided up into N spanwise elements, where the Nth element is at the rotor tip, then the differential calculus form of Equations 7.5 and 7.6 are

$$T_y^{i-1} = T_y^i + \frac{1}{2}\left(p_y^{i-1} + p_y^i\right)\left(x^i - x^{i-1}\right) \; ; i = N, N-1, \ldots, 2, \tag{7.19}$$

and

$$T_z^{i-1} = T_z^i + \frac{1}{2}\left(p_z^{i-1} + p_z^i\right)\left(x^i - x^{i-1}\right) \; ; i = N, N-1, \ldots, 2. \tag{7.20}$$

Similarly, Equations 7.7 and 7.8 take the form

$$M_y^{i-1} = M_y^i - T_z^i\left(x^i - x^{i-1}\right) - \left(\frac{1}{6}p_z^{i-1} + \frac{1}{3}p_z^i\right)\left(x^i - x^{i-1}\right)^2\ ;\ i = N, N-1, \ldots, 2, \tag{7.21}$$

and

$$M_z^{i-1} = M_z^i - T_y^i\left(x^i - x^{i-1}\right) - \left(\frac{1}{6}p_y^{i-1} + \frac{1}{3}p_y^i\right)\left(x^i - x^{i-1}\right)^2\ ;\ i = N, N-1, \ldots, 2. \tag{7.22}$$

Here we note that p_y and p_z in these equations have units of force per spanwise segment width, for example N/m, and where $x^i - x^{i-1}$ is the spanwise segment width for example with units of meters. Thus p_y and p_z are equivalent to dF_n/dr and dF_t/dr that are obtained from Equations 4.51 and 4.52 through BEM theory.

The deflections in the rotor blade are then found from

$$u_y^{i+1} = u_y^i - \theta_z^i\left(x^{i+1} - x^i\right) + \left(\frac{1}{6}\kappa_z^{i+1} + \frac{1}{3}\kappa_z^i\right)\left(x^{i+1} - x^i\right)^2\ ;\ i = 1, 2, \ldots, N-1, \tag{7.23}$$

and

$$u_z^{i+1} = u_z^i - \theta_y^i\left(x^{i+1} - x^i\right) + \left(\frac{1}{6}\kappa_y^{i+1} + \frac{1}{3}\kappa_y^i\right)\left(x^{i+1} - x^i\right)^2\ ;\ i = 1, 2, \ldots, N-1, \tag{7.24}$$

where the angular deformations, θ_y^i and θ_z^i are found from

$$\theta_y^{i+1} = \theta_y^i + \frac{1}{2}\left(\kappa_y^{i+1} + \kappa_y^i\right)\left(x^{i+1} - x^i\right);\ i = 1, 2, \ldots, N-1, \tag{7.25}$$

and

$$\theta_z^{i+1} = \theta_z^i + \frac{1}{2}\left(\kappa_z^{i+1} + \kappa_z^i\right)\left(x^{i+1} - x^i\right)\ ;\ i = 1, 2, \ldots, N-1, \tag{7.26}$$

where κ_z and κ_y are found from Equations 7.11–7.14.

Following the sample distributed load distribution on the rotor blade previously shown in Figure 7.1, the boundary conditions on the shear force are

$$\begin{aligned} T_y^N &= 0, \\ T_z^N &= 0, \\ T_y^1 &= \sum_i^N (R^i), \\ T_z^1 &= \sum_i^N (L^i). \end{aligned} \tag{7.27}$$

The boundary conditions on the moments are

$$\begin{aligned} M_y^N &= 0, \\ M_z^N &= 0, \\ M_y^1 &= \sum_i^N (L^i)(x^i), \\ M_z^1 &= \sum_i^N (R^i)(x^i). \end{aligned} \tag{7.28}$$

Finally assuming a rigid rotor support, the boundary conditions on the displacements are

$$\begin{aligned} u_y^1 &= 0, \\ u_z^1 &= 0. \end{aligned} \tag{7.29}$$

For reference, a uniformly loaded fixed-free cantilever beam of length L, with a uniform loading amplitude p, has a deflection distribution of

$$u_z(x) = -\frac{px^2}{24EI}\left(6L^2 - 4Lx + x^2\right), \tag{7.30}$$

so that

$$u_z(L) = -\frac{pL^4}{8EI}. \tag{7.31}$$

7.6 Rotor Vibration Modes

Rotor vibration is an important aspect of horizontal axis wind turbines because the long blades are partially elastic structures that are continually subjected to unsteady and cyclic loads that can excite a natural vibratory response. The presence of the vibrations can result in large deformations of the rotor blades that could result in material fatigue and structural failures.

Cantilever beam structures like that of a horizontal axis wind turbine rotor blade, exhibit natural vibration eigenmodes. An eigenmode is a vibrational state of an oscillatory system in which the frequency of vibration is the same for all elements. The frequencies of the eigenmodes of a system are known as its eigenfrequencies.

Starting for example with Equation 7.5, for a free vibration state, without any external loads,

$$\frac{dT_z}{dx} = m(x)\frac{d^2u_z(x)}{dt^2}. \tag{7.32}$$

For harmonic oscillation of the system, the displacement would be given by

$$u(t) = A(x)\sin(\omega t), \tag{7.33}$$

where ω is the associated eigenfrequency, and $A(x)$ is the eigenfunction. Therefore

$$\frac{d^2u_z(x)}{dt^2} \propto -\omega^2 u, \tag{7.34}$$

so that

$$\frac{dT_z}{dx} = -m(x)\omega^2 u_z(x), \tag{7.35}$$

and similarly

$$\frac{dT_y}{dx} = -m(x)\omega^2 u_y(x). \tag{7.36}$$

Comparing Equations 7.5 and 7.35, and Equations 7.6 and 7.36, it is apparent that eigenmodes can be found using the static beam equations that incorporate the external loads, namely

$$p_z = m(x)\omega^2 u_z(x), \tag{7.37}$$

and

$$p_y = m(x)\omega^2 u_y(x). \tag{7.38}$$

The solution of Equations 7.37 and 7.38 will lead to the lowest eigenfrequency mode, which for a cantilevered beam is known as the first flapping mode. Since the deflections, $u_z(x)$ and $u_y(x)$ in the equations are not known a priori, an iterative solution approach is necessary. The process [1] is as follows.

1. Start with uniform spanwise loading, p, in the z and y directions and calculate the deflections, u_z and u_y, based on static beam bending theory with a given EI based on material and cross-section geometry. Next calculate the eigenfrequency at the rotor tip using

$$\omega^2 = \frac{p_z^N}{u_z^N m^N}. \tag{7.39}$$

   where N is the tip location of the radially segmented blade.

2. Compute the new loading at all of the discrete spanwise locations, $i = 1, 2, \ldots, N$, as

$$p_z^i = \omega^2 m^i \frac{u_z^i}{\sqrt{(u_z^N)^2 + (u_y^N)^2}}, \tag{7.40}$$

   and

$$p_y^i = \omega^2 m^i \frac{u_y^i}{\sqrt{(u_z^N)^2 + (u_y^N)^2}}. \tag{7.41}$$

3. Recompute ω as in Step 1 using the loading, p_z^i and p_y^i, and apply that to obtain the next loading distribution.

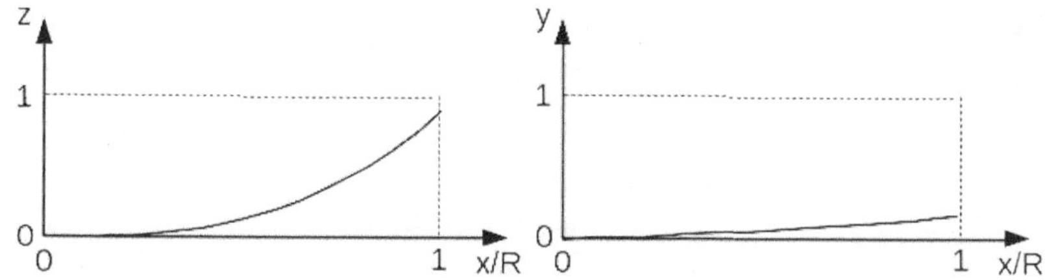

FIGURE 7.8
Deflection amplitude distribution for the first bending (flapping) eigenmode, u^{1f}, of a cantilevered beam that is representative of a HAWT rotor blade.

4. Repeat the procedure until the eigenfrequency converges to a constant value.
5. With the known value of ω, calculate the deflections at all of the discrete spanwise locations to the first flapping eigenmode shape, u_z^{1f} and u_y^{1f}. In this notation, the superscript "1f" refers to the first flapping mode. An example of the deflection amplitude distribution for the first flapping eigenmode is shown in Figure 7.8.

The procedure to determine the first edgewise (y-direction) eigenmode is similar to the procedure used in determining the first flapping mode, with one difference. In order for the iterative procedure to converge on the first edgewise mode, it is necessary that the motion of the first flapping mode be subtracted off, thus

$$u_z^{1e} = u_z - C_1 u_z^1 f, \tag{7.42}$$

and

$$u_y^{1e} = u_y - C_1 u_y^1 f, \tag{7.43}$$

where C_1 is a constant that is found by enforcing an orthogonality constraint between the u_z and u_y motions that is given by

$$\int_0^R u_z^{1f}(x)m(x)u_z^{1e}(x)dx + \int_0^R u_y^{1f}(x)m(x)u_y^{1e}(x)dx = 0. \tag{7.44}$$

Combining Equations 7.42–7.44 produces the equation for the C_1, namely

$$C_1 = \frac{\int_0^R u_z^{1f}(x)m(x)u_z(x)dx + \int_0^R u_y^{1f}(x)m(x)u_y(x)dx}{\int_0^R u_z^{1f}(x)m(x)u_z^{1f}(x)dx + \int_0^R u_y^{1f}(x)m(x)u_y^{1f}(x)dx}. \tag{7.45}$$

The previous iterative procedure outlined for the first flapping mode, is similarly followed to obtain the first edgewise eigenmode. However at each iteration, the returned displacements, u_z and u_y, will have the displacements of the first flapping mode subtracted off according to Equations 7.42 and 7.43,

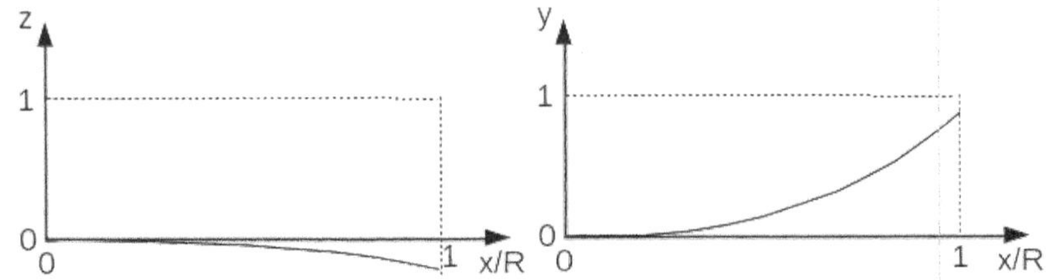

FIGURE 7.9
Deflection amplitude distribution for the first edgewise bending eigenmode, u^{1e}, of a cantilevered beam that is representative of a HAWT rotor blade.

with C_1 found through Equation 7.49. Convergence of the solution again corresponds to a reaching a constant eigenfrequency, ω. With the known value of ω, the deflections at all of the discrete spanwise locations, u_z^{1e} and u_y^{1e}, can be determined. Figure 7.9 shows an example of the deflection amplitude distribution for the first edgewise eigenmode.

The second flapping eigenmode is found by a similar procedure to that for the other two eigenmodes. In this case, it is necessary to subtract of the deflections of both the first flapping and first edgewise eigenmodes to properly achieve convergence of the iterative procedure to the second flapping eigenmode. Therefore at each iteration, the returned displacements, u_z and u_y, will have the displacements of the first flapping mode and the first edge mode subtracted off. This is achieved in the following equations:

$$u_z^{2f} = u_z - C_1 u_z^1 f - C_2 u_z^1 e, \tag{7.46}$$

and

$$u_y^{2f} = u_y - -C_1 u_y^1 f C_2 u_y^1 e, \tag{7.47}$$

where C_1 is again given by Equation 7.49, and C_2 was again found from the orthogonality condition

$$\int_0^R u_z^{1e}(x)m(x)u_z^{2f}(x)dx + \int_0^R u_y^{1e}(x)m(x)u_y^{2f}(x)dx = 0, \tag{7.48}$$

which when combined with Equations 7.46 and 7.47 gives

$$C_2 = \frac{\int_0^R u_z^{1e}(x)m(x)u_z(x)dx + \int_0^R u_y^{1e}(x)m(x)u_y(x)dx}{\int_0^R u_z^{1e}(x)m(x)u_z^{1e}(x)dx + \int_0^R u_y^{1e}(x)m(x)u_y^{1e}(x)dx}. \tag{7.49}$$

Convergence of the iterative process again is signified by reaching a constant eigenfrequency, ω. With the known value of ω, the deflections at all of the discrete spanwise locations, u_z^{2f} and u_y^{2f}, can be determined. Figure 7.10 shows an example of the deflection amplitude distribution for the second flapping eigenmode.

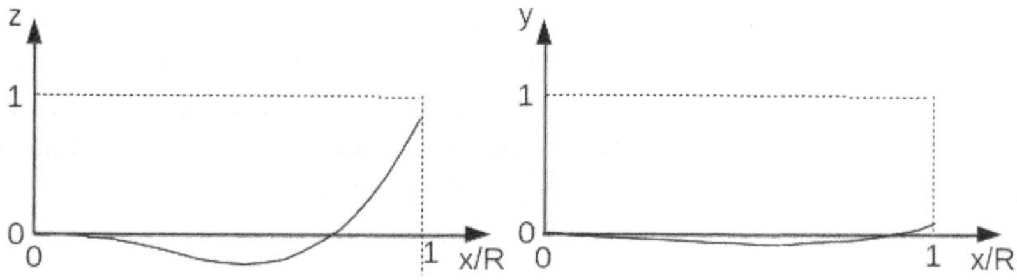

FIGURE 7.10
Deflection amplitude distribution for the second flapping eigenmode, u^{2f}, of a cantilevered beam that is representative of a HAWT rotor blade.

7.7 Design for Extreme Conditions

As mentioned at the start of this chapter, wind turbines are designed to operate for a minimum of 20 year. Over this amount of time, wind turbines are exposed to a broad range of wind conditions. The structural design has to account for the upper extremes in the steady and unsteady wind speeds which result in extremes in the steady and unsteady aerodynamic loads on the rotor.

A standard that is utilized in estimating extreme wind loads on structures such as buildings and bridges is to estimate the maximum wind speed based on a 10 minute averaged mean value, $\bar{V}$, plus three standard deviations, 3σ, of the probability distribution of wind speeds over a long period of time, for example over a month to a year. The wind speed statistics near the site of a wind turbine or wind farm can be obtained from the nearest airport. As a standard, this is measured at an elevation of 10 m. As presented in Chapter 2 for the atmospheric boundary layer, the wind speed varies as the natural log of elevation so that

$$\bar{V}(z) = \bar{V}(10)\frac{\ln(z/z_0)}{\ln(10/z_0)}, \tag{7.50}$$

where z_0 is the roughness height at the location where the velocity measurement was taken. If the extreme wind speed is

$$V_{3\sigma} = \bar{V} + 3\sigma, \tag{7.51}$$

then accounting for different elevations

$$V_{3\sigma}(z) = V_{3\sigma}(10)\frac{\ln(z/z_0)}{\ln(10/z_0)}. \tag{7.52}$$

For a Gaussian distribution of wind speeds, σ is the root-mean-square (r.m.s.) of the wind speed time series and has units of velocity (such as m/s).

The technical criteria for certification of wind turbines in Denmark provides an approach for the design of the internal structural elements of a wind turbine rotor to withstand extreme wind conditions. These standards are embodied in the document, DANSK DS 472 E-2008 [2]. This states that the loads on the wind turbine rotor be computed based on the wind speed at the $2/3R$ location with the rotor at the top-dead-center position. Therefore, the elevation used in determining the wind speed is

$$z = z_{\text{hub}} + \frac{2}{3}R. \tag{7.53}$$

The aerodynamic load on the rotor is

$$L(r) = \frac{1}{2}\rho\bar{V}^2 C_l c(r), \tag{7.54}$$

where C_l is the lift coefficient, and $c(r)$ is the local chord dimension. The units of $L(r)$ are force/meter-span. The DS 472 standard is to use $C_l = 1.5$. This is a nominal maximum lift coefficient for section shapes typically used with HAWT rotors.

As a test case, we consider a 1.5 MW wind turbine with a hub height of 65 m. and a rotor radius of 38 m. A smooth terrain is assumed with a roughness height of $z_0 = 0.010$ m. The rotor chord is assumed to be constant along the span, and equal to the mean aerodynamic chord of 1.3 m. The mean wind speed at the nearest airport was reported to be 27 m/s. The wind speed temporal distribution was Gaussian with an r.m.s. variation of 10% of the mean velocity, or 2.7 m/s.

Based on the listed conditions, the 3σ wind speed at the elevation of the 2/3R location in the top-dead-center position is

$$V_{3\sigma}(z) = [27 + (3)(2.7)]\frac{\ln(90.3/0.01)}{\ln(10/0.01)} = 46.3\,\text{m/s}. \tag{7.55}$$

The aerodynamic load on the section of the rotor at the 2/3R location, where the air density is taken as 1.28 kg/m^3, is then

$$L(R) = \frac{1}{2}(1.28)(46.2)^2(1.5)(1.3) = 2{,}673.4\,\text{N/m}. \tag{7.56}$$

We estimate the shear force and bending moment on the rotor at a near-root location of $r = 1$m. To accomplish this, we assume that the aerodynamic force acting at $r = 2/3R$ are acting over the entire span of the rotor. The shear force is then

$$T = \int_r^R L(r)dr = L(2/3R)(r)]_{r=1}^{r=38} = 98{,}916.2\,\text{N}. \tag{7.57}$$

The bending moment at $r = 1$ m. on the rotor is

$$M = \int_r^R rL(r)dr = \frac{1}{2}L(2/3R)(r^2)]_{r=1}^{r=38} = 1{,}928{,}865.1\,\text{N-m}. \tag{7.58}$$

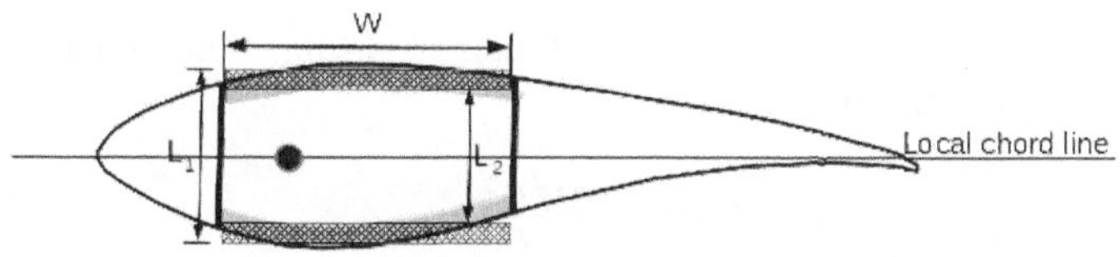

FIGURE 7.11
Simplified internal structure of a HAWT rotor designed to resist bending moments extreme wind loads.

The shear load and bending moment at $r = 1$ m. on the rotor represents the maximum estimated condition on which the internal structure is designed. As an example of this process, we take a simplified internal structure to the rotor section as shown in Figure 7.5. This simplified internal structure is shown in Figure 7.11. The structural elements are shown by the two cross-hatched rectangular strips that run parallel with the chord line of the airfoil section. These strips are meant to represent the thicker skin spar caps that follow the contour of the rotor section. The strips in our simplified version are located an equal distance on either side of the local chord line. The vertical spacing between the two strips is L_2. The strips are of equal thickness which is then $T = L_1/2 - L_2/2$.

For this simplified structure geometry, and assuming that the principle bending axis coincides with the local chord line, the bending moment of inertia is

$$I = \frac{1}{12} W (L_1^3 - L_2^3). \tag{7.59}$$

Substituting for L_2 in terms of the strip thickness t

$$I = \frac{1}{12} W (L_1^3 - [L_1 - 2T]^3). \tag{7.60}$$

In order to withstand the loads at the root span location of the rotor, the thickness to chord is large, nominally a thickness-to-chord ratio, $t/c = 0.35$. For this case study with a mean chord length of 1.3 m, the section maximum thickness is 0.46 m. The spar caps are generally a thickening of a portion of the skin of the rotor section. As a result that thickness, T, is generally much less than the thickness of the section, t. As a result, using the present notation in Figure 7.11,

$$T/t \ll 1. \tag{7.61}$$

As a result, Equation 7.60 simplifies to

$$I \simeq \frac{2}{3} W L_1^2 T. \tag{7.62}$$

For pure bending due to a positive lift component on the rotor, the lower strip will be under tension, and the upper strip will be under compression.

We will consider failure to be due to tensile loading. The tensile stress, σ_t, due to bending for the lower strip is

$$\sigma_t = \frac{ML_1/2}{I}, \tag{7.63}$$

where σ_t has units of force/area. To prevent failure of the structure, the tensile stress in the strip needs to be less the ultimate stress for the material, namely

$$\sigma_t < \sigma_{tu}. \tag{7.64}$$

Combining Equations 7.63 and 7.62, and applying the ultimate stress criteria (Equation 7.64) one obtains a relation for the minimum thickness of the structural strips, T_{min}, namely

$$T_{\text{min}} = \frac{M}{\frac{4}{3}WL_1\sigma_{tu}}. \tag{7.65}$$

Following the sample case study, if L_1 is taken as the section maximum thickness, then $L_1 = 0.46$ m. Based the cap strip design that was shown in Figure 7.5, $W/c = 0.35$ so that $W = 0.46$. Finally, the HAWT rotors are generally fabricated as glass-epoxy composite. Data for a 55% glass fiber volume glass-epoxy composite give an ultimate tensile stress, $\sigma_{tu} = 1,100$ MPa. Substituting these values into Equation 7.65, we obtain the minimum thickness of the structural spar caps that are required to withstand the aerodynamic bending loads at the $r = 1$ m. rotor location, namely

$$T_{\text{min}} = 0.006215\,\text{m} = 6.215\,\text{mm}. \tag{7.66}$$

We note that $T_{\text{min}}/t = 0.0135$ which substantiates the simplification used to obtain Equation 7.62.

Examples like this are useful to understand and refine the structural design of horizontal wind turbine rotors. This particular example only included the aerodynamic load. Flapping vibration of the rotor would produce additional loading that the structural design would also need to address. The steps for this are however similar to the present example, starting with the calculation of the maximum bending moment.

Problems

1. The BEM approach was used to determine the aerodynamic load distribution along a wind turbine rotor having geometric characteristics given below.

   Number of blades, $B = 3$

Tip speed ratio, $\lambda = 7$

Blade radius, $R = 4.953$ m.

Rated wind speed, $V_\infty = 11.62$ m/s

Rotor section shape, NACA 4415 airfoil

$C_l = 0.368 + 0.0942\alpha$

$C_d = 0.00994 + 0.000259\alpha + 0.0001055\alpha^2$

The angle of attack, α, has units of degrees

Rotor $\theta_{cp} = -2°$

r/R	$r(m)$	c(m)	θ_T	dF_n/dr (N/m)	dF_t/dr (N/m)
0.1	0.495	0.411	45	88.25	96.13
0.2	0.991	0.455	25.6	270.16	147.74
0.3	1.486	0.384	15.7	480.24	169.88
0.4	1.981	0.311	10.4	678.23	176.15
0.5	2.477	0.259	7.4	784.27	166.57
0.6	2.972	0.223	4.5	1176.41	192.36
0.7	3.467	0.186	2.7	1290.12	178.59
0.8	3.962	0.167	1.4	1517.54	178.04
0.9	4.458	0.137	0.4	1615.93	170.44
1.0	4.953	0.107	0.00	1553.83	153.81

Consider the structural properties of the rotor. For the NACA 4415 airfoil, the thickness-to-chord, t/c is 0.15. If we consider a box beam main spar that is internal to the rotor, as shown in Figure 7.5, with a width, $w = 0.35c$ and a height, $h = t = 0.15c$, the bending moment of inertia in the z (upward) direction is $I = wh^3/12$ or $I = (0.35c)(0.15c)^3/12$ or $I = 9.844 \times 10^{-5}c^4$. The following table summarizes the bending moment of inertia for the nine sections along the rotor.

r/R	$r(m)$	c(m)	$t(m)$	$I(m^4) \times 10^{-8}$
0.1	0.495	0.411	0.06165	280.89
0.2	0.991	0.455	0.06825	421.91
0.3	1.486	0.384	0.05760	214.04
0.4	1.981	0.311	0.04665	92.09
0.5	2.477	0.259	0.03885	44.30
0.6	2.972	0.223	0.03345	24.34
0.7	3.467	0.186	0.02790	11.78
0.8	3.962	0.167	0.02505	7.66
0.9	4.458	0.137	0.02055	3.46
1.0	4.953	0.107	0.01605	1.29

The modulus of elasticity for a thin glass/epoxy structure is $E_{xx} =$ 41 GPa. Therefore perform the following:

(a) Based on the lift force generated on each segment, calculate the shear force T_z for each segment along the rotor.
(b) Calculate and plot the rotor deflection, u_z, as a function of the radial position on the rotor. Assume that the principle bending axis is the airfoil section mean chord line so that $\nu = 0$.

2. The wind turbine used in the calculations in Section 7.7 on "Design for Extreme Conditions", had a rotor radius of 36 m. and a hub height of 65 m. The manufacturer has version of the same wind turbine with a slightly larger rotor radius of 39 m. and a higher hub height of 80 m. The wind turbine is intended to be located at the same site, where the mean wind speed is 27 m/s, and the temporal distribution is Gaussian, with an r.m.s. variation of 10%. Based on these conditions,

   (a) Determine $V_{3\sigma}$ at the hub height of the wind turbine. How does this compare to the smaller manufacturer version?
   (b) Estimate the shear force and bending moment of the rotor at the 1 m. radial location from the hub. How does this compare to the smaller manufacturer version?
   (c) Assume that the internal structure of the rotor blade is identical to the smaller version. Determine the minimum thickness, T, of the structural strips to prevent failure due to tensile loading. How does this compare to the smaller manufacturer version?

3. A wind turbine rotor such as shown schematically in Figure 7.4 has a rotor rotation rate of $\Omega = 0.159\ \text{s}^{-1}$ (or a frequency of 1 Hz.), and is yawing at an angular rate of $\omega = 0.175\ \text{s}^{-1}$ (or a velocity of $10°$/s). The polar moment of inertia of the rotor is $J = 13,558$ kg-m^2. The rotor weighs 1,459 kg and is 3.05 m from the bearing support that holds the main rotor shaft.

   (a) What is the load on the bearing when the turbine is not yawing?
   (b) What are the load on the bearing when the turbine is yawing?

4. Blade bending moments are being measured on a research wind turbine on a day in which the wind speed at the hub height is 9.0 m/s. A wind shear results in higher wind speed of 12.0 m/s at the top of the blade tip path and lower wind speeds of 6.0 m/s at the bottom of the blade tip path. A wind direction change results in a cross-wind of 0.60 m/s. The 24 m. diameter turbine starts with a flap hinge angle, $\beta_0 = 0.05$ radians and, at the moment that measurements of are being made, the rate of change of the flap hinge angle is $d\beta_{1s}/dt = 0.01$ rad/s.

   The rotor has a fixed rotation rate of $\Omega = 0.159\ \text{s}^{-1}$ (or a frequency of 1 hz.) and is 3.05 m/s from the yaw axis (bearing support). The blade pitch angle is $\theta_{cp} = 0.05$ radians ($2.86°$).

(a) Describe how the blade angle of attack would vary during one complete rotor revolution.

(b) List all of the effects these conditions have on the load on the bearing.

(c) Estimate the load due to the yawing motion of the rotor.

5. A new three-blade HAWT design having a down-wind rotor configuration (rotor is downwind of the tower) is proposed. The downwind rotor is to have a rotor diameter of 12.2 m. Each blade is 5.36 m. long, with a thickness of 7.62 cm. and an average width of 20.3 cm. The blades will be made of wood, which can be assumed to have a modulus of elasticity of 1.38×10^{10} Pa, and a weight density of 6,280 N/m. The rotor is designed to rotate at 120 rpm.

   (a) What are the pros and cons of having the rotor downwind of the tower?

   (b) If one models the rotor blades as a rectangle with the average width of the rotor and the same thickness and length, what is the frequency of the first bending mode?

   (c) For the proposed design, is there any possibility of exciting the first bending mode during blade rotation?

References

1. M. O. L. Hansen, *Aerodynamics of Wind Turbines*, 2nd Edition, Earthscan, London, 2008.
2. "Recommendation for Technical Approval of Wind Turbines", Design Basis for Offshore Wind Turbines, EFP-1363/99-0007, Denmark, December, 1999.

8

Wind Farms

Wind farms are a cluster of wind turbines that are located at a site to generate electricity. In the literature, wind farms are also sometimes referred to as a "plant", "array" or a "park". The first onshore wind farm was installed in 1980 on the shoulder of Crotched Mountain in southern New Hampshire, USA. It consisted of 20 wind turbines with rated power of 30 kW each, giving a combined capacity of 0.6 MW. The first offshore wind farm was built in 1991 off of the north coast of the Danish Island Lolland. It consisted of 11, 450 kW turbines that gave it a combined capacity of 4.95 MW.

The trend in the development of wind farms has been toward increased size and numbers of wind turbines that provide an overall larger power capacity. Typical modern wind farms consist of hundreds of wind turbines with multi-megawatt rated power that provide a total capacity of hundreds of megawatts. Photographs of modern onshore and offshore wind farms are shown in Figure 8.1. The multi-disciplinary nature and evolution towards larger size, smarter control, and more advanced capabilities of wind turbines has resulted in a more complex process of wind farm design. Often design objectives are constrained by aspects such as economic factors, operation and maintenance, environmental impact, and human factors play a significant role in the wind farm design.

Among all of the potential design objectives, one of the most critical is the arrangement of the wind turbines. The goal in this case is to determine the positions of the wind turbines within the wind farm to maximize or minimize some objective function(s). Examples include maximizing the energy production, or minimizing the cost, or environmental factors, under such constraints as finite wind farm size, noise emission standards, or initial investment limits. As a result, wind farm design optimization is a complex multi-objective problem that lacks an analytic formulation.

Different approaches toward wind farm design optimization have been proposed. These started with simplified formulations that ranged from an array of equally spaced turbines, to unequally spaced turbines, to a staggered grid arrangement. More complex arrangements have resulted from designs that evolved from randomly searched options using Monte Carlo methods and genetic algorithms.

 DOI: 10.1201/9781003611486-8

FIGURE 8.1
Photographs of modern onshore and offshore wind farms.

Source: General Electric Renewable Energy.

8.1 Wind Turbine Wake Effects

When a wind turbine extracts energy from the wind, it produces a cone-shaped wake of slower moving turbulent air. A remarkable photograph [1] that illustrates the wakes produced by wind turbines in an offshore wind farm is shown in Figure 8.2. The wakes of the wind turbines in the farm are made visible by a low level fog cover. This shows the long extent of the wakes of the most upwind wind turbines, that extends many rotor diameters downstream. Careful examination of the photograph reveals other downwind wind turbines that are completely engulfed in the wakes of the upwind turbines. The central issue is the impact this has on the power generated by the downwind turbines, and ultimately that of the wind farm as a whole.

An analytic wake model for a wind turbine was first proposed by Jenson [2] and subsequently used by others [3]. This model was developed by considering that momentum is conserved within the wake, and that the wake region expands linearly in the downstream direction away from the wind turbine. A schematic representation of the wake model is shown in Figure 8.3. With this, the upwind turbine rotor is designated by the thick black line and has a radius, r_0. The approaching wind is assumed to be uniform with a velocity of U_∞. At a distance, x, downstream of the wind turbine, the wind velocity is u. The wake radius is assumed to grow linearly with downstream distance according to

$$r_1 = \alpha x + r_r, \tag{8.1}$$

FIGURE 8.2
Photograph showing the wakes from wind turbines made visible by low level fog over an an offshore wind farm [1].

where α is the wake entrainment constant, also known as the wake decay constant. The entrainment constant has been determined in experiments to be

$$\alpha = \frac{0.5}{\ln\left(\frac{z}{z_0}\right)}, \tag{8.2}$$

where z is the wind turbine hub height, and z_0 is the surface roughness height at the site. The term r_r is called the equivalent *downstream rotor radius* and is given as

$$r_r = r_0\sqrt{\frac{1-a}{1-2a}}. \tag{8.3}$$

where r_0 is the rotor physical radius, and a is again the axial induction factor. The r_r being a function of the axial induction factor represents the blockage presented by the rotor actuator disk, which then affects the wake width.

If i is designated as the position of the wind turbine producing the wake, and j is the downstream position that is affected by the wake, then the wind speed at position j is given by

$$u_j = U_\infty(1 - u_{\mathrm{def_{ij}}}), \tag{8.4}$$

where $u_{\mathrm{def_{ij}}}$ is the *wake velocity deficit* induced on position j by an upstream wind turbine at position i.

The wake deficit can be computed through the following relation, namely

$$u_{\mathrm{def_{ij}}} = \frac{2a}{\left(1 + (\frac{\alpha}{r_r})x_{ij}\right)^2}, \tag{8.5}$$

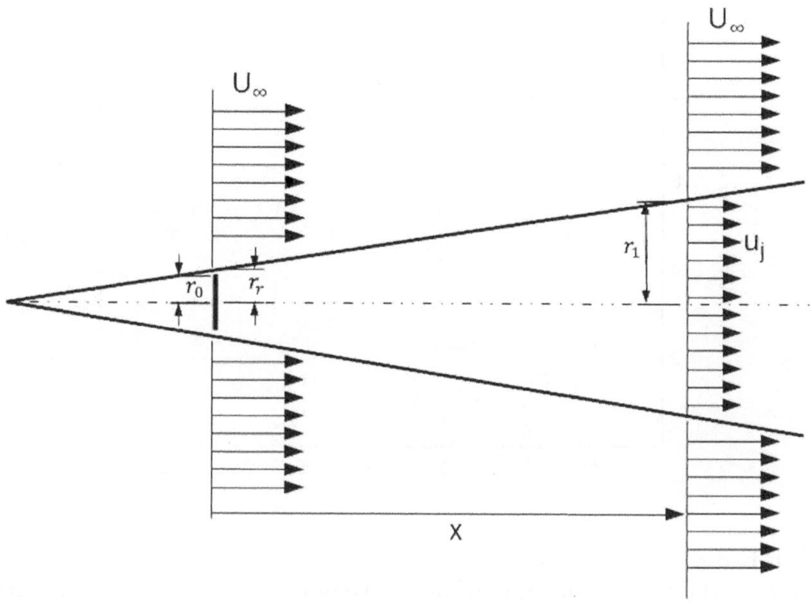

FIGURE 8.3
Schematic drawing of wind turbine wake model.

where a again is the axial induction factor which is related to the wind turbine thrust coefficient, C_T as

$$a = 0.5\left(1 - \sqrt{1 - C_T}\right), \tag{8.6}$$

and x_{ij} is the downstream distance between positions i and j. We note that for Betz efficiency, $a = 1/3$.

As an example, Figure 8.4 shows the change in the wind velocity, u, with increasing distance in the wake of an ideal, $a = 1/3$, upstream rotor. Note that it takes a downstream distance of more than 40 rotor diameters to recover the wind speed that is upstream of the rotor. The standard spacing in wind farms is 5 diameters!

In order to account for multiple wind turbines in which the wakes can intersect and affect a downstream turbine, the velocity deficit is the sum of the deficits produced by each wind turbine, namely

$$u_{\text{def}}(j) = \sqrt{\sum_{i \in W(j)} u^2_{\text{def}_{\text{ij}}}}, \tag{8.7}$$

where $W(j)$ is the set of upstream turbines affecting position j in the wake. The velocity deficit, $u_{\text{def}}(j)$ is then used in Equation 8.4 in place of $u_{\text{def}_{\text{ij}}}$ to compute u_j.

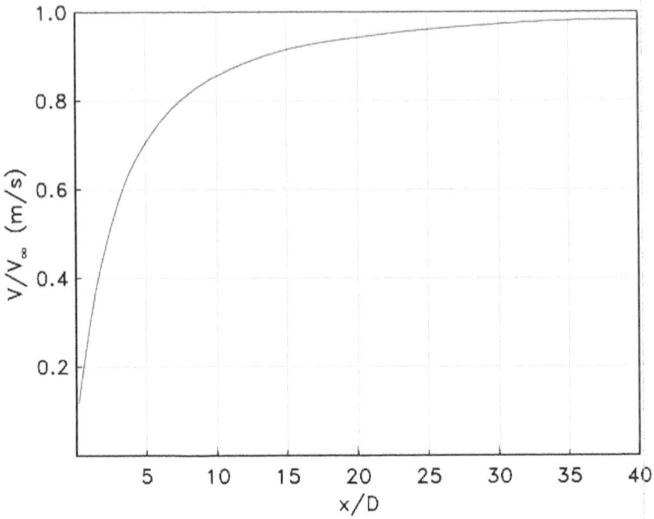

FIGURE 8.4
Velocity on the wake centerline of an upstream ideal, $a = 1/3$, wind turbine based on the wake model equations.

Example 8.1

Consider the arrangement of three wind turbines in the following schematic in which wind turbine C is in the wakes of turbines A and B.

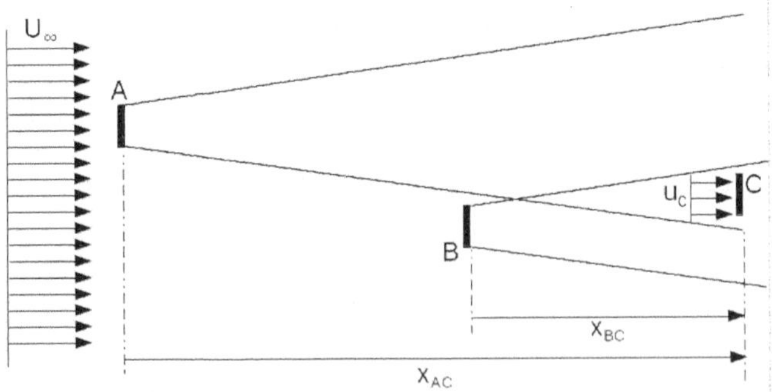

Given the following:

- $U_\infty = 12\,\mathrm{m/s}$
- $x_{AC} = 500\,\mathrm{m}$
- $x_{BC} = 200\,\mathrm{m}$

- $z = 60\,\text{m}$
- $z_0 = 0.3\,\text{m}$
- $r_0 = 20\,\text{m}$
- $C_T = 0.88$

Compute the total velocity deficit, $u_{\text{def}}(C)$ and the velocity at wind turbine C, namely u_C.

(1.) We note that wind turbine C is in the wakes of both wind turbines A and B. We also note that although down-wind of wind turbine A, wind turbine B is not in the wake of wind turbine A, so that its approaching wind speed is U_∞.

Then based on the previous equations, $u_{\text{def}_{\text{AC}}} = 0.0902$ and $u_{\text{def}_{\text{BC}}} = 0.2324$. Based on Equation 8.7, $u_{\text{def}}(C) = 0.2493$, and thus the wind speed approaching wind turbine C has been reduced by 24.93% as a result of being in the wakes of turbines A and B.

2. The wind velocity approaching wind turbine C is then

$$U_C = U_\infty\,(1 - u_{\text{def}}(C)) = 9.008\,\text{m/s}. \tag{8.8}$$

This example highlights a very important property of multiple wake combinations, namely that the total velocity deficit depends most on the closest wind turbine that generates a wake.

The power generated by any one of the wind turbines is

$$P_j \propto a_j u_j^3, \tag{8.9}$$

where where a_j is the inflow induction for the wake-affected turbine and u_j is the wind velocity approaching the wake-affected turbine. The total power generated by all of the wind turbines is

$$P_{\text{tot}} \propto \sum_{i \in W(j)} a_j u_{ij}^3, \tag{8.10}$$

where $W(j)$ is the set of turbines with inflow induction factors, a_j and approaching velocities u_{ij}. The wind farm efficiency is then defined as

$$\eta = \frac{P_{\text{tot}}}{N \cdot P_{\text{iso}}}, \tag{8.11}$$

where P_{iso} is the power produced by an isolated wind turbine under the same inflow velocity, U_0.

The wind farm efficiency can be used as a metric of merit for evaluating different arrangements and spacing of wind turbines within a wind farm. This leads to a discussion of optimum wind farm designs.

8.2 Wind Farm Design Optimization

In an optimization of a wind farm, one might seek to maximize the power with respect to the initial cost of the wind turbines purchased for the wind farm. This example requires a cost model [4] such as

$$\mathrm{Cost_{tot}} = N_t \left(\frac{2}{3} + \frac{1}{3} e^{-0.00174 N_t^2} \right), \tag{8.12}$$

where N_t is the number of turbines installed. Note that the cost per turbine decreases as N_t increases, thus reflecting the "economy of scale".

The objective function for the optimization process could then be

$$\mathrm{Obj} = \frac{1}{P_{\mathrm{tot}}} w_1 + \frac{\mathrm{Cost_{tot}}}{P_{\mathrm{tot}}} w_2, \tag{8.13}$$

where w_1 and w_2 are weighting coefficients where $w_1 + w_2 = 1$.

An optimization scheme might start with a conventional wind turbine pattern for a wind farm such as that shown in Figure 8.5. This shows wind turbines in wind parks as they are usually spaced, with somewhere between 5 and 9 rotor diameters apart in the prevailing wind direction, and between 3 and 5 diameters apart in the direction perpendicular to the prevailing winds. Various optimization schemes would seek to add or remove wind turbines in the grouping while seeking to improve the cost model object function.

An optimization study was conducted to examine the potential of optimized patterns of wind turbines [5]. The results are presented in Figure 8.6. This shows the impact of site area and number of wind turbines on wind farm efficiency. It considers either 64, 5 MW turbines or 106, 3 MW turbines. The total power installed is similar for the two cases. Each case is solved by imposing a predefined geographical extension (or site area) of the wind farm, which is equivalent to imposing a predefined density of installed power, that is, the smaller the area, the higher the power density.

In Figure 8.6, the open blue symbols represent the results obtained by the "rule of thumb" pattern shown in Figure 8.5. The open square symbols are for the wind farm with 106 wind turbines. The open triangle symbols are for the wind farm with 64 wind turbines. We notice that when using the rule of thumb pattern with either 106 or 64 turbines, as the site area increases (power density decreases), the efficiency of the wind farm increases. This trend is highlighted by the two lines in the figure.

The red filled symbols in Figure 8.6 represent the results obtained by seeking an optimum pattern [5]. The red filled square symbols correspond to the wind farm with 106 wind turbines. The red filled triangle symbols correspond to the wind farm with 64 wind turbines. Keeping the power density nearly constant, the optimization process was observed to improve the efficiency for the 64 turbine wind farm. In contrast to that, the optimized wind farm with

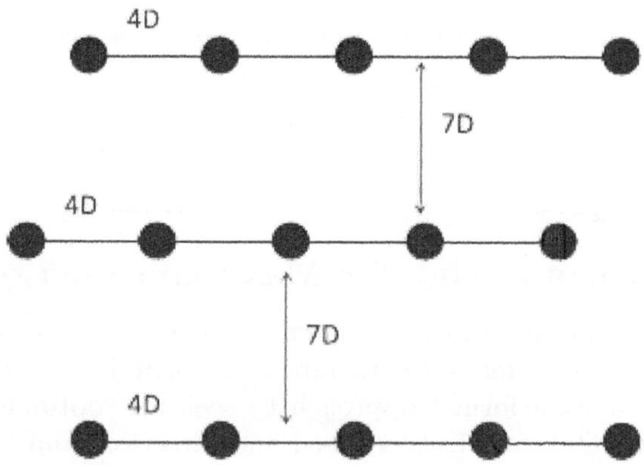

FIGURE 8.5
Rule of thumb pattern of wind turbines in a wind farm. The predominant wind direction is from bottom to top.

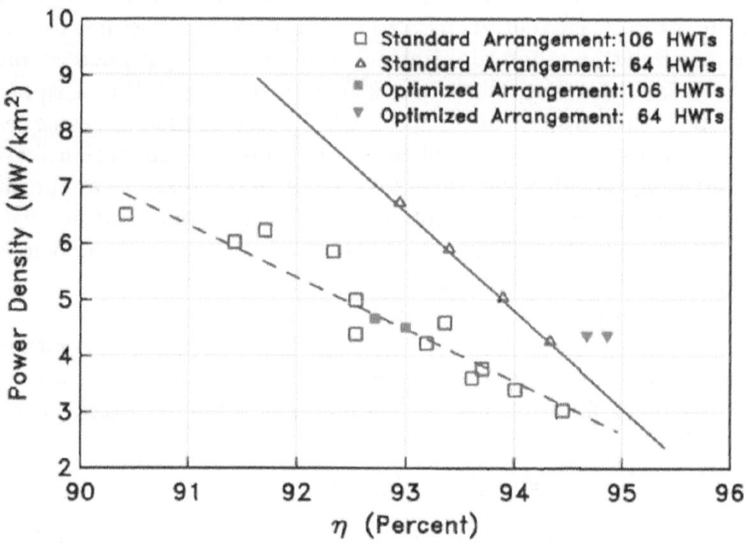

FIGURE 8.6
Impact of site area and number of wind turbines on wind farm efficiency [5].

106 turbines followed the general trend of the "standard" arrangement, and therefore showed no improvement in efficiency. Thus, the potential improvement over the rule of thumb pattern is more evident if the turbines are fewer and larger. This may be a physical limit of optimized wind farm design, or a product of the optimization method which clearly is more complex as the

number of wind turbines increases. Further research is needed to answer this question.

8.3 Wind Farm Design for Maximum Energy Capture

The previous section illustrates the potential benefits that can come by optimizing the pattern of wind farm turbines to maximize the energy capture. This section presents a formal approach to seek an "optimum" layout that considers the probabilities of wind speed and direction, and the power curve (cut-in, rated and cut-out wind speeds) of the wind turbines in the wind farm.

In addition to the previously cited [4,5], formal methods of wind farm layout optimization has been addressed by others [6,7]. Grady et al. [6] and Mosetti et al. [4], in particular, used a genetic algorithm to seek layouts that would minimize a weighted sum of wind energy and turbine costs. Lackner and Elkinton [3] presented a general framework to optimize the offshore wind turbine layout. Mora Castro et al. [7] also used a genetic algorithm to maximize an economic function, that took into account turbine parameters and wind turbine locations. In that, the wind farm was represented as a square grid. Unfortunately, their optimization approach did not include the wake loss. The following presents an approach [8] to a constrained optimization for a wind farm layout that considers the wake loss, wind characteristics at the site, and wind turbine performance. Presented with this is a double-objective evolutionary algorithm used to solve the system of performance equations. This utilizes the wake model presented in this chapter as well as wind probability and turbine performance models developed in Chapters 2 and 4.

The approach will assume that any two wind turbines are separated from each other by at least four rotor diameters. Thus for a rotor radius, R, any two wind turbines at locations (x_i, y_i) and (x_j, y_j) must satisfy the spatial constraint that

$$(x_i - x_j)^2 + (y_i - y_j)^2 \geq 64r_0^2. \tag{8.14}$$

The wind farm terrain is assumed to be homogeneous with a uniform terrain elevation and surface roughness. The wind turbines are considered to be identical with the same power curves, $P = f(U)$, where U is the wind speed at rotor hub height.

As presented in Chapter 2, the wind speed is assumed to follow a Weibull probability distribution

$$p(U) = \frac{k}{c}\left(\frac{U}{c}\right)^{k-1} e^{-(U/c)^k}, \tag{8.15}$$

where k and c are Weibull coefficients.

It is assumed that the wind velocity is a continuous function of the wind direction, θ, $0° \leq \theta \leq 360°$. In terms of the Weibull distribution, then $k = k(\theta)$ and $c = c(\theta)$.

The effect of energy capture as a function of wind direction of a single wind turbine can be expressed as

$$E(P,\theta) = \int_0^\infty f(U)p(U,k(\theta),c(\theta))dU. \tag{8.16}$$

Substituting for the Weibull probability

$$E(P,\theta) = \int_0^\infty f(U)\frac{k(\theta)}{c(\theta)}\left(\frac{U}{c(\theta)}\right)^{k(\theta)-1} e^{-(U/c(\theta))^{k(\theta)}}dU. \tag{8.17}$$

Note that although the wind speed integration upper limit is infinity, in reality it is the cut-out wind speed above which the wind turbine is not producing power.

Integrating Equation 8.17 with respect to the wind direction, θ, gives

$$E(P) = \int_0^{360} p_\theta(\theta)d\theta \int_0^\infty f(U)\frac{k(\theta)}{c(\theta)}\left(\frac{U}{c(\theta)}\right)^{k(\theta)-1} e^{-(U/c(\theta))^{k(\theta)}}dU. \tag{8.18}$$

Following Equation 8.4,

$$u_{\text{def}_{\text{ij}}} = 1 - \frac{u_j}{U_\infty}. \tag{8.19}$$

And again from Equation 8.5,

$$u_{\text{def}_{\text{ij}}} = \frac{2a}{\left(1+(\frac{\alpha}{r_r})x_{ij}\right)^2}, \tag{8.20}$$

where x_{ij} is the downstream distance between positions i and j. From before, r_r is the equivalent *downstream rotor radius* that represents the blockage presented by the rotor actuator disk and given as

$$r_r = r_0\sqrt{\frac{1-a}{1-2a}}. \tag{8.21}$$

where r_0 is the rotor physical radius, and a is again the axial induction factor.

As before, a can be put in terms of the wind turbine thrust coefficient, C_T namely

$$a = 0.5\left(1-\sqrt{1-C_T}\right). \tag{8.22}$$

Incorporating these into $u_{\text{def}_{\text{ij}}}$ gives

$$u_{\text{def}_{\text{ij}}} = \frac{1-\sqrt{1-C_T}}{\left(1+\alpha\left(\frac{x_{ij}}{r_r}\right)\right)^2}, \tag{8.23}$$

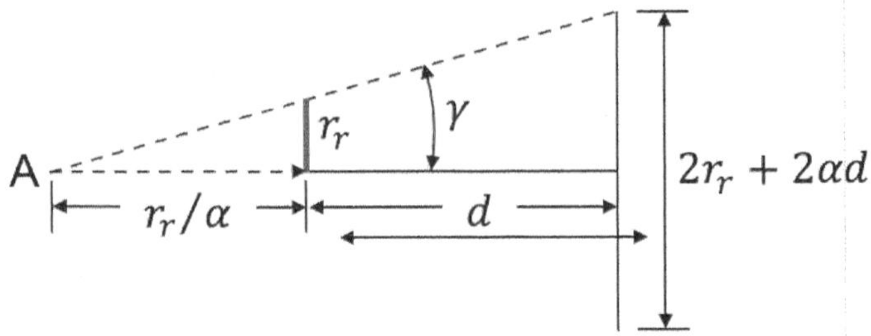

FIGURE 8.7
Schematic illustration of the half cone of a wind turbine wake.

where as before, α is the wake entrainment constant, also known as the wake decay constant given as

$$\alpha = \frac{0.5}{\ln\left(\frac{z}{z_0}\right)}, \tag{8.24}$$

and of which z is the wind turbine hub height, and z_0 is the surface roughness height at the site.

For multiple wind turbines in which the wakes of upstream wind turbines can intersect and affect a downstream turbine, the velocity deficit is the sum of the deficits produced by each upstream wind turbine, namely

$$u_{\text{def}}(j) = \sqrt{\sum_{i=1, i\neq j}^{N} u^2_{\text{def}_{\text{ij}}}}, \tag{8.25}$$

where $W(j)$ is the set of upstream turbines affecting position j in the wake. The velocity deficit, $u_{\text{def}}(j)$ is then used in Equation 8.19 in place of $u_{\text{def}_{\text{ij}}}$ to compute u_j.

Figure 8.7 illustrates a half cone of the wake formed by a wind turbine located at position (x, y). The wake spreading half angle, γ, is a parameter where based on Equation 8.1 corresponds to $\tan^{-1}(\alpha)$, where again, α is the wake entrainment coefficient. The distance between the labeled point A and the center point of the rotor is r_r/α.

Kusiak and Song [8] derived parameter, β_{ij}, that relates how the wind direction affects the wake impingement on a downstream wind turbine caused by an upstream wind turbine. The parameter has the following form

$$\beta_{ij} = \cos^{-1}\left[\frac{(x_j - x_i)\cos\theta + (y_j - y_i)\sin\theta + r_r/\alpha}{\sqrt{(x_j - x_i + \frac{r_r}{\alpha}\cos\theta)^2 + (y_j - y_i + \frac{r_r}{\alpha}\sin\theta)^2}}\right]. \tag{8.26}$$

Note that r_r was substituted for r_0 in the original expression [8] to reflect the effective radius to be consistent with the previous formulations.

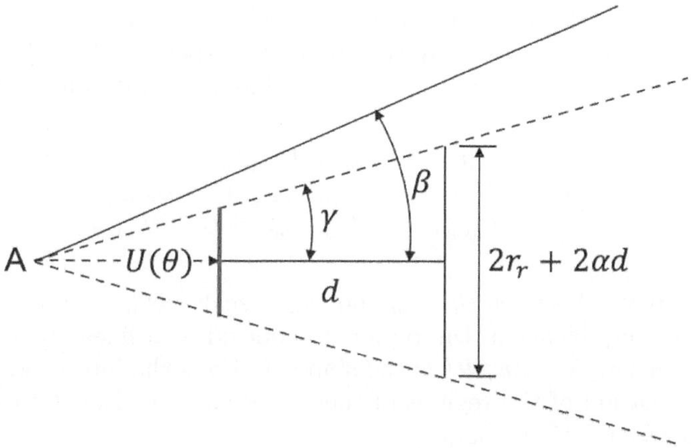

FIGURE 8.8
Schematic illustration of a wind turbine inside the cone of an upstream wind turbine for a wind vector angle, θ, and the resulting wind direction parameter, β.

The parameter, $\beta_{ij} = \beta(\theta, x_i, y_i, x_j, y_j)$ is the angle used to determine whether turbine j is in the cone of upstream turbine, i, given the wind direction, θ. This is illustrated in Figure 8.8. If the downstream wind turbine, j, is inside the wake cone of upstream wind turbine, i, the effective wake development distance between the wind turbines given by Kusiak and Song [8] is

$$d_{ij} = \mid (x_j - x_i)\cos\theta + (y_j - y_i)\sin\theta \mid . \tag{8.27}$$

The effect of the wind direction is incorporated into Equation 8.25 as

$$u_{\text{def}}(j) = \sqrt{\sum_{i=1, i\neq j, \beta_{ij}<\gamma}^{N} u^2_{\text{def}_{\text{ij}}}}. \tag{8.28}$$

Now $u_{\text{def}}(j)$ is a function of the wind direction, θ, and all of the turbine locations. With regard to the Weibull probability model, only the Weibull coefficient, c, is affected by the wake loss [3]. This is expressed as

$$c_j(\theta) = c(\theta)(1 - u_{\text{def}}(j)), \;\; j = 1, \ldots, N. \tag{8.29}$$

Therefore Equation 8.18 as expressed as

$$E(P_j) = \int_0^{360} p_\theta(\theta) d\theta \int_0^{\infty} f(U) \frac{k(\theta)}{c_j(\theta)} \left(\frac{U}{c_j(\theta)}\right)^{k(\theta)-1} e^{-(U/c_j(\theta))^{k(\theta)}} dU, \tag{8.30}$$

represents the expected power from a *single* turbine in the wind farm.

In Equation 8.30, $f(U)$ represents the power curve for the wind turbines. This can be simply expressed by the three wind speed regions: below cut-in, between cut-in and rated, and between rated and cut-out. Therefore

$$f(U) = \begin{array}{ccl} 0 & : & U < U_{\text{cut-in}} \\ \lambda U + \eta & : & U_{\text{cut-in}} \leq U \leq U_{\text{rated}} \\ P_{\text{rated}} & : & U_{\text{cut-out}} > U > U_{\text{rated}}. \end{array} \tag{8.31}$$

The turbine power between $U_{\text{cut-in}}$ and U_{rated} scales as U^3. However in Equation 8.31 for simplification, the power is modeled as a linear function of the wind speed, where $\lambda = dP/dU$ is the slope, and η is the intercept.

The application of the regions of the power curve to Equation 8.30 yields the three power relations, namely,

$$E(P_j) = \begin{array}{l} \lambda \int_{U_{\text{cut-in}}}^{U_{\text{rated}}} U \left(\frac{k(\theta)}{c_j(\theta)}\right) \left(\frac{U}{c_j(\theta)}\right)^{k(\theta)-1} e^{-(U/c_j(\theta))^{k(\theta)}} dU \\ \eta \int_{U_{\text{cut-in}}}^{U_{\text{rated}}} \left(\frac{k(\theta)}{c_j(\theta)}\right) \left(\frac{U}{c_j(\theta)}\right)^{k(\theta)-1} e^{-(U/c_j(\theta))^{k(\theta)}} dU \\ P_{\text{rated}} \int_{U_{\text{rated}}}^{U_{\text{cut-out}}} \left(\frac{k(\theta)}{c_j(\theta)}\right) \left(\frac{U}{c_j(\theta)}\right)^{k(\theta)-1} e^{-(U/c_j(\theta))^{k(\theta)}} dU. \end{array} \tag{8.32}$$

A digitized approach is used to solve the integral equations. This involves a discretization of the wind speed and direction based on their probability. The wind direction is discretized into N_θ equal width bins corresponding to wind directions $\theta_1, \theta_2, \ldots, \theta_{N_\theta}$ where $0° < \theta_1 \leq \theta_2 \leq \theta_{N_\theta} < 360°$ where $\theta_0 = 0°$ and $\theta_{N_{\theta+1}} = 360°$. Each bin is associated with the relative frequency, w_i, of the wind direction (wind rose), where $0 \leq w_i \leq 1$, where $i = 0, \ldots, N_\theta$. An example of a wind direction probability distribution is shown in Figure 8.9.

The wind speed can be similarly discretized into $N_U + 1$ bins, where $U_1, U_2, U_3, \ldots,, U_{N_U}$, and $U_{\text{cut-in}} < U_1 \leq U_2 \leq U_3 \leq \ldots \leq U_{N_U}$, and which $U_0 = U_{\text{cut-in}}$ and $U_{N_U} = U_{\text{rated}}$.

Following this, the integral

$$\int_{U_{\text{cut-in}}}^{U_{\text{rated}}} U \left(\frac{k(\theta)}{c_j(\theta)}\right) \left(\frac{U}{c_j(\theta)}\right)^{k(\theta)-1} e^{-(U/c_j(\theta))^{k(\theta)}} dU, \tag{8.33}$$

can be approximated as

$$\sum_{i=1}^{N_U+1} \left[e^{-(U_{i-1}/c_j(\theta))^{k(\theta)}} - e^{-(U_i/c_j(\theta))^{k(\theta)}} \right] (U_{i-1} + U_i)/2. \tag{8.34}$$

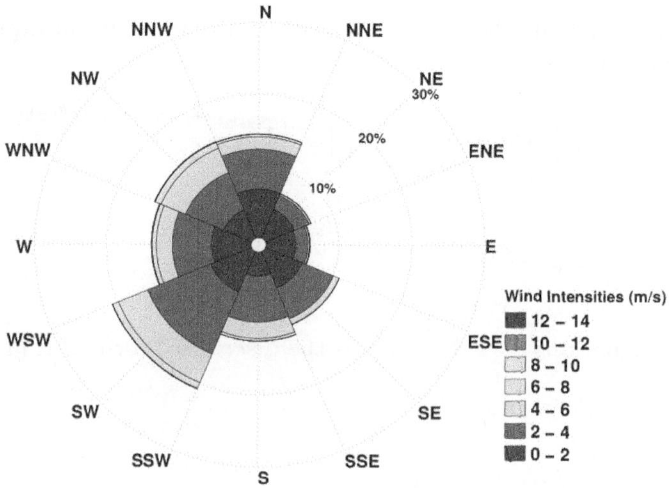

FIGURE 8.9
Example of a wind direction probability (wind rose) map.

Incorporating this into the formulation for the expected power,

$$\begin{aligned} E(P_j) = \int_0^{360} p_\theta(\theta) \left(\lambda \sum_{i=1}^{N_U+1} \left[e^{-(U_{i-1}/c_j(\theta))^{k(\theta)}} - e^{-(U_i/c_j(\theta))^{k(\theta)}} \right] \right. \\ \left. ((U_{i-1} + U_i)/2) \right) d\theta \\ + P_{\text{rated}} \int_0^{360} p_\theta(\theta) e^{-(U_{\text{rated}}/c_j(\theta))^{k(\theta)}} d\theta \\ + \int_0^{360} p_\theta(\theta) \eta \left(e^{-(U_{\text{cut-in}}/c_j(\theta))^{k(\theta)}} - e^{-(U_{\text{rated}}/c_j(\theta))^{k(\theta)}} \right) d\theta, \end{aligned} \tag{8.35}$$

which can be further simplified as

$$\begin{aligned} E(P_j) = \lambda \sum_{i=1}^{N_U+1} \left(\frac{U_{i-1} + U_i}{2} \right) \int_0^{360} p_\theta(\theta) \left[e^{-(U_{i-1}/c_j(\theta))^{k(\theta)}} - e^{-(U_i/c_j(\theta))^{k(\theta)}} \right] d\theta \\ + P_{\text{rated}} \int_0^{360} p_\theta(\theta) e^{-(U_{\text{rated}}/c_j(\theta))^{k(\theta)}} d\theta \\ + \eta \int_0^{360} p_\theta(\theta) \left(e^{-(U_{\text{cut-in}}/c_j(\theta))^{k(\theta)}} - e^{-(U_{\text{rated}}/c_j(\theta))^{k(\theta)}} \right) d\theta. \end{aligned} \tag{8.36}$$

The integral quantity in the second line of Equation 8.36 can be discretized as

$$P_{\text{rated}} \sum_{l=1}^{N_\theta+1} \left[(\theta_l - \theta_{l-1}) w_{l-1} e^{-(U_{\text{rated}}/c_j(\frac{\theta_l+\theta_{l-1}}{2}))^{k(\frac{\theta_l+\theta_{l-1}}{2})}} \right]. \tag{8.37}$$

In a similar fashion, the integral quantity in the third line of Equation 8.35 can be discretized as

$$\eta \sum_{l=1}^{N_\theta+1} \left[(\theta_l - \theta_{l-1}) w_{l-1} \left(e^{-\left(\frac{U_{\text{cut-in}}}{c_j} \frac{\theta_l+\theta_{l-1}}{2}\right)^{k\left(\frac{\theta_l+\theta_{l-1}}{2}\right)}} \right.\right.$$
$$\left.\left. - e^{-\left(\frac{U_{\text{rated}}}{c_j} \frac{\theta_l+\theta_{l-1}}{2}\right)^{k\left(\frac{\theta_l+\theta_{l-1}}{2}\right)}} \right) \right]. \tag{8.38}$$

Combining Equations 8.35–8.38, yields the discretized form of Equation 8.32, namely

$$\begin{aligned} E(P_j) = \lambda \sum_{i=1}^{N_U+1} \left(\frac{U_{i-1}+U_i}{2} \right) & \\ \sum_{l=1}^{N_\theta+1} \left[(\theta_l - \theta_{l-1}) w_{l-1} \left(e^{-\left(\frac{U_{i-1}}{c_j}\Theta\right)^{k(\Theta)}} - e^{-\left(\frac{U_i}{c_j}\Theta\right)^{k(\Theta)}} \right) \right] & \\ + P_{\text{rated}} \sum_{l=1}^{N_\theta+1} \left[(\theta_l - \theta_{l-1}) w_{l-1} e^{-\left(\frac{U_{\text{rated}}}{c_j} \frac{\theta_l+\theta_{l-1}}{2}\right)^{k\left(\frac{\theta_l+\theta_{l-1}}{2}\right)}} \right] & \\ + \eta \sum_{l=1}^{N_\theta+1} \left[(\theta_l - \theta_{l-1}) w_{l-1} \left(e^{-\left(\frac{U_{\text{cut-in}}}{c_j}\Theta\right)^{k(\Theta)}} - e^{-\left(\frac{U_{\text{rated}}}{c_j}\Theta\right)^{k(\Theta)}} \right) \right], & \end{aligned} \tag{8.39}$$

where

$$\Theta = \frac{\theta_l + \theta_{l-1}}{2}. \tag{8.40}$$

Equation 8.39 can then be used to maximize the energy capture of the wind farm such that

$$\max \sum_{j=1}^{N} E(P_j), \tag{8.41}$$

is subject to the turbine spacing constraint that

$$(x_i - x_j)^2 + (y_i - y_j)^2 \geq 64 r_0^2,\ i, j = 1 \ldots N, i \neq j, \tag{8.42}$$

where r_0 is the physical radius of the wind turbines.

A wind farm design that maximizes wind energy capture ultimately minimizes the wake effects among turbines. The optimum placement of the wind turbines to achieve this, particularly when the number is large, is complex. Therefore the following section introduces a maximizing method that is based on an evolutionary genetic algorithm.

8.3.1 Evolutionary Genetic Algorithm

The object is to find the optimum locations, $(x_i, y_i), i = 1, N$, of wind turbines that minimize wake interactions and subsequently maximize the wind farm energy capture.

With the evolutionary optimization formulation, the jth individual is defined as (z^i, σ^i), where z^i and σ^i are two vectors with $2N$ entries, defined as $z^i = (x_1^i, y_1^i, \ldots, x_N^i, y_N^i)^T$ and $\sigma^i = (\sigma_{1,x}^i, \sigma_{1,y}^i, \ldots, \sigma_{N,x}^i, \sigma_{N,y}^i)^T$.

The pair (x_1^i, y_1^i) is the position of the first turbine, while the pair $(\sigma_{1,x}^i, \sigma_{1,y}^i)$ is used to mutate the position of the first turbine. The pair (x_N^i, y_N^i) is the position of the Nth wind turbine on which $(\sigma_{N,x}^i, \sigma_{N,y}^i)$ acts to mutate its position. The vector, σ^i, is taken as the zero mean standard deviation. The initial population, σ^i, $i = 1, \ldots, \mu_{\text{Child}}$, is generated by uniformly sampling from the range, $(\sigma_{\text{lower}}, \sigma_{\text{upper}})$, where the subscripts refer to the lower and upper bounds for the standard deviation vector.

An intrinsic element in an iterative optimization process is formulating the object function. In this case, it is maximizing the expected energy capture, namely

$$\text{Obj} = \max \left[\sum_{i=1}^{N} E(P_j) \right]^{-1}, \tag{8.43}$$

subject to $(x_i)^2 + (y_i)^2 \leq r_{\text{farm}}^2$, and $(x_i - x_j)^2 + (y_i - y_j)^2 \geq 64 r_0^2$

8.3.1.1 Iterative Process behind Evolutionary Algorithms

The following are steps in an iterative evolutionary algorithm.

1. Formulate the object function(s) which in this case is $\max \left[\sum_{i=1}^{N} E(P_j)\right]^{-1}$.
2. Identify decision variables. In this case they are the wind turbine locations, $(x_i, y_i), i = 1, N$.
3. Identify the problem dimension as the total number of decision variables, which in this case is $2N$.
4. A solution generated by an evolutionary algorithm is referred to as a "chromosome/individual" which is made up of genes. Following the selection of decision variables and problem dimension, the choice of a suitable type (e.g. integer, real, binary, etc.) is important. In this case, it could be realistic to make the (x_i, y_i) locations to be either integer (within the nearest meter) or real (within the nearest rounded fraction of a meter). This will obviously depend on the size of the wind farm (radius in this case). For example with $r_{\text{farm}} = 500$ m, integer positions are suitable.
5. Impose a boundary constraint, which is the range of the search space. In this case, the search space is bounded to be within the wind farm boundary.

TABLE 8.1
Functional Evaluation of Initial Population

	Initial Random Population				Feval	Remarks
Solution 1	x_1^1	y_1^1	x_2^1	y_2^1	$\lvert[\sum_{i=1}^{2} E(P_j^1)\rvert$	-
Solution 2	x_1^2	y_1^2	x_2^2	y_2^2	$\lvert[\sum_{i=1}^{2} E(P_j^2)\rvert$	-
Solution 3	x_1^3	y_1^3	x_2^3	y_2^3	$\lvert[\sum_{i=1}^{2} E(P_j^3)\rvert$	Best
Solution 4	x_1^4	y_1^4	x_2^4	y_2^4	$\lvert[\sum_{i=1}^{2} E(P_j^4)\rvert$	-
Solution 5	x_1^5	y_1^5	x_2^5	y_2^5	$\lvert[\sum_{i=1}^{2} E(P_j^5)\rvert$	-
Solution 6	x_1^6	y_1^6	x_2^6	y_2^6	$\lvert[\sum_{i=1}^{2} E(P_j^6)\rvert$	-

The following describes the different stages of the optimization process. As an illustration, a wind farm consisting of two wind turbines, with positions $(x_i, y_i), i = 1, 2$ is considered.

Stage 1: Population Generation - Initial Solution Set

For the two wind turbines, four genes, (x_1, y_1, x_2, y_2) are randomly generated while satisfying the lower and upper limits of the boundary constraints:

$$(x_i)^2 + (y_i)^2 \leq r_{\text{farm}}^2, \tag{8.44}$$

and

$$(x_i - x_j)^2 + (y_i - y_j)^2 \geq 64r_0^2. \tag{8.45}$$

A chromosome thus generated is a vector comprised of, in this case, four genes. Chromosome refers to the solution/individual of the formulated optimization problem, while the collection of such chromosomes is referred to as a population. As an example, we will choose a population size that is $2N + 2$, or 6 in this case. Therefore the system size is 6 by 4. Table 8.1 shows the 6x4 array that will initially be randomly populated.

Stage 2: Function Evaluation: Feval

All the chromosomes of random population will go through a process known as a fitness evaluation to measure the quality of the Solution created by EA. The fitness evaluation of chromosome/solution is carried out by calculating the objective function value as Feval $= \lvert[\sum_{i=1}^{N} E(P_j)\rvert$. The Solution with the maximum Feval is then identified. In this example, this was arbitrarily selected to be that of Solution 3.

Stage 3: Parent Selection

The chromosomes are selected from the initial population to act as "parent" for reproduction, based on the fitness of the solution/individual. This tells how to choose individuals in the population that will create offspring for the next generation. The fittest solution will have a higher probability to be selected as a parent. The selection procedure consists of the following:

1. Compute the probability, Prob(i), where

$$\text{Prob}(i) = \frac{\text{Fit}(i)}{\sum_{i=1}^{\text{No. of solutions}} \text{Fit}(i)}, \tag{8.46}$$

where in this example, the number of solutions is 6, and

$$\text{Fit}(i) = \frac{1}{(1 + \text{Feval}(i))}. \tag{8.47}$$

2. The "parent" is then found using a "Roulette-wheel" (RW) process. This divides the RW into pie slices that correspond to the number of solutions (6 in our example). Each solution occupies a portion of the RW that is in proportion to its fitness value, that is larger fitness occupies a larger pie slice. Therefore the probability of choosing a chromosome depends only on the fitness.
3. The steps involving the RW selection process are:
   (a) Compute cumulative probability values for all the solutions, Cum(i).
   (b) Allot the pie segments for all of the solutions (6 in the example).
   (c) To arrange the chromosomes in the parent pool, "spin" the RW a number of solutions (6 in the example) times. This is equivalent to generating random numbers the number of solutions (6) times, namely Rand(i) $< 1, 2, \ldots, 6$. As an example consider the following:

$$\text{Rand}(\text{Place1}, 2, \ldots, 6) = 0.200, 0.285, 0.098, 0.812, 0.397, 0.50.$$

   (d) Fit the random numbers, Rand(i) $< 1, 2, \ldots, 6$, into the place on the RW based on where they fall between the range of cumulative probability, Prob(i), of each "pie slice".

As an example of the process, consider the following RW, where the size of the pie slices corresponds to the cumulative probability values for all the solutions, Cum(i). An example is shown in Figure 8.10.

From the previous step, Rand(Place(2) = 0.285 is seen to fall into the pie slice of individual/chromosome [9]. Therefore, individual/chromosome [9] will occur in Place(2) of

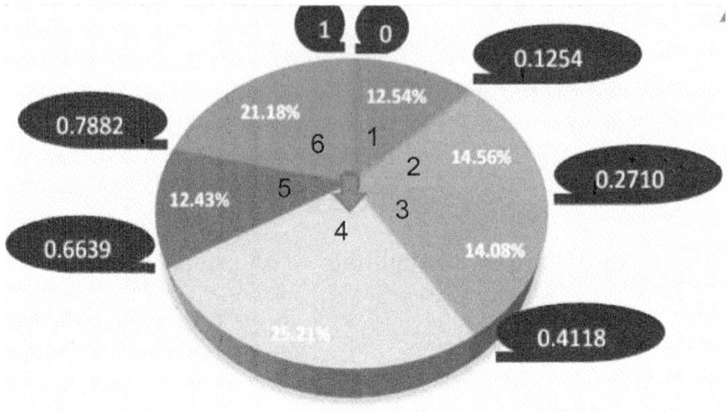

FIGURE 8.10
Example of roulette wheel with pie slices that correspond to the cumulative probability values, Cum(i), for all the solutions.

TABLE 8.2
Chromosomes in Mating Pool after RW Selection Process

Position	Chromosome	Population after Selection			
Place 1	II	x_1^2	y_1^2	x_2^2	y_2^2
Place 2	III	x_1^3	y_1^3	x_2^3	y_2^3
Place 3	I	x_1^1	y_1^1	x_2^1	y_2^1
Place 4	VI	x_1^6	y_1^6	x_2^6	y_2^6
Place 5	III	x_1^3	y_1^3	x_2^3	y_2^3
Place 6	IV	x_1^4	y_1^4	x_2^4	y_2^4

the parent/mating pool. Similarly, Rand(Place(1) $=$ 0.200 falls into the pie slice of individual/chromosome [10] and therefore individual/chromosome [10] will occur in Place(1). Place(3) falls into the pie slice of individual/ chromosome [1], therefore individual/chromosome [1] will occur in Place(3). Individual/chromosome [4] will occur in Place(4). Individual/chromosome [9] will occur in Place(5), and Individual/chromosome [2] will occur in Place(6). As a result, Table 8.1 is reordered as shown in Table 8.2.

8.3.1.1.4 Stage 4: Crossover

The crossover operation involves three steps: (1) selecting mating chromosomes, (2) determining cut point for crossover, and (3) updating the population.

For this, parent chromosomes from the parent pool that undergo the mating process are randomly selected. The number of mate solutions is decided using a crossover rate, p_c. Solution, (i), will become a parent, if random number, Rand(i), falls below the crossover rate.

As an example, assume a crossover rate of $p_c = 25\%$. Generate a set of random numbers below 1 of the population size (6 in the example). For example say,

$$\text{Rand}(1, 2, \ldots, 6) = 0.190, 0.249, 0.750, 0.005, 0.149, 0.320.$$

Those that fall below p_c are rand(1), rand(4), and rand(5). Thus, the three chromosomes/solutions [1–3] are selected for crossover operation.

Next, chromosome cut points are to be determined. For this, generate random numbers between 1 to the number of chromosomes minus 1 ($4 - 1 = 3$ in this case). For this example, assume, Cut(1) = 1, Cut(2) = 1 and Cut(3) = 2.

In the crossover operation, parent individuals get cut at the crossover point and their genes are interchanged. For the first, second, and third crossovers, the Parent genes are cut at positions 1, 1, and 2, respectively. This is shown for each crossover operation in the following. The cut points are denoted by the column line.

In the first cross-over operation, original Solution(1) replaces chromosomes in Solution(4) along Cut(1), and the result replaces the original Solution(1). Thus starting from Table 8.2,

1	x_1^2	y_1^2	x_2^2	y_2^2
4	x_1^6	y_1^6	x_2^6	y_2^6
1	x_1^2	y_1^6	x_2^6	y_2^6

In the second cross-over operation, original Solution(4) replaces chromosomes in Solution(5) along Cut(1), and the result replaces the original Solution(4). Again starting from Table 8.2,

4	x_1^6	y_1^6	x_2^6	y_2^6
5	x_1^3	y_1^3	x_2^3	y_2^3
4	x_1^6	y_1^3	x_2^3	y_2^3

In the third cross-over operation, original Solution(5) replaces chromosomes in Solution(1) along Cut(2), and the result replaces the original Solution(5). Again starting from Table 8.2,

TABLE 8.3
Chromosomes in Mating Pool before and after Crossover Process

Population after Selection				Population after Crossover			
x_1^2	y_1^2	x_2^2	y_2^2	x_1^2	y_1^6	x_2^6	y_2^6
x_1^3	y_1^3	x_2^3	y_2^3	x_1^3	y_1^3	x_2^3	y_2^3
x_1^1	y_1^1	x_2^1	y_2^1	x_1^1	y_1^1	x_2^1	y_2^1
x_1^6	y_1^6	x_2^6	y_2^6	x_1^6	y_1^3	x_2^3	y_2^3
x_1^3	y_1^3	x_2^3	y_2^3	x_1^3	y_1^3	x_2^2	y_2^2
x_1^4	y_1^4	x_2^4	y_2^4	x_1^4	y_1^4	x_2^4	y_2^4

5	x_1^3	y_1^3	x_2^3	y_2^3
1	x_1^2	y_1^2	x_2^2	y_2^2
5	x_1^3	y_1^3	x_2^2	y_2^2

Table 8.3 shows the full chromosome arrangement before (Table 8.2) and after crossover.

8.3.1.1.5 Stage 5: Mutation

The mutation operation is a small random sharp change in the chromosome necessary to obtain a new solution. Mutation is used to sustain population diversity and it generally has a low probability, p_m.

Mutation in EA refers to an "exploration" of search space. It has been proven that the mutation operation is crucial for the convergence of the algorithm, whereas crossover operation is not. The following is the process.

1. Calculate the total number of genes in the population. In the example that is $4 \cdot 6 = 24$ genes.

2. Calculate the number of mutable genes which is based on the mutation rate, p_m. For example, for a $p_m = 10\%$, the number of mutations is $0.1 \cdot 24 = 2.4$ or nominally 2 mutations.

3. Calculate gene positions found by generating two random numbers below (in this case) 24. For example, 12 and 18. Therefore for population after cross-over in Table 8.3, the mutable genes are Chromosome(3)-gene 4, y_2^1, and Chromosome(5)-gene2, y_1^3.

4. The values of the mutable genes are substituted with random numbers that satisfy the original boundary constraint(s) of decision variables/genes.

TABLE 8.4
Replacement Population for the Next Iteration

Population After Mutation				Next Generation Initial Population				Feval (Example)	Remarks
x_1^2	y_1^6	x_2^6	y_2^6	x_1^2	y_1^6	x_2^6	y_2^6	37	-
x_1^3	y_1^3	x_2^3	y_2^3	x_1^3	y_1^3	x_2^3	y_2^3	77	-
x_1^1	y_1^1	x_2^1	Rand	x_1^1	y_1^1	x_2^1	and	47	-
x_1^6	y_1^3	x_2^3	y_2^3	Rand	Rand	Rand	Rand	47	Replaced sol.
x_1^3	Rand	x_2^2	y_2^2	x_1^3	Rand	x_2^2	y_2^2	56	-
x_1^4	y_1^4	x_2^4	y_2^4	x_1^4	y_1^4	x_2^4	y_2^4	46	-

8.3.1.1.6 Stage 6: Survival Selection and Replacement Mechanism

Following the mutation operation, one iteration/generation of EA is over. A functional evaluation is again performed on the offspring for a survival selection.

From the functional evaluation of the population after mutation, it may be evident that the objective function value of the best solution is reducing in comparison with the minimum objective value of the initial random population. Thus the minimization objective is being met, and thus the process is converging towards a solution.

To execute the iteration process continuously, the population is to be revised at the end of each iteration. This is referred to as replacement mechanism, where at the end of each iteration, 80%–90% of the best solutions from the offspring population (four to five best children), and 20%–10% of the best solutions from the initial population (two to one random solution), are selected to form a new population for the next generation. The process of mutation and survival selection is illustrated in Table 8.4

8.3.1.1.7 Stage 7: Stopping the Iteration

The optimization process is repeated until the objective function from one iteration to the next stops changing within a given tolerance.

The following are results from a genetic algorithm based wind farm optimization that sought to maximize the captured wind energy. For this, the wind turbines had the following parameters: a rotor radius of $r_0 = 38.5$ m; a cut-in wind speed of 3.5 m/s; a rated wind speed 14 m/s; and a rated power of 1,500 kW. The linearly approximated power function curve had coefficients, $\lambda = 140.86$ and $\eta = -500$. The hub height was 80 m. The thrust coefficient CT was 0.8, and the spreading parameter, α, was taken to be 0.075 which is representative of a wind farm land site.

The wind speed was divided into $N_v = 20$ intervals of 0.5 m/s each. The wind direction was divided into $N_\theta = 23$ intervals of 15° each. The Weibull

TABLE 8.5
Wind Direction Probability Used in Optimized Wind Farm Design

Index (i)	θ (deg)	w_i	Index (i)	θ (deg)	w_i
1	0	0	13	180	0.01
2	15	0.01	14	195	0.01
3	30	0.01	15	210	0.01
4	45	0.01	16	225	0.01
5	60	0.01	17	240	0.01
6	75	0.20	18	255	0.01
7	90	0.01	19	270	0.01
8	105	0.01	20	285	0.01
9	120	0.01	21	300	0.01
10	135	0.01	22	315	0.01
11	150	0.01	23	330	0.01
12	165	0.01	24	345	0.01

coefficients were $c = 13$ m/s and $k = 2$. As previously explained, only the Weibull coefficient c was assumed to depend on wind direction as expressed in Equation 8.29. The probability of the wind direction used in the example is given in Table 8.5. of the

The wind farm is a cartesian grid with dimensions, $x = 2,000$ m by $y = 2,000$ m. The wind direction $\theta = 0$ is parallel to the y axis. The wind turbines have a minimum spacing corresponding to four-times the physical rotor radius, $4r_0$.

Figure 8.11 shows three examples of arrangements of two wind turbines that were found to maximize the captured wind power based on the wind direction probability given in Table 8.5. This is somewhat of a trivial case in which the optimum arrangement places the wind turbines out of each others wakes. This is similarly demonstrated with the arrangements of three wind turbines that is shown in Figure 8.12.

The optimization is not as obvious as the number of wind turbines increases. This is demonstrated in Figure 8.13 which includes arrangements of from two to seven wind turbines that maximize the captured power based on the wind direction probability given in Table 8.5. For example, with the larger number of wind turbines, the effect of the $\theta = 75°$ wind direction with probability $w_i = 0.2$ is apparent with a few pairs of wind turbines aligned along $\theta = 90°$ to capture the less prevalent wind direction. Another important note is that although the wind turbines are constrained to be on a uniformly spaced grid with spacing of four rotor radii, the optimum arrangement is not evenly spaced as is traditional in wind farms.

Figure 8.14 shows the wind farm expected power as a function of the number of wind turbines with optimized locations based on the wind direction probability. For each number of wind turbines, the optimization was repeated three times. With the fewer number of wind turbines, the repeated optimized power remained the same and thus the data points overlap in the plot. With the larger number of wind turbines, the optimized power varied slightly from one optimization run to the next and thus the data points show a small degree of scatter. The repeatability can be improved by involving more "genes" in the genetic algorithm.

Figure 8.14 shows a linear relation between the expected power and the number of wind turbines. Most importantly, the slope is very close to the ratio of the number of wind turbines. This linear scaling between the captured power with the number of wind farm turbines would not necessarily be expected with traditional turbine packing arrangements where wake losses combine.

Other uses of the genetic algorithm based wind farm optimization could consider the effect of the turbine radius on the degree of optimization. Another might investigate an optimized energy density, namely the captured power normalized by the wind farm area. Problem 5 in this chapter examines that question for a simple pair of wind turbines. Finally an optimization might examine a number of parameters such as rotor radius, hub height, site conditions, etc. on the cost of electricity (COE). This will be examined in Chapter 11.

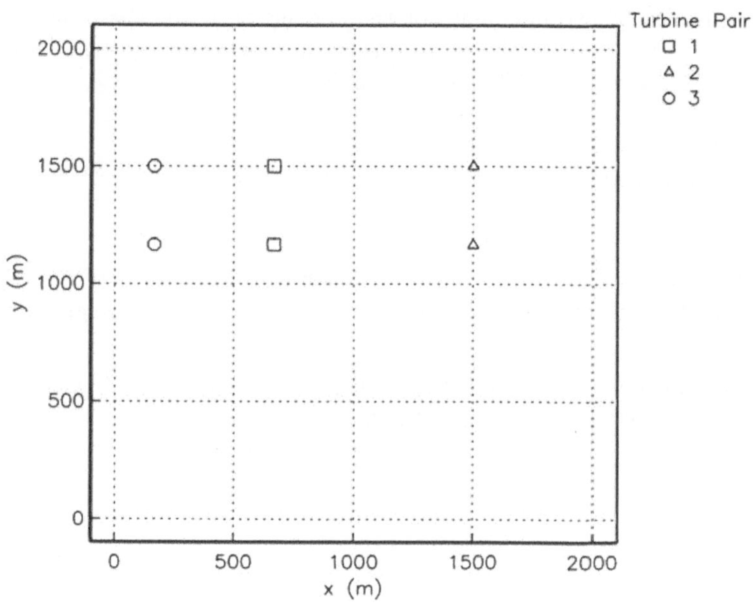

FIGURE 8.11
Three examples of arrangements of two wind turbines that maximize the captured power based on the wind direction probability given in Table 8.5. Wind $\theta = 75°$, where $\theta = 0°$ is along x-axis.

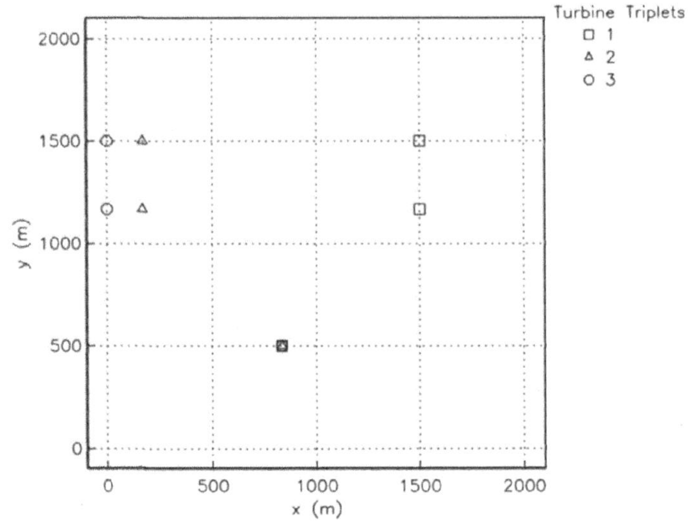

FIGURE 8.12
Three examples of arrangements of three wind turbines that maximize the captured power based on the wind direction probability given in Table 8.5. Wind $\theta = 75°$, where $\theta = 0°$ is along x-axis.

Problems

1. Consider a pair of wind turbines, each with rotor radii of $r_r = 20$ m, hub heights of $z = 60$ m. At the site the roughness height is $z_0 = 0.3$ m.

   For both wind turbines, the design torque coefficient is $C_T = 0.88$. The free-stream wind speed is $U_\infty = 12$ m/s.

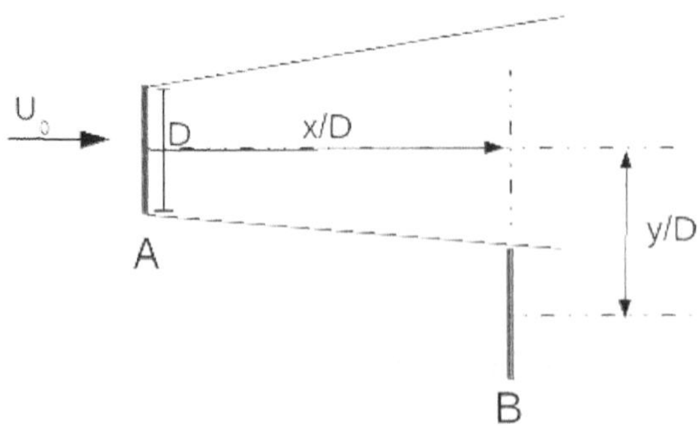

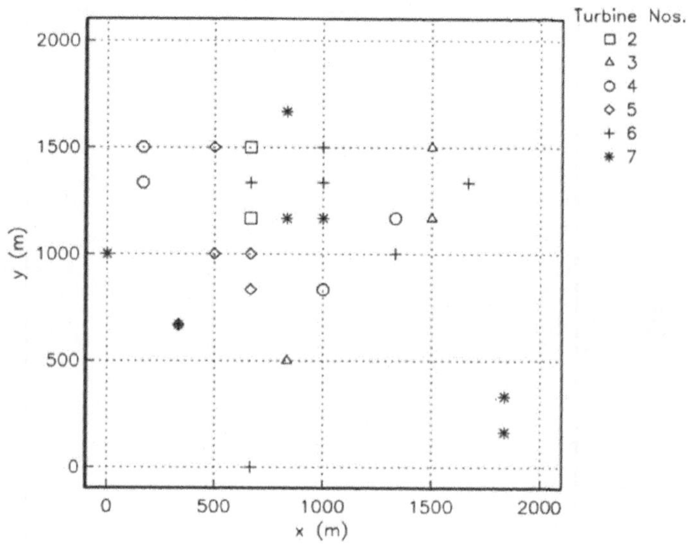

FIGURE 8.13
Examples of arrangements of from two to seven wind turbines that maximize the captured power based on the wind direction probability given in Table 8.5. Wind $\theta = 75°$, where $\theta = 0°$ is along x-axis.

(a) For this system, based on the analytic wake model described in this chapter, determine the spanwise distance, y/D of the downwind turbine (B) so that it is **not affected** by the wake of the upstream turbine (A) for positions of Turbine B of $x/D = 4$, 7 and 10.

(b) Repeat Part 1 with $C_T = 0.80$ for Turbine A.

(c) How did this lowering of the design torque coefficient affect the energy density (power-per-square-area) of the wind farm?

2. A wind farm is proposed to be built outside of Denver Airport. The airport wind data obtained at a $z_{\text{ref}} = 10$ m., gave a Weibull wind frequency distribution fit with coefficients of $k = 1.54$ and $C = 3.78$ m/s. The wind turbines in the proposed wind farm have a rated power of $P_{\text{rated}} = 1.5$ MW, a rotor radius of $R = 35$ m, and a hub height of $H = 65$ m. The rated wind speed for the wind turbines is $V_{\text{rated}} = 8$ m/s and the cut-out wind speed is $V_{\text{cut-out}} = 20$ m/s.

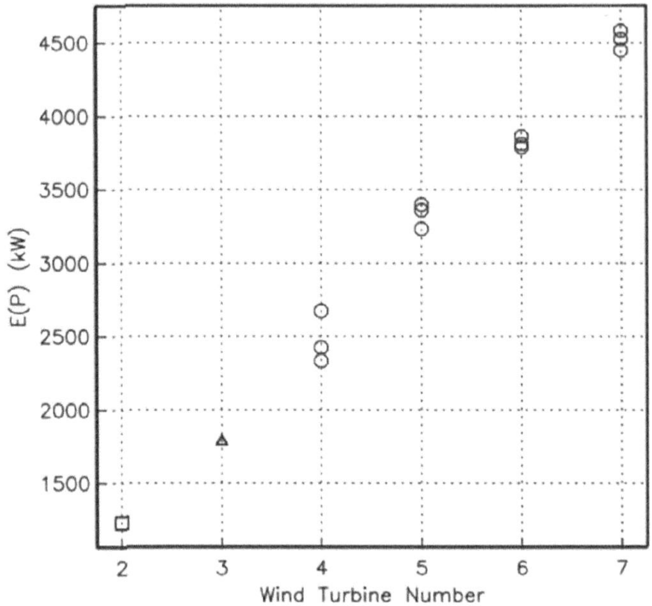

FIGURE 8.14
Wind farm expected power as a function of the number of wind turbines with optimized locations based on the wind direction probability.

(a) Based on the Weibull coefficients at $z_{\mathrm{ref}} = 10$ m., determine the new Weibull coefficients for the wind speed at the wind turbine hub height. Assume $n = 0.23$.

(b) What is the probability for the wind at the hub height to be between V_{rated} and $V_{\mathrm{cut-out}}$?

(c) We would like to estimate the effect of placing one of the wind turbines a distance of ten rotor radii (five rotor diameters) directly down-wind of another identical wind turbine in the wind farm. For this, assume that the wind turbines are optimal, with a Betz axial inflow induction factor of $a = 0.33$, the local surface roughness height is $z_0 = 0.01$ m, and the wind speed is $V_\infty = V_{\mathrm{rated}} = 8$ m/s.

  i. Under the stated conditions, determine the wind speed approaching the down-wind turbine.

  ii. What is the power generated by the downstream wind turbine in MW?

3. Four identical wind turbines are lined up in a row, 12 rotor diameters apart. The approaching wind speed is $U_\infty = 12\,\text{m/s}$. The wind direction is parallel to the long line of wind turbines. The thrust coefficient for each of the wind turbines is $C_T = 0.7$. The hub heights are $z = 60\,\text{m}$., and the roughness height is $z_0 = 0.3\,\text{m}$.
   (a) Determine the wind speed approaching the most downstream (fourth) wind turbine.
   (b) Plot the ratio of the power generated by each of the wind turbines with respect to the power generated by the most upstream turbine.
4. For the conditions in Problem 3, determine the minimum spacing of the wind turbines in the direction perpendicular to the wind direction so that none of the wind turbines are in the wakes of the others.
5. In a reconsideration of Problem 1, look at the effect of having a smaller rotor radius for the upstream wind turbine (A) on the wind farm energy density. Thus, keeping the rotor radius of the downstream wind turbine the same at $r_r = 20\,\text{m}$, reduce the rotor diameter of the upstream wind turbine by 20%. The hub heights will remain the same at $z = 60\,\text{m}$., as well as the same roughness height of $z_0 = 0.3\,\text{m}$. Again for both wind turbines, the design torque coefficient is $C_T = 0.88$, and the free-stream wind speed is $U_\infty = 12\,\text{m/s}$.
   (a) For this system, based on the analytic wake model described in this chapter, determine the spanwise distance, y/D of the downwind turbine (B) so that it is **not affected** by the wake of the upstream turbine (A) for positions of Turbine B of $x/D = 4$, 7 and 10.
   (b) How did reducing the upwind rotor radius affect the energy density (power-per-square-area) of the wind farm?
6. Write a genetic algorithm computer code to find the unknown coefficients, x, y, z, and u to the following equation

$$x + 2y + 3z + 4u = 30, \tag{8.48}$$

   where the range of unknown coefficients is 0–30. Is there only one set of coefficients? Explain.

7. Write a genetic algorithm computer code to determine best locations for a pair of wind turbines within 2,000 by 2,000 m wind farm that will maximize $E(P)$ based on the frequency of wind directions listed in the following table.

Index (i)	θ (deg)	w_i	Index (i)	θ (deg)	w_i
1	0	0	13	180	0.01
2	15	0.01	14	195	0.01
3	30	0.01	15	210	0.01
4	45	0.01	16	225	0.01
5	60	0.20	17	240	0.01
6	75	0.60	18	255	0.01
7	90	0.01	19	270	0.01
8	105	0.01	20	285	0.01
9	120	0.01	21	300	0.01
10	135	0.01	22	315	0.01
11	150	0.01	23	330	0.01
12	165	0.01	24	345	0.01

(a) For the conditions in the table with a predominant wind direction of $\theta = 75°$, list the grid turbine locations and the maximum $E(P)$.

(b) Repeat 1 where the prominent wind direction is 45°.

(c) Repeat 1 where the prominent wind direction is 90°.

References

1. U.S. Department of Energy, "Historic wind development in New England: The age of PURPA Spawns the "Wind Farm", October 9, 2008, Retrieved 24 April 2010.

2. N. Jensen, "A note on wind generator interaction", Risϕ DTU National Laboratory for Sustainable Energy, 1983.

3. M. Lackner and C. Elkinton, "An analytical framework for offshore wind farm layout optimization", *Wind Engineering*, 2007, 31:17–31.

4. G. Mosetti, C. Poloni, and D. Diviacco, "Optimization of wind turbine positioning in large wind farms by means of a Genetic algorithm", *Journal of Wind Enginering Industrial Aerodynamics*, 1994, 51:105–116.

5. R. Rivas, J. Clausen, and K. Hansen, "Solving the turbine positioning problem for large offshore wind farms by simulated annealing", *Wind Engineering*, 2009, 33:287–297.

6. S. A. Grady, M. Y. Hussaini, and M. M. Abdullah, "Placement of wind turbines using genetic algorithms", *Renewable Energy*, 2005, 30:259–270.

7. J. Mora Castro, J. M. Baron Calero, J. M. Santos Riquelme, and M. Payan Burgos, "An evolutive algorithm for wind farm optimal design", *Neurocomputing*, 2007, 70:2651–2658.
8. A. Kusiak and Z. Song, "Design of wind farm layout for maximum wind energy capture", *Renewable Energy*, 2010, 35:685–694.
9. Photographer is Christian Steiness. Courtesy of Vattenfall Wind Power, Denmark.
10. University of Massachusetts Amherst, "Wind energy center alumni and the early wind industry", 2010, Retrieved 24 April 2010.

9

Wind Turbine Acoustics

Wind turbines generate sound by both mechanical and aerodynamic sources. As the technology has advanced, wind turbines have become quieter; however, sound remains an important criterion used in the siting of wind farms. As a result, sound emission from wind turbines has been one of the more studied environmental impact areas in wind energy engineering. Although sound levels can be measured accurately, the perception of the acoustic impact of wind turbines on people is sometimes subjective. Thus in this case, the psychological aspect can be as important as the physical perception.

Acoustic "noise" is defined as any unwanted sound. Concerns about noise depend on:

1. the level of intensity, frequency, frequency distribution, and patterns of the noise source,
2. background sound levels,
3. the terrain between the emitter and receptor,
4. the nature of the receptor, and
5. the attitude of the receptor about the emitter.

The effects of noise on people can be classified into three general categories:

1. subjective effects including annoyance, nuisance, dissatisfaction,
2. interference with activities such as speech, sleep, and learning, and
3. physiological effects such as anxiety, tinnitus, or hearing loss.

In almost all cases, the sound levels associated with wind turbines, regardless of the size, produce effects only in the above categories (1) and (2). Modern wind turbines typically only produce noise effects in Category (1). Whether a sound is objectionable however depends on the type of sound (tonal, broad-band, low frequency, or impulsive), and the circumstances and sensitivity of the person (receptor) who hears it. Because of the wide variation in the levels of individual tolerance for noise, there is no completely satisfactory manner to measure the subjective effects of noise or the corresponding reactions of annoyance and dissatisfaction. With this background, the significant factors relevant to the potential environmental impact of wind turbine noise [1]

DOI: 10.1201/9781003611486-9

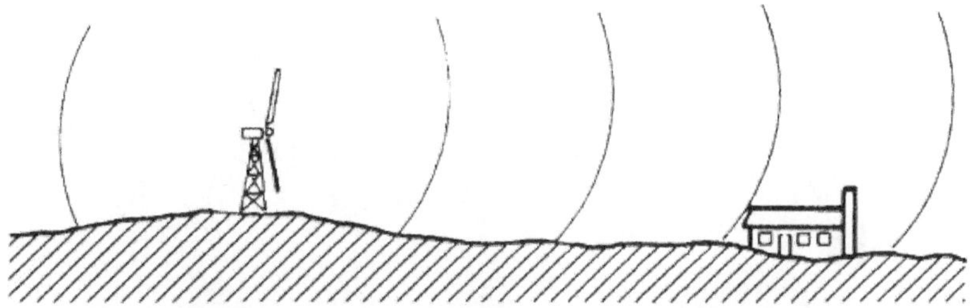

FIGURE 9.1
Schematic examples of wind turbine sound sources, propagation paths and receivers [1].

are illustrated in Figure 9.1. This includes sound sources, sound propagation paths, and sound receivers. A more detailed discussion of these is presented in the following sections.

9.1 Acoustics Fundamentals

Sound consists of pressure waves that travel through a medium. This is illustrated in Figure 9.2. Sound waves are characterized by their amplitude, wavelength, λ, frequency, ω and velocity, c, where

$$c = \omega\,\lambda, \tag{9.1}$$

where ω has units of sec^{-1}, and λ has units of length such as meters. The physical sound frequency is $f = \omega/2\pi$ with units of Hertz. The velocity of sound in air depends on the air density, which are functions of temperature, pressure, and humidity. For air at standard temperature and pressure, the speed of sound is approximately 340 m/s.

The intensity of sound is the average amount of sound power transmitted through a unit area in a specified direction. The unit of intensity is Watts/m^2.

Sound frequency denotes the "pitch" of the sound, and in many cases corresponds to notes on the musical scale, for example Middle C is 262 Hz. An octave is a frequency range between a sound having one frequency and another having twice that frequency. Octaves are often used to define ranges of sound frequency values. For example, the frequency range of human hearing corresponds to 10 Octaves, from about 20 Hz to 20 kHz.

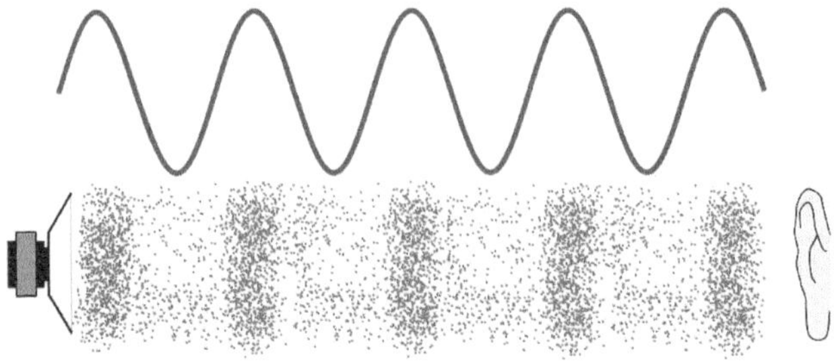

FIGURE 9.2
Schematic representation of a sound pressure wave.

Because of the five order of magnitude range of sound pressure to which the human ear responds, it is convenient to represent sound levels on a logarithmic scale [2]. Therefore sound *intensity*, I, is then represented as

$$I = 10\log_{10}(I/I_0), \tag{9.2}$$

where I has units of decibels (named after Alexander Graham Bell), and I_0 represents the lowest threshold of human hearing corresponding to $1 \times 10^{-12}\,\mathrm{W/m^2}$.

Because audible sound consists of pressure waves, sound *power* is also quantifiable by its relation to a reference pressure. The sound *power* level of a source, L_W, in units of decibels (dB), is given as

$$L_W = 10\log_{10}(P/P_0), \tag{9.3}$$

where P is equal to the sound power level in units of power density, and P_0 is the reference threshold sound power level, $P_0 = 2 \times 10^{-5}\,\mathrm{N/m^2}$.

It is also customary to measure the root-mean-square (r.m.s.) of the pressure fluctuations, p_{rms}, which has units of pressure. The sound *pressure* level in decibels is then defined as

$$L_P = 10\log_{10}(p_{\mathrm{rms}}/p_0)^2, \tag{9.4}$$

or

$$L_P = 20\log_{10}(p_{\mathrm{rms}}/p_0), \tag{9.5}$$

where p_0 in this case is again the reference threshold sound pressure level, $p_0 = 2 \times 10^{-5}\,\mathrm{N/m^2}$.

The human response to sounds measured in decibels has the following characteristics:

- Except under laboratory conditions, a change in sound pressure level of 2 dB cannot be perceived.

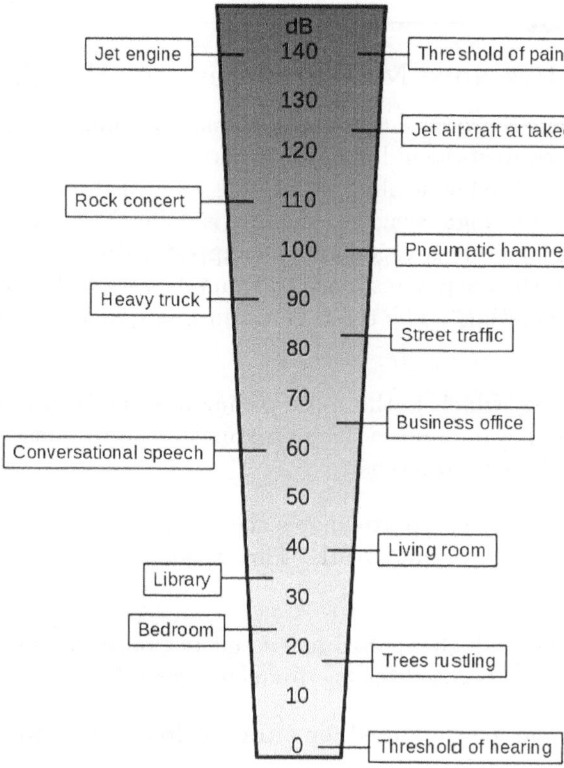

FIGURE 9.3
Examples of sound pressure levels that occur in different activities.

- Doubling the energy of a sound source corresponds to a 3 dB increase in the sound intensity level, or 6 dB increase in sound pressure level.
- Outside of the laboratory, a 3 dB change in sound intensity level is considered a barely discernible difference.
- A change in sound intensity level of 5 dB will typically result in a noticeable community response.
- A 3 dB increase in sound intensity level, or a 6 dB increase in sound pressure level, is equivalent to moving half the distance towards a sound source.
- The threshold of pain sound pressure level is 140 dB.

Figure 9.3 illustrates the relative magnitudes of common sounds on the decibel scale [3].

9.2 Sound Pressure Measurement and Weighting

Sound pressure levels are measured using sound level meters that consist of a microphone that converts sound pressure fluctuations into a voltage time series output that is calibrated in decibels. A sound level measurement that combines all frequencies into a single weighted reading is defined as a broadband sound level. Sound level meters are generally equipped with band-pass frequency filters that shape the output response to simulate human hearing. These are referred to as "weighting". The types of sound pressure level weighting are as follows:

- A-scale weighting, which is the most common scale for assessing environmental and occupational noise. It approximates the response of the human ear to sounds of medium intensity.
- B-scale weighting, which approximates the response of of the human ear for medium-loud sounds, around 70 dB. This weighting scale is not commonly used.
- C-scale weighting, which approximates the response of the human ear to loud sounds. It can be used for low-frequency sound.
- G-scale weighting, which is used for ultra-low frequency, infrasound.

A representation of the frequency response of the A, B and C-scale weighting [3] is shown in Figure 9.4.

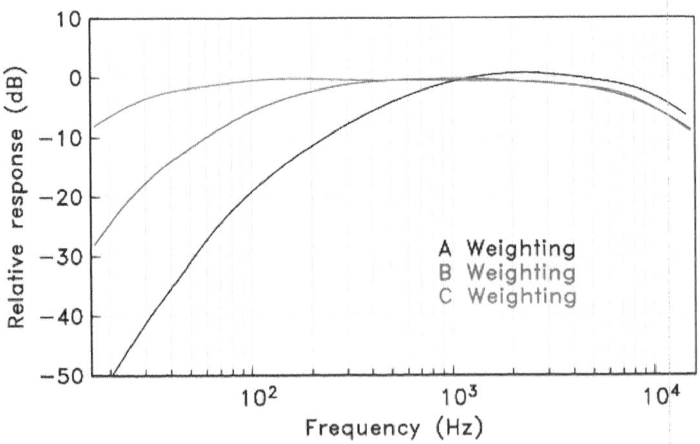

FIGURE 9.4
Frequency response curves for A, B, and C weighting scales.

Once the A-weighted sound pressure level is measured over a period of time, it is possible to determine a number of statistical descriptions of time-varying sound. Terms commonly used in describing environmental sound include [3]:

1. L_{10}, L_{50}, and L_{90}, which are the A-scale weighted sound levels that are exceeded 10%, 50%, and 90% of the time, respectively. During the measurement period, L_{90} is generally taken as the background sound pressure level.
2. Equivalent Sound Level, L_{eq}, which is the average A-scale weighted sound pressure level that gives the same total energy as the varying sound level during the measurement period of time. Also referred to as $L_{A_{eq}}$.
3. Day-Night Level, L_{dn}, which is the average A-scale weighted sound level during a 24 hours day, obtained after the addition of 10 dB to levels measured in the night time between 10 p.m. and 7 a.m.

9.3 dB Math

The logarithmic nature of sound intensity level requires care in determining the sound level from multiple sound sources. For example, consider two sound sources of 90 and 80 dB. To determine the sum of the two sound pressure levels in decibels, we first convert the decibel value into sound pressure, namely

$$90\text{dB} = 20\log\left(\frac{P'_{90}}{2\times10^{-5}\text{Pa}}\right) = 0.632\text{Pa}, \tag{9.6}$$

$$80\text{dB} = 20\log\left(\frac{P'_{80}}{2\times10^{-5}\text{Pa}}\right) = 0.200\text{Pa},$$

therefore

$$(90+80)\text{dB} = 20\log\left(\frac{0.832}{2\times10^{-5}\text{Pa}}\right) = 92.38\text{dB}.$$

9.4 Low Frequency and Infrasound

Low frequency sound consists of pressure fluctuations that can be heard near the lowest end of the frequency response of the human ear, namely from 10 to 200 Hz. Infrasound is pressure fluctuations at frequencies that are below

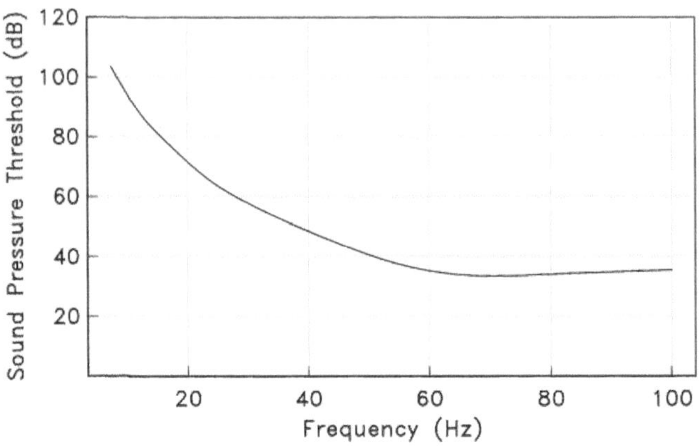

FIGURE 9.5
Perception threshold of the human ear for low frequency sound.

the common limit of the human ear. This is generally considered to be below 20 Hz.

Infrasound is always present in the environment and stems from many sources including ambient air turbulence, ventilation units, waves on the seashore, distant explosions, traffic, aircraft, and other machinery. Infrasound propagates farther, with lower levels of dissipation, than higher frequencies.

Some characteristics of the human perception of infrasound and low frequency sound are [4]

1. Frequencies in the 2–100 Hz range are perceived as a mixture of auditory and tactile sensations.
2. Because of the poorer low frequency response of the human ear (A-scale weighting), such lower frequencies must be of a higher magnitude (dB) to be perceived. For example, the threshold of hearing at 10 Hz is approximately 100 dB as shown in Figure 9.5
3. Tonality cannot be perceived below around 18 Hz.
4. Infrasound may not appear to be coming from a specific location because of its long wavelength.

The primary human response to perceived infrasound is annoyance, with resulting secondary effects. Annoyance levels typically depend on other characteristics of the infrasound, including intensity, variations with time, such as impulses, loudest sound, periodicity, etc. Infrasound has three annoyance mechanisms:

1. A feeling of static pressure.
2. Periodic masking effects in medium and higher frequencies.

3. Rattling of doors, windows, etc. from strong low frequency components.

Human effects vary by the intensity of the perceived infrasound, which can be grouped into these approximate ranges:

1. 90 dB and below, where there is no evidence of adverse effects,
2. 115 dB, where fatigue, apathy, abdominal symptoms, and hypertension in some humans occurs,
3. 120 dB, which is the approximate threshold of pain at 10 Hz, and
4. 120–130 dB and above, where exposure for 24 hours causes physiological damage.

There is no reliable evidence that infrasound below the perception threshold produces physiological or psychological effects.

9.5 Wind Turbine Sound Sources

Wind turbines generate four types of sound characteristics: tonal, broadband, low frequency, and impulsive. Tonal sound is defined as sound that occurs at discrete frequencies. It is caused by components such as meshing gears, non-aerodynamic instabilities interacting with a rotor blade surface, or unstable flows over holes or slits or a blunt trailing edge [5]. Broadband sound is characterized by a broad spectrum of frequencies, generally greater than 100 Hz. It is often caused by the interaction of wind turbine blades with atmospheric turbulence. It is commonly described as a "swishing" or "whooshing" sound that accompanies the rotor rotation. This usually occurs rotor rotation cycle when it is retreating from top dead center position.

Low frequency sound occurs in the range from 20 to 100 Hz. It is primarily associated with rotors that are *downwind* of the tower support. This is the result of an interaction between the rotor wake and the support tower flow field. Figure 9.6 shows an example of the type of interaction that occurs, where the rotor plane cuts through the unsteady wake vortex street produced by the tower, resulting in "bursts" of sound observed in the time traces from a microphone [1].

Finally, impulsive sound consists of short acoustic impulses or thumping sounds that vary in amplitude with time. It is again associated with the interaction between the rotor wake and the support tower flow field with rotors that are downwind of the tower support.

The sources of sound from a wind turbine can be separated into two types: mechanical and aerodynamic. Mechanical sounds come from components such

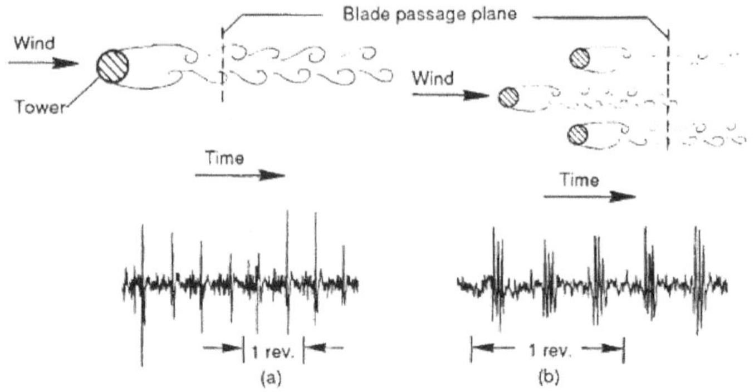

FIGURE 9.6
Example of the type of interaction that occurs, when the rotor plane cuts through the unsteady wake vortex street produced by the tower, resulting in "bursts" of sound observed in the time traces from a microphone [1].

as the gear box, generator, yaw drives, cooling fans, and other auxiliary equipment. The sound from these components is generally associated with the rotation of the rotor and therefore is mostly tonal in nature. The transmission path can be air-borne or structure-borne, namely it is emitted directly into the air, or is transmitted along structural elements of the wind turbine. A summary of wind turbine aerodynamic sound mechanisms is given in Table 9.1 [5].

Aerodynamic sources originate from the flow of air around the blades. This is typically the largest component of wind turbine acoustic emissions. There are numerous mechanism for aerodynamic sound generation on the rotor. These are illustrated in Figure 9.7. These aerodynamic sound sources are generally divided into three groups [5]:

1. *Low frequency sound* that is generated when the rotating blade encounters localized flow deficiencies (wakes) due to the flow around a tower, wind speed changes, or wakes shed from other blades.
2. *Inflow turbulence sound* that results from unsteady aerodynamic loading (pressure fluctuations) caused by the passage of turbulent wind gusts.
3. *Airfoil self noise* that results from air flowing along the surface of the airfoil. This includes trailing-edge noise, tip noise, stall or flow separation noise, laminar boundary layer noise, blunt trailing edge noise, and noise from holes, slits, and intrusions. These can be either tonal or broadband noise.

Figure 9.8 provides the scaling of the sound power level with the characteristic velocity and lengths of the wind turbine rotor. For inflow turbulence sound, the sound level scales with the local velocity to the fourth power, V^4, the

TABLE 9.1
Wind Turbine Aerodynamic Sound Mechanisms [5]

Indication	**Mechanism**	**Main Characteristics**
Low-frequency sound		
Steady thickness noise steady loading noise	Rotation of blades or rotation of lifting surfaces	Frequency is related to blade passing frequency, not important at current rotational speeds
Unsteady loading noise	Passage of blades through tower velocity deficit or wakes	Frequency is related to blade passing frequency, small in cases of upwind rotors, though possibly contributing in case of wind farms
Inflow turbulence sound	Interaction of blades with atmospheric turbulence	Contributing to broadband noise; not yet fully quantified
Airfoil self-noise		
Trailing-edge noise	Interaction of boundary layer turbulence with blade trailing edge	Broadband, main source of high frequency noise (770 Hz $\leq f \leq$ 2 kHz)
Tip noise	Interaction of tip turbulence with blade tip surface	Broadband; not fully understood
Stall, separation noise	Interaction of turbulence with blade surface	Broadband
Laminar boundary layer noise	Non-linear boundary layer instabilities interacting with the blade surface	Tonal, can be avoided
Blunt trailing edge noise	Vortex shedding at blunt trailing edge	Tonal, can be avoided
Noise from flow over holes, slits and intrusions	Unstable shear flows over holes and slits, vortex shedding from intrusions	Tonal, can be avoided

nose radius squared, σ^2, and linearly with the length of the blade element and chord. This is usually a broadband source, and not fully quantified.

The airfoil trailing-edge self noise scales as V^5, and linearly with the wake width, δ. This is usually broadband in nature. With a blunt trailing edge. Finaly, the self noise scales as $V^{5.3}$, and linearly with the trailing-edge thickness, t. This is usually tonal in nature.

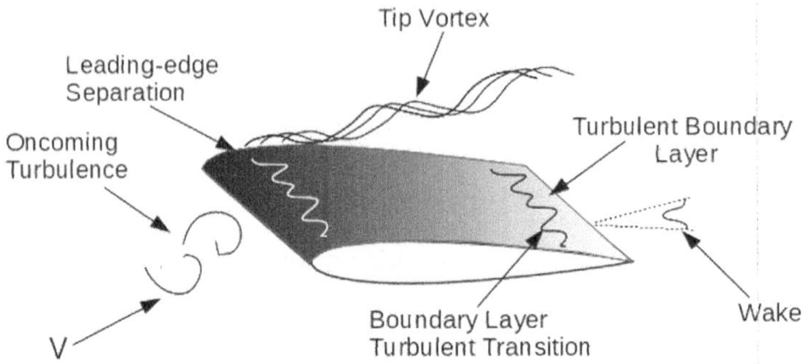

FIGURE 9.7
Mechanisms for sound generation due to the air flow over the turbine rotor.

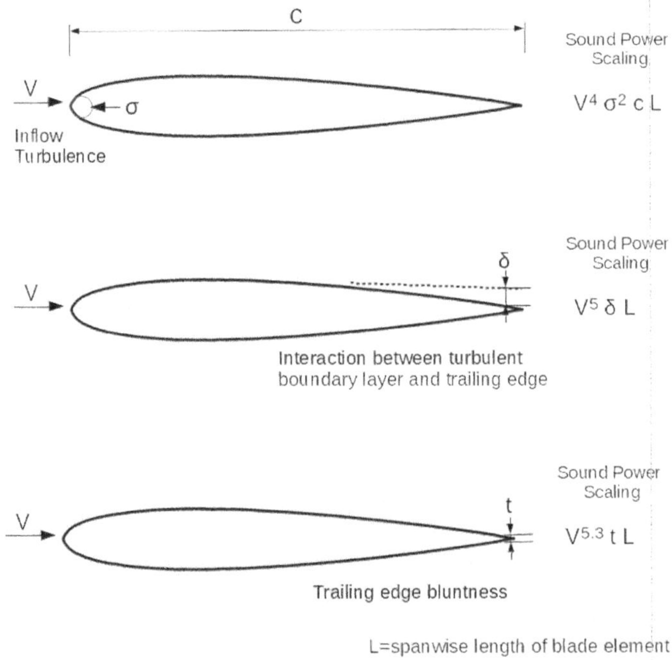

FIGURE 9.8
Sound level power scaling for different aerodynamic sound source mechanisms on the turbine rotor [6].

In addition to the mechanisms for aerodynamic sound generation, the sound generated from the rotor plane is directional. This is illustrated in Figure 9.9 which shows the sound pressure levels measured in a 360° plane around a wind turbine [1,7]. The wind is from the 0° vector, The rotor plane is perpendicular to the wind direction. This illustrates that the azimuthal sound

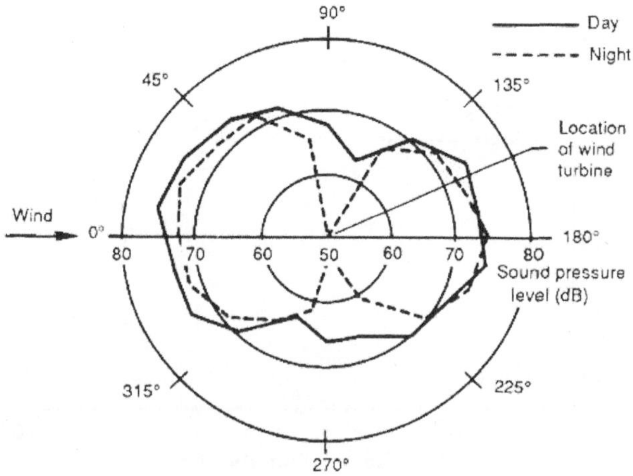

FIGURE 9.9
Sound pressure level azimuthal radiation pattern for a wind turbine [1,7].

level distribution forms two lobes with maxima on the upwind and downwind locations from the rotor plane (0° and 180° vectors), and minima on the edges of the rotor plane (90° and 270° vectors).

Efforts to reduce aerodynamic sounds have included the use of lower tip speed ratios, lower blade angles of attack, upwind rotor designs, variable speed operation and most recently, the use of specially modified blade trailing edges. This is reflected in the data in Figure 9.10 which shows the trends in sound pressure levels as a function of rotor diameter for different generations of wind turbines [11]. In general, sound pressure levels increases logarithmically with the rotor diameter. The earlier generation wind turbines, circa 1980s, were considerably louder than modern generation turbines. The improvements reflect a better understanding and control of the sound sources.

9.6 Sound Propagation

In order to predict the sound pressure level at a distance from a source with a known power level, one must determine how the sound waves propagate. In general, as sound propagates without obstruction from a point source, the sound pressure level decreases. The initial energy in the sound is distributed over a larger and larger area as the distance from the source increases. Thus assuming spherical propagation, the same energy that is distributed over a square meter at a distance of 1 m from a source, is distributed over 10,000 m

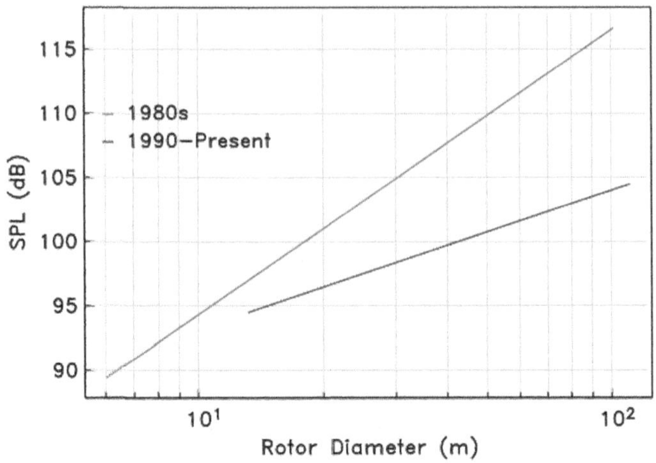

FIGURE 9.10
Trends in sound pressure levels as a function of rotor diameter for different generations of wind turbines [11].

at a distance of 100 m away from the source. With spherical propagation, the sound pressure level is reduced by 6 dB per doubling of distance.

This simple model of spherical propagation must be modified in the presence of reflective surfaces and other disruptive effects. As illustrated in Figure 9.11, if the source is on a perfectly flat and reflecting surface, then the sound level would be 3 dB higher at a given distance than what would be predicted with hemispherical spreading [8]. Thus, the development of an accurate sound propagation model generally must include

1. source characteristics, for example directivity, height, etc.,
2. the distance of the source from the observer,
3. ground effects, for example reflection and absorption of sound on the ground which depend on the source height, the terrain cover, the ground properties, and the sound frequency,
4. blocking of the sound by obstructions and uneven terrain,
5. weather effects, for example wind speed, change of wind speed or temperature with elevation,
6. prevailing wind direction which can cause differences in sound pressure levels between upwind and downwind positions, and
7. the shape of the land whereby certain land forms can focus sound.

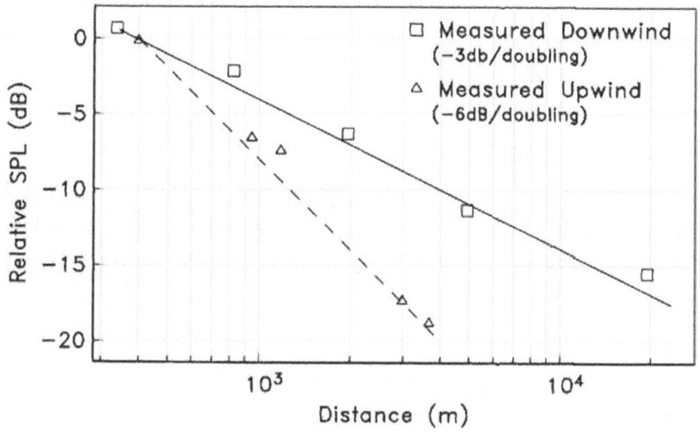

FIGURE 9.11
Example of the effect of wind on the propagation of low frequency rotational harmonic noise from a large-scale HAWT.

These effects are embodied in the illustration in Figure 9.12.

For estimation purposes, a simple model based on the more conservative assumption of hemispherical sound propagation over a reflective surface, including air absorption is often used, namely

$$L_{W_p} = L_{W_s} - 10\log_{10}\left(2\pi R^2\right) - \alpha R, \tag{9.7}$$

where L_{W_s} is the sound power level (dB) measured at the sound source, L_{W_p} is the propogated sound power level (dB) measured at a radial distance R from a sound source, and α is the frequency-dependent sound absorption coefficient with units of dB per distance. An estimate of the broadband sound absorption coefficient is, $\alpha = 0.005\,\text{dB/m}$. For multiple wind turbines, the total sound power level can be calculated by summing up the sound power levels due to each turbine at a specific location using the dB math previously discussed.

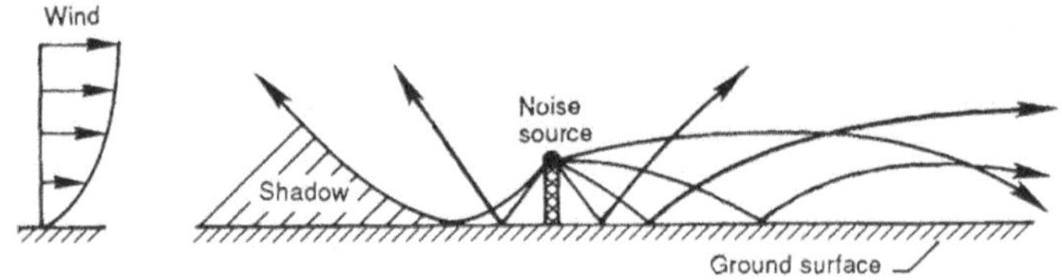

FIGURE 9.12
Example of the effects of wind-induced refraction on acoustic rays radiating from an elevated source [9].

Example 9.1

An example of the sound that might be propagated from a single wind turbine is shown in the illustration below. The wind turbine hub height is 50 m. The sound power level measured at the hub height is 102 dB(A).

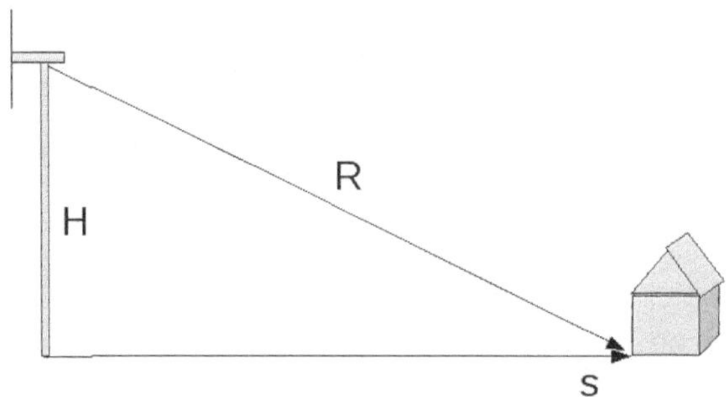

Assuming that the sound radiates in a spherical pattern, and that the atmospheric sound absorption coefficient is $\alpha = 0.005\,\text{dB/m}$, we wish to determine the sound power level along the ground, measured from the base of the wind turbine tower to the location of the house which is 1,000 m. from the base of the wind turbine.

1. The sound propogates in a radial pattern along the radial vector, R. The sound power level along R is given by Equation 9.7. However, our interest is in the sound power level along the ground with a distance s, measured from the base of the wind turbine tower. The relation between s and R is

$$R = \left(s^2 + H^2\right)^{1/2}, \tag{9.8}$$

which can be substituted into Equation 9.7 to give an expresssion for the sound power level on the ground, namely

$$L_{W_p} = L_{W_s} - 10\log_{10}\left(2\pi\left(s^2 + H^2\right)\right) - \alpha\left(s^2 + H^2\right)^{1/2}. \tag{9.9}$$

Based on this, the next table gives the decibel level on the ground at greater distances from the base of the wind turbine.

Ground Distance (m)	dB Level
0	59.79
100	52.49
200	46.70
300	42.84
400	39.89
500	37.48
600	35.41
700	33.59
800	31.93
900	30.41
1,000	29.00

These have been plotted in Figure 9.13.

2. The atmospheric sound absorption coefficient can depend on temperature and relative humidity. As shown in the following table, it is most sensitive to temperature.

Relative Humidity (%)	Temperature (°F)	α (dB/m)
20	70	0.006
40	70	0.005
60	70	0.005
80	70	0.005
90	70	0.006
20	20	0.014
20	30	0.018
20	40	0.015
20	50	0.011
20	60	0.008
20	70	0.006
20	80	0.006
20	90	0.006

If the conditions were such that the sound absorption coefficient were the highest value in the table, $\alpha = 0.018$, the sound power level at $s = 1,000$ m would be lowered by nearly a factor of two to 15.98 dB.

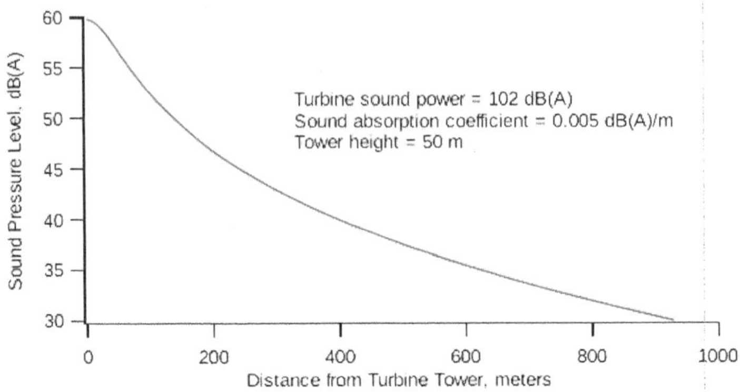

FIGURE 9.13
Sound pressure as a function of distance the wind turbine example problem with $\alpha = 0.005\,\text{dB/m}$.

9.7 Background Sound

The ability to hear a wind turbine in a given installation also depends on the ambient sound level. When the background sounds and wind turbine sounds are of the same magnitude, the wind turbine sound gets lost in the background. Ambient baseline sound levels will be a function of things such as local traffic, industrial sounds, farm machinery, barking dogs, lawnmowers, children playing and the interaction of the wind with ground cover, buildings, trees, power lines, etc. It will vary with the time of day, wind speed and direction, and the level of human activity.

The most likely sources of wind-generated sounds are interactions between the wind and vegetation. A number of factors affect this. For example, the total magnitude of wind-generated sound depends more on the size of the windward surface of the vegetation than the foliage density or volume. The sound level and frequency content of wind generated sound also depends on the type of vegetation. For example, sounds from deciduous trees tend to be slightly lower and more broadband than that from conifers, which generate more sounds at specific frequencies. The equivalent A-weighted broadband sound power level generated by wind in foliage has been shown [10] to vary as

$$L_{A,eq} \propto \log_{10}(U), \tag{9.10}$$

where U is the local wind speed. Thus the wind-generated contribution to background sound tends to increase fairly rapidly with wind speed.

Sound levels from large modern wind turbines during constant speed operation tend to increase more slowly with increasing wind speed than ambient wind generated sound. As a result, wind turbine noise is more commonly a concern at lower wind speeds.

TABLE 9.2
ISO 1996-1971 Recommendations for Community Noise Limits

Location	Daytime - db(A) 7 AM–7 PM	Evening - db(A) 7 PM–11 PM	Night - dB(A) 11 PM–7 AM
Rural	35	30	25
Suburban	40	35	30
Urban residential	45	40	35
Urban mixed	50	45	40

9.8 Noise Standards

At the present time, there are no common international noise standards or regulations for sound pressure levels. In most countries, however, noise regulations define upper bounds for the noise to which people may be exposed. These limits depend on the country and may be different for daytime and nighttime.

In the U.S., although no federal noise regulations exist, the U.S. Environmental Protection Agency (EPA) has established noise guidelines. Most states do not have noise regulations, but many local governments have enacted noise ordinances to manage community noise levels. Table 9.2 lists ISO 1996-1971 recommendations for community noise limits.

9.9 Wind Turbine Project Noise Assessment

When a new wind turbine (or wind farm) is proposed near a sensitive receptor site, a noise assessment study is appropriate. Such studies will typically contain the following four major parts:

1. An estimation or survey of the existing ambient background noise levels.
2. The prediction (or measurement) of noise levels from the turbine(s) at and near the site.
3. The identification of a model for sound propagation.
4. Comparing calculated sound pressure levels from the wind turbines with background sound pressure levels at the locations of concern.

The following is an example of the steps in assessing the noise anticipated from a wind turbine at a new site [11].

9.9.1 Ambient Background Levels

Ambient sound levels vary widely and are important for understanding the noise as well as complying with ambient-based regulations. Background sound pressure levels should be measured for the specific wind conditions under which the wind turbine will be operating.

9.9.2 Source Sound Levels

In order to calculate noise levels measured at different distances, the reference sound levels need to be determined. The reference sound level is the acoustic power being radiated at the source and is not the actual sound pressure level as heard at ground level. Reference sound levels can be obtained from manufacturers and independent testing agencies.

9.9.3 Sound Propagation Model

Sound propagation is a function of the source sound characteristics (directivity, height), distance, air absorption, reflection, and absorption by the ground and nearby objects and weather effects such as changes of wind speed and temperature with height. One could assume a conservative hemispherical spreading model or spherical propagation in which any absorption and reflection are assumed to cancel each other out. More detailed models could be used that include the effects of wind speed and direction since sound travels more easily in the downwind direction; however, a conservative model will assume that all directions are downwind at some time.

9.9.4 Comparison of Calculated Sound Levels with Baseline Sound Levels

Calculated wind turbine sound levels do not include the additional background ambient sound levels. As an example of one state's regulations, the mathematical relationship governing the addition of dB(A) levels require that if the turbine sound level is no more than 9.5 dB(A) above the ambient noise level, then the total noise levels will be within 10 dB(A) of the ambient sound level. If the ambient sound level is 45 dB(A), then the turbine can generate no more than 54.5 dB(A) at locations of concern. For the example problem presented earlier, this would require that a sensitive receptor site be no closer than about 75 m. from the wind turbine.

9.10 Wind Turbine Sound Mitigation Techniques

While understanding the sources of sound emanating from horizontal wind turbines that was presented in Section 9.5, this section presents rotor design methods to interfere with the source mechanisms, thereby reducing their acoustic signature and levels.

Problems

1. The sound level standard for night time in a community is 30 dB(A). Therefore it is desired to make a **conservative** estimate of the closest distance, s, that a wind turbine can be located near the community. The wind turbine has the following characteristics: a height, H, of 30 m. and a generated Sound Pressure Level of 110 dB.

   (a) Determine the minimum distance, s, that a wind turbine can be located near the community assuming zero atmospheric absorption, namely $\alpha = 0$.
   (b) Repeat Part (a) using an standard atmospheric absorption, $\alpha = 0.005$ dB/m.
   (c) Repeat Part (a) using an standard atmospheric absorption, $\alpha = 0.015$ dB/m.
   (d) Based on these results, what would be your conservative recommendation to the community?

2. A community wishes to replace an existing 600 kW wind turbine with one rated at 2.5 MW. The higher powered wind turbine has a sound pressure level in the audible range that is 5 dB higher than the lower powered wind turbine, which had a sound pressure level of 80 dB.

   If the sound level standard for night time in the community is 30 dB, and the original wind turbine had a height, H, of 50 m.

   (a) What was the required distance, s, assuming atmospheric absorption of $\alpha = 0.005$ dB/m, that the older wind turbine needed to be placed from the community to meet the night time sound level standard?
   (b) If the higher powered wind turbine is placed at the same location as the old lower-powered wind turbine, what would its height, H, need to be to meet the night time dB sound level standard?
   (c) Does increasing the hub height in Part (b) seem like a reasonable solution to reducing the sound level reaching the community? Can you suggest other options that could reduce the sound level that would not involve changes in the wind turbine?

3. Describe five design features of a wind turbine that affect the generated sound power level. List these in order of importance, focusing on those that, if modified, have the potential to lower the generated sound.

4. The community decides to add an additional wind turbine to double the generated power. They wish to know the impact this will have on the sound levels. The two wind turbines generate a sound power level of 102 dB(A). They are placed side-by-side with a hub center distance of 50 m. Their hub heights are 50 m.

    (a) With the two wind turbines operating, determine the sound power level at a home in the community that is located 300 m. down-wind of the wind turbines on the centerline between the two wind turbines. Use an atmospheric absorption of α = 0.005 dB/m.

    (b) How much does the sound power level at the home change if only one of the wind turbines was operating?

5. In Problem 4, the community is located downwind of the wind turbine.

    (a) What is the impact on the sound level if the community were located directly *upwind* of the wind turbine?

    (b) Repeat Problem 4 but with the community being located *upwind* of the wind turbine.

References

1. H. Hubbard and K. Shepherd, "Wind turbine acoustics," NASA Technical Paper 3057 DOE/NASA/20320-77, 1990.
2. L. Beranek and I. Ver, *Noise and Vibration Control Engineering: Principles and Applications*, Wiley, New York, 1992.
3. Bruel and Kjaer Instruments, Skodsborgvej 307 DK-2850 Nærum, Denmark.
4. A. Moorhouse, D. Waddington, and M. Adams, "Proposed criteria for the assessment of low frequency noise disturbance", Contract no NANR45, University of Salford, UK, February 2005.
5. S. Wagner, R. Bareib, and G. Guidati, *Wind Turbine Noise*, Springer, Berlin, 1996.
6. F. Grosveld, "Prediction of broadband noise from horizontal axis wind turbines", *AIAA Journal of Propulsion and Power*, 1985, 1:4.
7. K. Shepherd, W. Willshire, and H. Hubbard, "Comparison of measured and calculated sound pressure levels around a large horizontal axis wind turbine generator, NASA TM 1000654, 1988.

8. W. Willshire and W. Zorumski, "Low frequency acoustic propogation in high winds", *Proceedings of Noise-Con 87*, Noise Control Foundation, New York, June 1987.
9. K. Shepherd and H. Hubbard, "Sound propagation studies for a large horizontal axis wind turbine", NASA CR-172564, March 1985.
10. O. Fegeant, "On the masking of wind turbine noise by Ambient noise", *Proceedings of European Wind Energy Conference*, Nice, France, March 1–5, 1999.
11. A. Rogers, J. Manwell, and S Wright, "Wind turbine acoustic noise", Renewable Energy Research Laboratory, Department of Mechanical and Industrial Engineering, University of Massachusetts at Amherst, June 2002, amended January 2006, pp. 21–23.

10

Wind Energy Storage

Most electricity in the U.S. is produced at the same time it is consumed, and suppliers bring plants on and off line depending on demand. Peak-load plants, usually fueled by natural gas, run when demand surges, often on hot days when consumers run air conditioners.

In contrast to electric power plants, wind generated power cannot be guaranteed to available when demand is highest. As an example, Figure 10.1 shows an hourly electric power demand time series over a two week period for a city in the Northern U.S. The hourly electric power demand is relatively periodic on a 24 hours cycle with the peak demand occurring in the daylight hours. In contrast, the wind power generation is not periodic or correlated to the demand cycle. When the wind energy is available, it can help to accommodate the demand. However in this example, it is unable to provide all of the electric energy demand because of the intermittent production. A solution to this involves methods to store the energy captured from the wind and regenerating this energy in the form of electricity to match the consumer demand cycles.

There are many methods of energy storage that are being used, developed, or theorized that can apply to wind energy. These include electro-chemical energy storage such as batteries, chemical storage such as electro-hydrogen

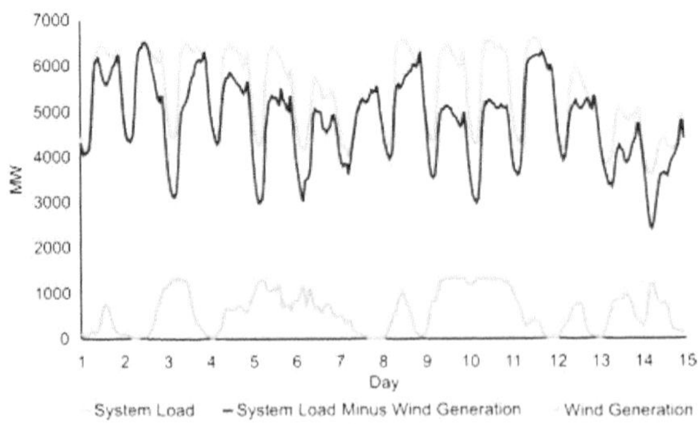

FIGURE 10.1
Example of a two week period of system loads, system loads minus wind generation, and wind generation [1].

DOI: 10.1201/9781003611486-10

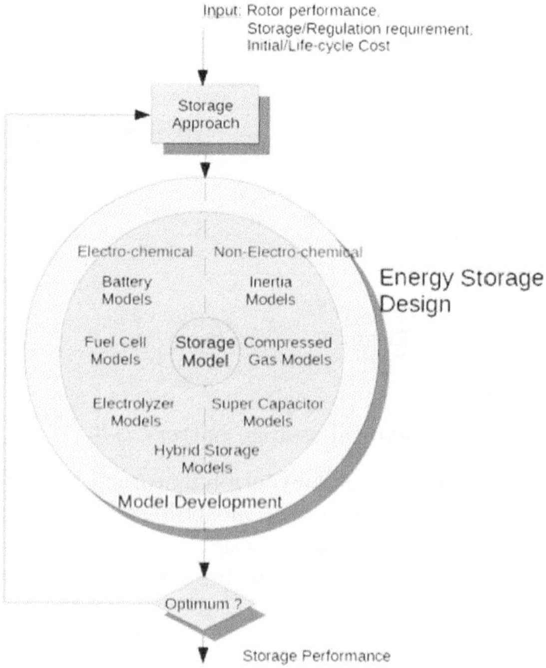

FIGURE 10.2
Wind turbine energy storage optimization flow chart.

generation, gravitational potential energy storage such as pumped-storage hydroelectric, electrical potential storage such as electric capacitors, latent heat storage such as phase-change materials, and kinetic energy storage such as flywheels. Some of these methods provide only short-term energy storage, whereas others can provide very long-term storage. Other important aspects of energy storage are the maximum discharge rate and the number of possible charge-discharge cycles. Figure 10.2 illustrates a wind turbine energy storage flow chart that could be used to evaluate and optimize potential methods. Comprehensive reviews of energy storage methods from the perspective of wind energy are given in references [2–4] listed at the end of this chapter.

10.1 Electro-Chemical Energy Storage

Rechargeable batteries or "storage batteries" are the most common form of electric storage devices. They come in a large range of sizes and power capacities. Battery systems with storage power levels totaling megawatts are being used to stabilize electric power in some portions of the distribution grid.

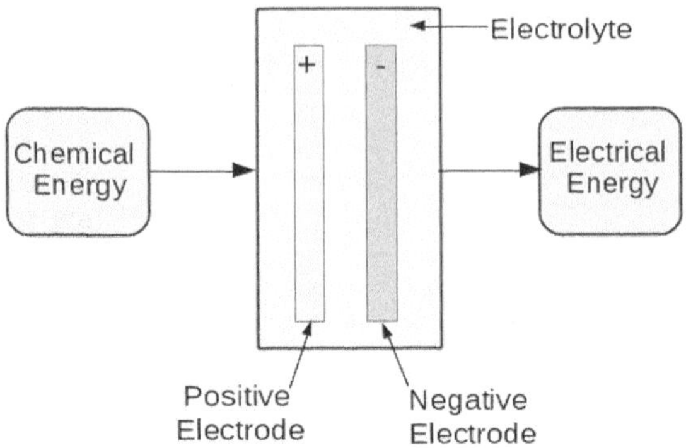

FIGURE 10.3
Illustration of an electro-chemical storage battery cell.

There are three main types of conventional storage batteries that are used extensively today: lead-acid batteries, nickel-based batteries, and lithium-based batteries. Each has a common design consisting of cells made up of positive and negative electrodes that are immersed in an electrolyte. This is illustrated in Figure 10.3.

10.1.1 Lead-Acid Batteries

Lead-acid batteries are the oldest type of rechargeable battery and most commonly used. They are based on a chemical reactions involving lead dioxide, which forms the cathode electrode, and sulfuric acid which acts as the electrolyte. The rated voltage of a lead-acid cell is 2 V. The typical energy density is around 30 W-h/kg, with a power density of approximately 180 W/kg.

Lead-acid batteries have a high energy efficiency between 80% and 90%. They are easy to install and have relatively low maintenance and initial investment costs. In addition, the self-discharge rates for lead-acid batteries are very low, approximately 2% of the rated capacity per month at 25°C. This makes them ideal for long-term storage applications.

The limiting factors for for lead-acid batteries are (1) the low cycle life and (2) battery operational lifetime. The typical lifetime of lead acid batteries are between 1,200 and 1,800 charge/discharge cycles, or approximately 5–15 years of operation. The cycle life is negatively affected by the depth of discharge and temperature. Attempts to fully discharge the battery can be particularly damaging to the electrodes, thus reducing lifetime. Regarding temperature levels, although high temperatures, up to the 45°C upper limit for battery operation, may improve battery performance in terms of higher capacity, it can reduce total battery lifetime as well as the battery energy efficiency.

10.1.2 Nickel-Based Batteries

Nickel-based batteries mainly consist of nickel-cadmium (NiCd), nickel-metal hydride (NiMH) and nickel-zinc (NiZn). All three types use the same material for the positive electrode, and an electrolyte which is a combination of nickel-hydroxide and an aqueous solution of potassium-hydroxide and some lithium-hydroxide. For the negative electrode, the NiCd type uses cadmium-hydroxide. The NiMH battery uses a metal alloy, and the NiZn battery uses zinc-hydroxide.

The rated voltage per cell for these batteries is 1.2 V (1.65 V for the NiZn type). The typical maximum energy density is higher than that for lead-acid batteries. Typically, values are 50 W-h/kg for the NiCd, 80 W-h/kg for the NiMH and 60 W-h/kg for the NiZn. Typical operational life and cycle life of NiCd batteries is also superior to that of the lead-acid batteries. At deep discharge levels, typical lifetimes range from 1,500 to 3,000 cycles. The NiMH and NiZn batteries have a lesser cycle life, being similar to or lower than that of lead-acid batteries.

NiCd and the rest of the nickel-based batteries have several disadvantages compared to the lead-acid batteries in terms of industrial use or for use in supporting renewable energy power systems. Generally, the NiCd battery is the only one of the three types of nickel-based batteries that is commercially used for industrial UPS applications such as in large energy storage for renewable energy systems. However, the NiCd battery may cost up to ten times more than the lead-acid battery. In addition, the energy efficiency for the nickel-based batteries are lower than for the lead-acid batteries. The NiMH batteries have energy efficiency between 65% and 70%, whereas the NiZn batteries have an 80% efficiency. Another area where NiCd batteries are inferior to lead-acid batteries is the self-discharge rate, which can reach more than 10% of rated capacity per month.

10.1.3 Lithium-Based Batteries

Lithium technology batteries consist of two main types: lithium-ion and lithium-polymer. Their advantage over the NiCd and lead-acid batteries is a higher energy density and energy efficiency, lower self-discharge rate, and extremely low required maintenance. Lithium-ion cells have a nominal voltage of about 3.7 V. The energy density of lithium-ion batteries ranges from 80 to 150 W-h/kg, while that of lithium-polymer ranges from 100 to 150 W-h/kg. The energy efficiency of both range from 90% to 100%. The power density of lithium-ion cells range from 500 to 2,000 W/kg, while that for lithium-polymer cells ranges from 50 to 250 W/kg.

The self-discharge rate for lithium-ion batteries is very low, with a maximum of 5% per month. Its battery lifetime can reach more than 1500 charge cycles. However, the lifetime of a lithium-ion battery is temperature dependent, being worse at high temperatures. The battery lifetime can also

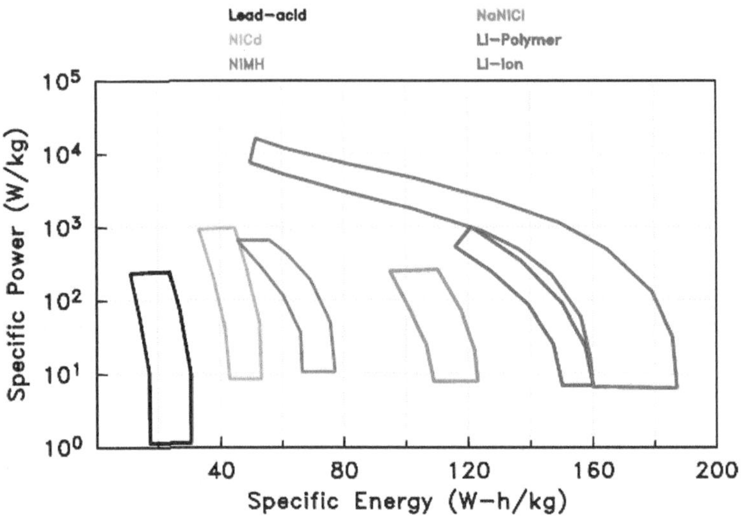

FIGURE 10.4
Specific power versus specific energy for types of electro-chemical storage batteries [3,4].

be severely shortened by deep discharges. This makes lithium-ion batteries unsuitable for use in back-up applications where they may become completely discharged.

Lithium-ion batteries are also fragile and require a protection circuit to maintain safe operation. The protection circuit limits the peak voltage of each cell during charge and prevents the cell voltage from dropping too low on discharge. In addition, the cell temperature is monitored to prevent temperature extremes. These precautions are necessary to eliminate the possibility of metallic lithium plating occurring due to overcharge.

Lithium-polymer battery lifetime can only reach about 600 charge cycles. Its self-discharge dependents on temperature, but it has been reported to be around 5% per month. Compared to the lithium-ion battery, the lithium-polymer battery requires a much narrower operation temperature range that avoids lower temperatures. Overall, lithium-polymer batteries are lighter, and safer, with a minimum self-inflammability compared to lithium-ion batteries. The cost of lithium-based batteries is between \$900 and \$1,300 kW-h. Figure 10.4 provides a graphical comparison of the specific power, W/kg, versus specific energy, W-h/kg, for the types of storage batteries discussed in this section.

10.1.4 Additional Electro-Chemical Storage Technologies

In addition to the three types of batteries described in the previous section, a few additional types also exist, although they are not as widely used. These are the sodium-sulfur (NaS) battery, the Redox-flow battery, and the metal-air battery.

10.1.5 Sodium Sulfur Batteries

The NaS battery consists of liquid (molten) sulfur at the positive electrode and liquid (molten) sodium at the negative electrode as active materials separated by a solid beta alumina ceramic electrolyte. The electrolyte allows only the positive sodium ions to go through it and combine with the sulfur to form sodium polysulphides. During discharge, sodium gives off electrons, while positive Na+ ions flow through the electrolyte and migrate to the sulfur container. The electrons flow in the external circuit of the battery producing about 2 V and then through the electric load to the sulfur container. The electrons react with the sulfur to form S cations, which then forms sodium polysulfides after reacting with sodium ions. As the cell discharges, the sodium level drops. This process is reversible as charging causes sodium polysulfides to release the positive sodium ions back through the electrolyte to recombine as elemental sodium. Once running, the heat produced by charging and discharging cycles is enough to maintain operating temperatures and no external heat source is required to maintain this process. The process produces temperatures in the range of 300°C–350°C.

NaS batteries are highly energy efficient (89%–92%) and are made from inexpensive and non-toxic materials. However, the high operating temperatures and the highly corrosive nature of sodium make them suitable only for large-scale stationary applications. NaS batteries are currently used in electricity grid related applications such as peak shaving and improving power quality.

10.1.6 Redox Flow Battery

A flow battery is a type of rechargeable battery where recharge-ability is provided by two chemical components dissolved in liquids contained within the system and separated by a membrane. Ion exchange (providing flow of electrical current) occurs through the membrane while both liquids circulate in their own respective space. Cell voltage is chemically determined and ranges in practical applications from 1.0 to 2.2 V. A schematic of the process is shown in Figure 10.5.

A flow battery is technically both a fuel cell and an electro-chemical accumulator cell (electro-chemical reversibility). It offers significant advantages such as no self-discharge and no degradation for deep discharge. Commercial applications of most flow batteries are appealing only for long duration stationary energy storage, such as back up grid power for emergency, since increasing a system's overall energy capacity (measured in MW-h) basically requires only an increase in the size of its liquid chemical storage reservoirs.

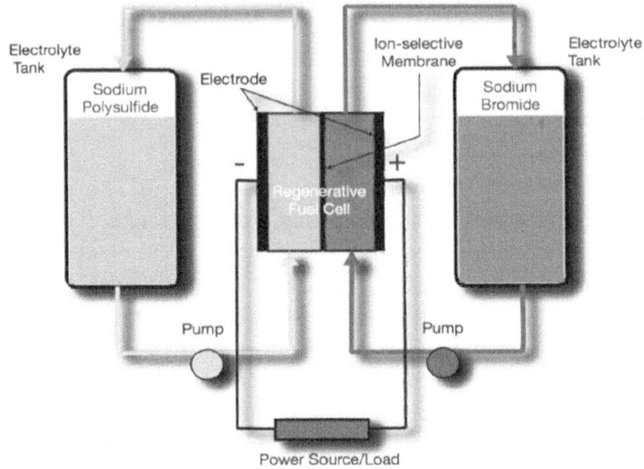

FIGURE 10.5
Schematic drawing of a flow battery [4].

10.1.7 Metal-Air Battery

A metal-air battery is an electro-chemical cell that uses an anode made from pure metal and an external cathode of ambient air, typically with an aqueous electrolyte. Metal-air technology offers high energy density (compared to lead-acid batteries) and long shelf life while promising reasonable cost levels. However, tests have shown that the metal-air batteries suffer from limited operating temperature range and a number of other technical issues not least of which is the difficulty in developing efficient, practical fuel management systems and cheap and reliable bi-functional electrodes.

10.2 Supercapacitor Storage

Supercapacitors (or ultracapacitors) are very high surface area activated capacitors that use a molecule-thin layer of electrolyte as the dielectric to separate charge [5]. The supercapacitor resembles a regular capacitor except that it offers very high capacitance in a small package. Supercapacitors rely on the separation of charge at an electric interface that is measured in fractions of a nanometer, compared with micrometers for most polymer film capacitors. Energy storage is by means of static charge rather than of an electro-chemical process inherent to the battery. Figure 10.6 shows an illustration of a super capacitor.

Depending on the material technology used for the manufacture of the electrodes, supercapacitors can be categorized into electro-chemical double

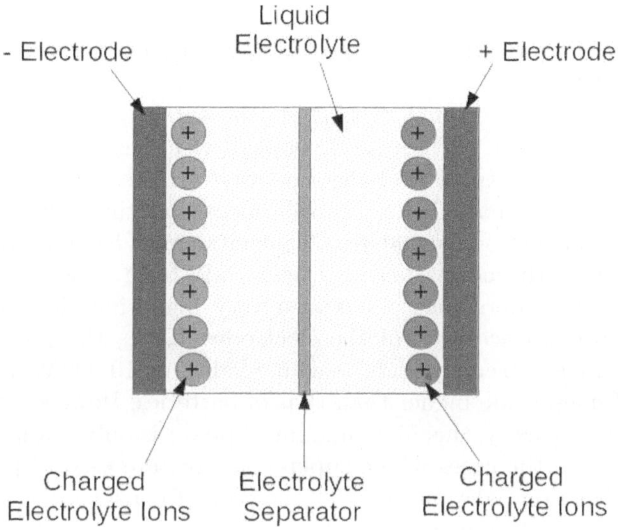

FIGURE 10.6
Schematic of a super capacitor.

layer supercapacitors (ECDL) and pseudo-capacitors. ECDL super-capacitors are currently the least costly to manufacture and are the most common type of supercapacitor.

The ECDL supercapacitors have a double-layer construction consisting of carbon-based electrodes immersed in a liquid electrolyte, which also contains the separator. Porous active carbon is usually used as the electrode material. The electrolyte is either organic or aqueous. The organic electrolytes use usually acetonitrile and allow nominal voltage of up to 3 V. Aqueous electrolytes use either acids or bases (H_2SO_4, KOH), but the nominal voltage is limited to 1 V. During charging, the electrically charged ions in the electrolyte migrate towards the electrodes of opposite polarity due to the electric field between the charged electrodes created by the applied voltage. Thus, two separate charged layers are produced.

Although, similar to a battery, the double-layer capacitor depends on electrostatic action. Since no chemical action is involved the effect is easily reversible with minimal degradation in deep discharge or overcharge and the typical cycle life is hundreds of thousands of cycles. Reported cycle life is more than 100,000 cycles at 100% depth of discharge. The limiting factor in terms of lifetime may be the years of operation with reported lifetimes reaching up to 12 years [6].

One limiting factor of supercapacitors is the high self-discharge rate that is much higher than batteries, reaching a level of 14% of nominal energy per month. However, the fact that no chemical reactions are involved means that supercapacitors can be easily charged and discharged in seconds, thus being much faster than batteries. In addition, no thermal heat or hazardous substances are released during discharge. Energy efficiency is very high, ranging from 85% to 98%.

Compared to conventional capacitors, the supercapacitors have a significantly larger electrode surface area coupled with a much thinner electrical layer between the electrode and the electrolyte. These two attributes mean that supercapacitors have higher capacitances and therefore higher energy densities than conventional capacitors. Capacitances of 5,000 Farads have been reported, along with energy densities up to 5 W-h/kg. The current carrying capability of the supercapacitors is also very high since it is directly proportional to the surface area of the electrodes. Thus, the power density of supercapacitors is extremely high, reaching values of 10,000 W/kg, which is a few orders of magnitude higher than that of batteries. However, as a result of their low energy density, this high amount of power is only available for a very short duration. In the cases where supercapacitors are used to provide power for prolonged periods of time, it is at the cost of considerable added weight and bulk due to its low energy density.

The cost of supercapacitors is a significant issue for their use in industrial applications. The cost, which is estimated to be about $20,000/kW-h, is significantly higher than that of well-established storage technologies such as lead-acid batteries.

Currently, the high power storage ability of supercapacitors, together with the fast discharge cycles, makes them ideal for use as temporary energy storage, for capturing and storing the energy from regenerative braking, and for providing a booster charge in response to sudden power demands. One approach is to combine supercapacitors with conventional storage batteries in a load sharing arrangement, in which the batteries provide power only during the longer duration loads, and the supercapacitors only handle peak loads.

10.3 Hydrogen Storage

Hydrogen is also being developed as an electrical power storage approach. Electricity is used with water to make hydrogen gas through the process of electrolysis. Approximately 50 kW-h of electric energy is required to produce a kilogram of hydrogen. Given this yield, the cost of the electricity used to make hydrogen is clearly crucial to make it a viable storage approach. For example at $0.03/kW-h, which is a common off-peak high-voltage line rate in the United States, hydrogen produced by electrolysis costs approximately $1.50 per kilogram. Figure 10.7 provides a schematic of the elements involved in the use of electricity for hydrogen production and possible storage.

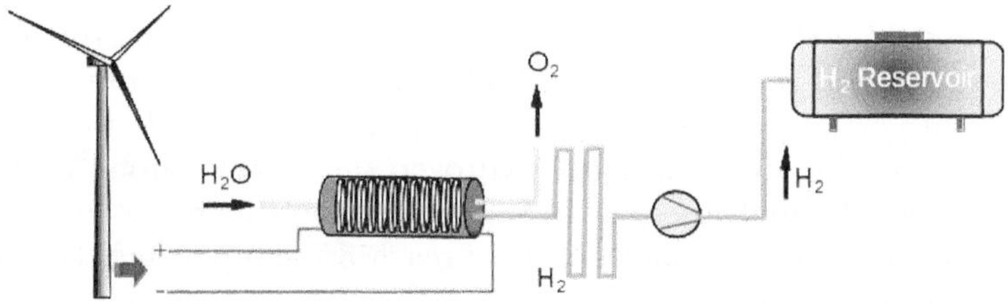

FIGURE 10.7
Illustration of the elements in the use of electricity for hydrogen production and possible storage.

The two most mature methods of hydrogen storage are hydrogen pressurization and the hydrogen adsorption in metal hydrides. Pressurized hydrogen technology relies on materials that are impermeable to hydrogen and mechanical stable under pressure. Currently steel tanks can store hydrogen at 200–250 bar, but present a very low ratio of stored hydrogen per unit weight. Storage capability increases with higher pressures, but stronger materials are then required. Storage tanks with aluminum liners and composite carbon fibre/polymer containers are being used to store hydrogen at 350 bar providing a higher ratio of stored hydrogen per unit weight of up to 5%. In order to reach higher storage capabilities, higher pressures in the range of 700 bar are needed.

The use of metal hydrides offers an excellent alternative to pressurized storage. Metal hydrides, such as MgH_2, $NaAlH_4$, $LiAlH_4$, LiH, $LaNi_5H_6$, $TiFeH_2$ and palladium hydride, with varying degrees of efficiency, can be used as a storage medium for hydrogen, often reversibly. Some of these are easy-to-fuel liquids at ambient temperature and pressure, others are solids that can be turned into pellets. These materials have good energy density by volume, although their energy density by weight is often worse than the leading hydrocarbon fuels.

Most metal hydrides bind with hydrogen very strongly. As a result, high temperatures around 120°C are required to release their hydrogen content. However, metal hydride compounds have some disadvantages. Typically, they exhibit rather low mass absorption capacities (except magnesium hydrides) and do require thermal management system. This is because the absorption of hydrogen is an exothermic reaction (releases heat) while desorption of hydrogen is endothermic. Heating and cooling of the metal hydrides is achieved through a water-based heat exchanger. Absorption/desorption kinetics are however very fast in most hydrides, thus allowing for fast hydrogen storage and release.

Liquid hydrogen storage technology use is currently limited. This is due to the properties and cost of the materials used in the manufacturing of the container/tank and the extreme temperatures that are required for such storage, around −253°C. Storage containers require specific internal liners that are surrounded by thermal insulators in order to maintain the required temperature and avoid any evaporation. The whole process is quite inefficient since a large portion of electric energy is used in the initial stage of hydrogen liquefaction. In addition, liquid hydrogen tanks suffer from leaks caused by unavoidable thermal losses that lead to pressure increases in the tanks. Hydrogen self-discharge of the tank may reach 3% daily, which translates to a 100% self-discharge in 1 month.

10.4 Mechanical Energy Storage Systems

Mechanical storage systems involve the conversion of electric energy into potential or kinetic energy. It includes pumped storage hydroelectricity, compressed air storage, and flywheel energy storage [7]. These are each discussed in the following sections.

10.4.1 Pumped Storage Hydroelectricity

Pumped storage hydroelectricity is a method of storing and producing electricity to supply high peak demands by moving water between reservoirs at different elevations. The principle is that during times of low electricity demand, the excess generation capacity is used to pump water into a reservoir at a higher elevation. When the electric demand is higher, the water is released back into the lower reservoir. In doing so, the water is run through a turbine that generates electricity. In this process, a reversible turbine/generator acts as both a pump and a turbine. Figure 10.8 illustrates the process. Some facilities use abandoned mines as the lower elevation reservoir, but many use the height difference between two natural bodies of water or artificial reservoirs.

Worldwide, pumped storage hydroelectricity is the largest form of grid energy storage available, accounting for more than 99% of bulk storage capacity, representing approximately 127,000 MW. Taking into account evaporation losses from the exposed water surface and conversion losses in the pump, turbine and piping, approximately 70%–85% of the electrical energy used to pump the water into the elevated reservoir can be regained. The approach is currently the most cost-effective means of storing large amounts of electrical energy on an operating basis. However, capital costs and the presence of appropriate geography are critical decision factors.

Pumped storage systems have a relatively low energy density so that they require either a very large body of water or a large variation in elevation. In some locations, this occurs naturally. In others instances, one or both bodies of water have been man-made. They can be economical in use because they can flatten out load variations on the power grid, permitting thermal power stations such as coal-fired and nuclear plants to provide base-load electricity

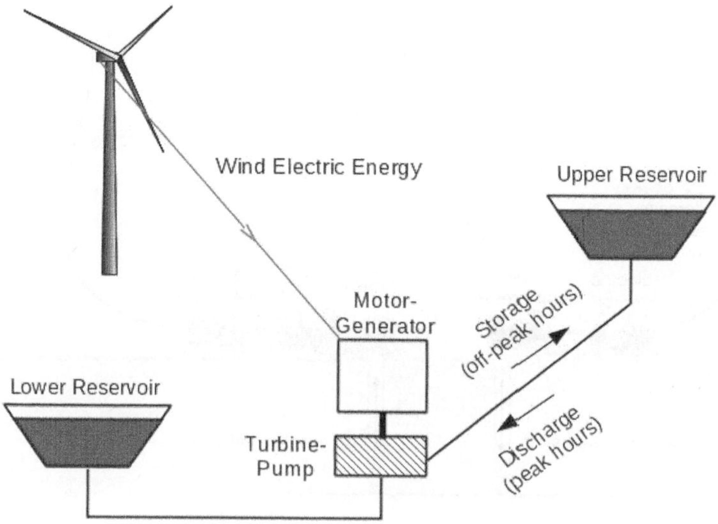

FIGURE 10.8
Illustration of pumped storage hydroelectric power plant.

at peak efficiency, and reducing the need for peak-load power plants that use costly fuels. Pumped storage plants, like other hydroelectric plants, can respond to load changes within seconds.

10.4.2 Compressed Air Storage

Compressed Air Storage (CAES) is another method of storing electric energy during off-peak demand and to be used later when the demand is higher. In this case, the electric energy is used to compress air where it is stored, most often, in underground reservoirs. An illustration of a compressed air storage power plant is shown in Figure 10.9.

There are many geologic formations that can be used for the underground reservoirs. These include naturally occurring aquifers, solution-mined salt caverns and constructed rock caverns. In general, rock caverns are about 60% more expensive to mine than salt caverns for CAES purposes. Aquifer storage is by far the least expensive method and is therefore used in most of the current locations.

The components making up a basic CAES power plant are shown in Figure 10.10. These include

1. a motor/generator that employs clutches to provide for alternate engagement to the compressor or turbine power train,
2. an air compressor that may require two or more stages, intercoolers and aftercoolers to reduce moisture in the compressed air, and to increase the power plant efficiency, and

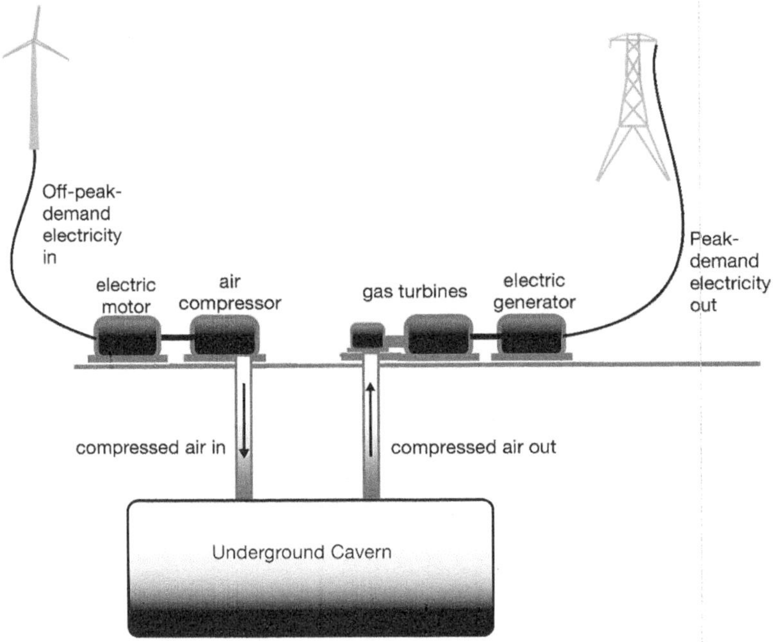

FIGURE 10.9
Illustration of compressed air storage power plant [3].

3. high and low pressure turbines and a recuperator to again increase the power plant efficiency.

During off-peak demand, the excess electric power drives an electric motor that powers the air compressor. This often involves multiple staged compressors in which inter-stage heat exchangers are used to remove heat resulting from compressing the air. This heat can be stored and utilized in a combined or recuperated cycle to improve the plant efficiency. An example of this arrangement is shown in Figure 10.11. The air is typically pressurized to about 75 bar.

When the demand is high, the compressed air is released to pass through the turbine. Prior to this, the air is heated by passing it through a recuperator. This is a heat exchanger that makes use of the stored heat that was released during the air compression. In some CAES power plants, fuel is injected into the air and heated further in a combustor. The hot gas then expands through the turbine, which is connected to the electric generator. The electric generator is a synchronous machine that can be operated as a motor or generator. In the former, it drives the compressor. In the latter, it is driven by the turbine to generate electricity. In the combined or recuperated cycle configurations, waste heat from the turbines is used for inter-stage turbine heating or in the recuperator.

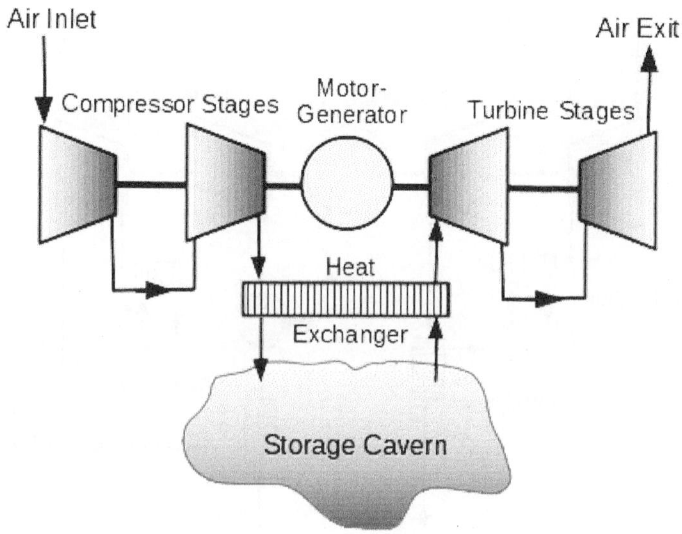

FIGURE 10.10
Components of a basic compressed air storage power plant.

CAES systems can be used as large-scale power plants. Apart from the pumped storage hydroelectric system, no other storage method has a storage capacity that is as large as the CAES. Typical capacities are from 50 to 300 MW. As a result of having very small losses over time, the storage period is the longest of the other systems, easily storing energy for more than a year.

Fast start-up is also an advantage of CAES power plant, with a start-up time of about 9 minutes in an emergency, and about 12 minutes under normal conditions. By comparison, conventional combustion turbine peak-load plants typically require 20–30 minutes for a normal start-up.

10.4.3 Flywheel Storage

Flywheel storage uses a mass rotating about an axis to store energy mechanically in the form of kinetic energy. Energy supplied to an electric motor is used to accelerate the flywheel to its design rotation speed. Once it is rotating, it is in effect a mechanical battery. The energy stored is

$$E = \frac{1}{4} M r^2 \omega^2 \ \sim \ M \nu^2, \tag{10.1}$$

where M is the mass of the flywheel, r is the radius of the flywheel, ω is the rotation rate, and ν is the linear velocity at the outer rim of the flywheel. The stored energy can be retrieved by slowing down the flywheel through a decelerating torque that would be imparted by the generator, which is a

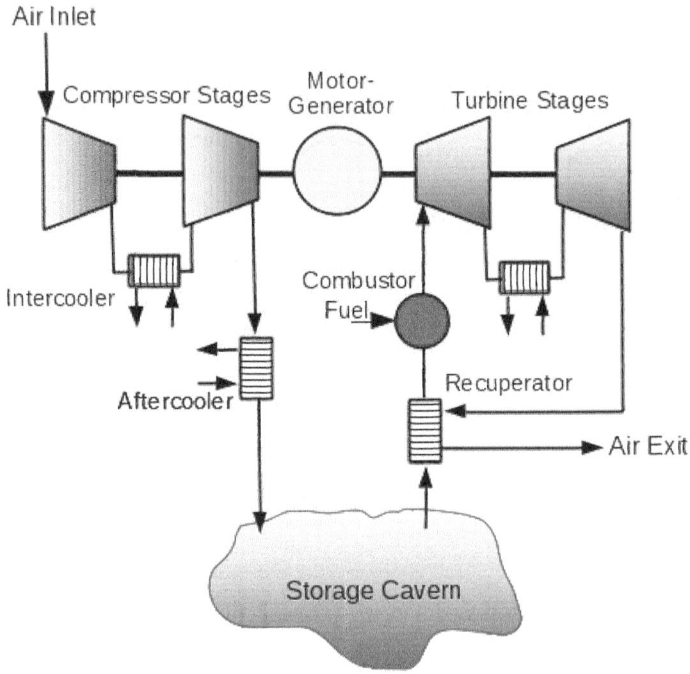

FIGURE 10.11
Recuperated cycle representation of a compressed air storage power plant.

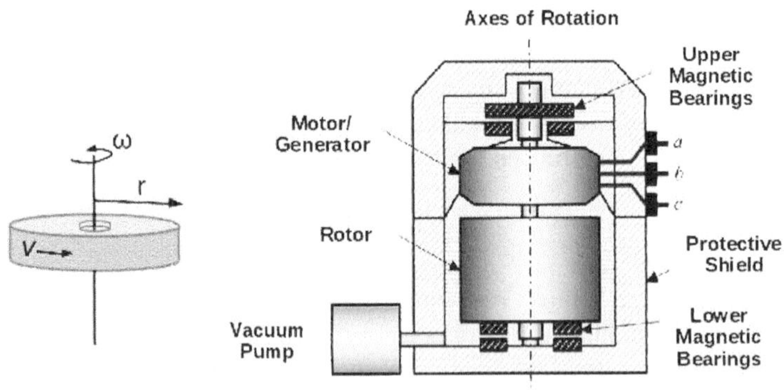

FIGURE 10.12
Illustration of a flywheel energy storage system.

synchronous machine that can operate as both a generator and motor. Figure 10.12 shows an illustration of a flywheel storage system.

The flywheel can be either low-speed, with operating speeds up to 6,000 rpm, or high-speed with operating speeds up to 50,000 rpm. Low-speed flywheels usually consist of steel rotors and conventional bearings. These can achieve specific energy of approximately 5 W-h/kg. High-speed flywheels use advanced composite materials for the rotor along with ultra-low friction bearing assemblies. These light-weight and high-strength composite rotors can achieve specific energies of 100 W-h/kg. These lightweight flywheel designs also come up to speed in a matter of minutes, rather than the hours needed to recharge a battery. The enclosure for high-speed flywheel systems are either evacuated or filled with helium to reduce aerodynamic losses and rotor stresses.

The main advantage of flywheel storage systems is their high charge and discharge rate. Their energy efficiency is typically around 90% at rated power. Their operation lifetime is estimated to be 20 years.

Their main disadvantages are their high cost and the relatively high standing loss. The self-discharge rate for flywheel systems are approximately 20% of the stored capacity per hour. Thus they are not a suitable device for long-term energy storage.

A summary [8] of all of the discussed electric power storage systems with regard to power rating and discharge time is presented in Figure 10.13. This indicates that the pumped storage hydroelectric and CAES systems combine the highest power rating and discharge power capabilities. The groupings of the different technologies suggest regions of applications that relate to "energy management", "bridging power" and "power quality".

It is evident from this that batteries are the dominant technology to be used when continuous energy supply is paramount, while technologies such as flywheel and super-capacitors are better suited for power storage applications where very brief power supply is required such as in uninterrupted power supply requirements. Lithium-ion batteries are becoming increasingly important and have several advantages over the traditional lead acid batteries. Fuel cell performance is constantly improving in terms of reliability and investment cost, while some types (e.g. SOFC) can provide very high efficiency in the context of combined heat and power (CHP) applications. However, the future expansion in use of fuel cells remains tied to the high-cost hydrogen production and storage processes. Finally, pumped storage and CAES technologies are best suited to very high power, high investment cost generation applications to be used in the transmission system.

Tables 10.1 and 10.2 provide other statistics about the different types of storage. Table 10.1 compares the different electric energy storage types in terms of the capital costs for the plant and storage method, the amount of power "capital" that can be stored, and the Operation and Maintenance (O&M) costs for each of the storage approaches. The CAES and pumped hydro-electric have the lowest capital costs for storage since they generally make use of natural formations in the land (caverns or hills). Their plant costs are comparable to the others, but their storage capability is significantly larger than the other three approaches listed.

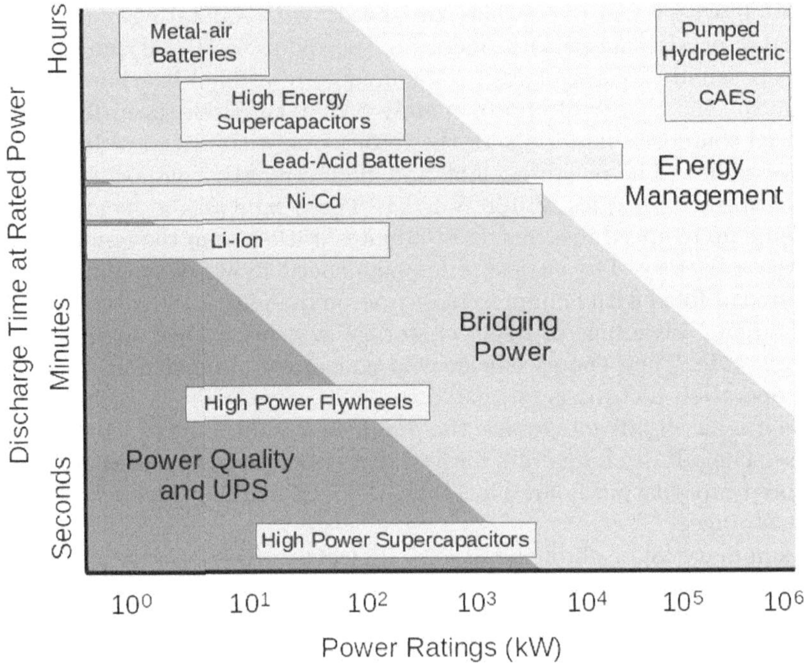

FIGURE 10.13
Summary comparison of different electric power storage systems with regard to power rating and discharge rate.

Table 10.2 provides a comparison of the efficiency, the time over which full power can be provided, and the level of full power for the different storage approaches. The flywheel and super conducting magnet are highly efficient. However, they cannot deliver full power for more than approximately a second. Batteries are the least efficient as well as also being limited in the time over which they can provide maximum power. The CAES and pumped hydroelectric are again the best in terms of time at which they can provide full power, ranging on the order of minutes. In addition, the maximum power is as much as three orders of magnitude larger than the other three storage approaches.

10.5 CAES Case Study

It is instructive to investigate in more detail a Compressed Air Electric Storage (CAES) power plant from a thermodynamic point of view. Figure 10.14 shows a thermodynamic representation of a typical CAS power plant. The plant

TABLE 10.1
Capital Costs of Installed Storage

Type	Storage Capital Cost (\$/kW-h)	Plant Capital Cost (\$/kW)	Storage (MW-h)	O&M \$/kW/yr)
CAES	> 3	> 425	5–100 K	1.35
Pumped hydro	> 10	> 600	> 20 K	4.3
Flywheel	300 to 25 K	280 to 360	0.0002 to 500	7.5
Super-cond. mag.	500 to 72 K	300	0.0002–100	1
Battery	1–15	500–1,500	0.0002–2	-

TABLE 10.2
Efficiency and Hours at Full Power of Installed Storage

Type	Efficiency (%)	Time @ Full Power	Power (MW)
CAES	> 70	1–10 minutes	0.5–2,700
Pumped Hydro	> 70	10 seconds to 4 minutes	300–1,800
Flywheel	90–93	< 1 seconds	0.001–1
Super-conducting mag.	95	< 1 seconds	0.001–2
Battery	59	< 1 seconds	0.01–3

consists of a series of three compressors with temperature inter-coolers in between, and an after-cooler that removes the last amount of heat before the compressed air is stored in an underground reservoir. The compressors are driven by an electric motor that we presume would receive its power from a wind turbine.

When the stored compressed air is released, it is warmed by passing through a recuperator. It is then injected with fuel and combusted to heat the air to the highest point in the cycle. The heated air is expanded through a series of three turbines. Heat is added between each turbine stage. The final exhaust air from the turbine passes through the recuperator before being exhausted to the atmosphere. The turbines drive the generator, which is an asynchronous device that doubles as the compressor motor.

The thermodynamic cycle so depicted is known as an Erickson cycle. In the representation in Figure 10.14, the portion of the cycle denoted by the numbers represent the following:

1–3 the "charging mode" where the electric motor compresses the air using power either from the wind or from the grid at low demand periods of time, and

3–7 the "discharge mode" in which the compressed air is expanded through the turbines to drive the electric generator during peak demand periods of time.

The efficiency of the thermodynamic cycle is

$$\eta_{th} \equiv \frac{w_t}{\frac{w_c}{\eta_{ex}} + q_f}, \tag{10.2}$$

where w_t is the specific work done by the gas turbine, w_c is the specific work done on the compressor, q_f is the specific heat from combustion, η_{ex} is the

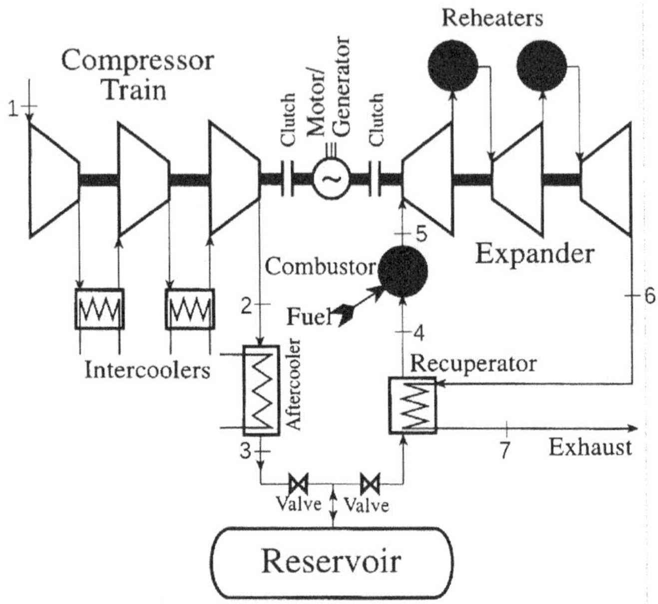

FIGURE 10.14
Thermodynamic representation of a CAES power plant.

external efficiency of the base load power plant, that is, the wind turbine efficiency or the efficiency of any other source of electricity used to power the electric motor for the air compression.

Note that when no heat of combustion is supplied, $q_f = 0$, and the efficiency of such an adiabatic CAES system is

$$\eta_{th}^{ad} = \beta\eta_{ex} = \frac{w_t}{w_c}\eta_{ex}. \tag{10.3}$$

In this case, $\beta = w_t/w_c$ is a relevant index of performance.

How do we improve the thermodynamic efficiency of a CAES power plant? To address that, we start by considering air to be an ideal gas, and process such as compression or expansion to be polytropic, namely

$$pV^k = \text{Constant}, \tag{10.4}$$

so that

$$\frac{T_2}{T_1} = \left(\frac{P_2}{P_1}\right)^{\frac{k-1}{k}}, \tag{10.5}$$

where for air, $k = 1.4$.

For the compression stage, the work is given as

$$w_c = \frac{C_p T_1}{\eta_c \eta_{\text{elm}}} n \left(\sigma_c R^{1/n} - 1 \right), \tag{10.6}$$

where

$$\begin{aligned}
r_{mt} = T_5/T_1 &= \text{maximum temperature ratio} \\
r_{st} = T_3/T_1 &= \text{storage temperature ratio} \\
\eta_c &= \text{compressor efficiency} \\
\eta_{\text{elm}} &= \text{electro-mechanical efficiency} \\
\eta_t &= \text{turbine efficiency} \\
R = T_2/T_1 &= \text{terminal isentropic temperature ratio} \\
\sigma &= \text{pressure losses factor} \\
n - 1 &= \text{number of intercoolers} \\
m - 1 &= \text{number of reheaters}
\end{aligned}$$

The subscripts t and c respectively refer the the turbine and compressor. Note that with regard to the number of intercoolers and reheaters, "none" corresponds to n and m equal to 1.

The "energy storage effectiveness", $\beta = w_t/w_c$ is then given as

$$\beta = \frac{\eta_t \eta_c \eta_{\text{elm}} r_{mt}\, m \left(1 - \sigma_t/R^{1/m}\right)}{n \left(\sigma_c . R^{1/n} - 1\right)}. \tag{10.7}$$

Now the economics of a CAES power plant depends on the instantaneous price of electricity, which in turn depends on the instantaneous price of electricity which depends on the instantaneous demand. As a model for the cost of electricity, $P(t)$ consider

$$P(t) \simeq A_0 + A_1 N(t) + A_2 N(t), \tag{10.8}$$

where $N(t)$ is the time variation in the electric power demand, and the A's are best-fit coefficients that relate the cost of electricity to the demand. An example of the electric power demand and corresponding consumer price of electricity over a 24 hours period is shown in Figure 10.15 [9].

Once the instantaneous price function, $P(t)$, has been evaluated for a given power demand curve, then charging and discharging price functions, C_{ch}, and C_d, respectively, can be developed. The price functions depend on the duration of the charging and discharging, h_{ch} and h_d, respectively. An example of the charging and discharging price functions that corresponds to Figure 10.15 is shown in Figure 10.16.

10.5.1 Cost Function

The cost function is intended to provide a formula for estimating the costs associated with a CAES plant. The total cost is broken down into **fixed costs**

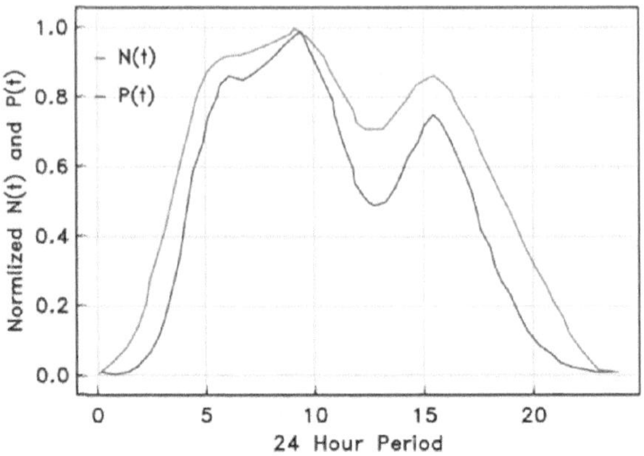

FIGURE 10.15
Example of the electric power demand and corresponding consumer price of electricity over a 24 hours period [9].

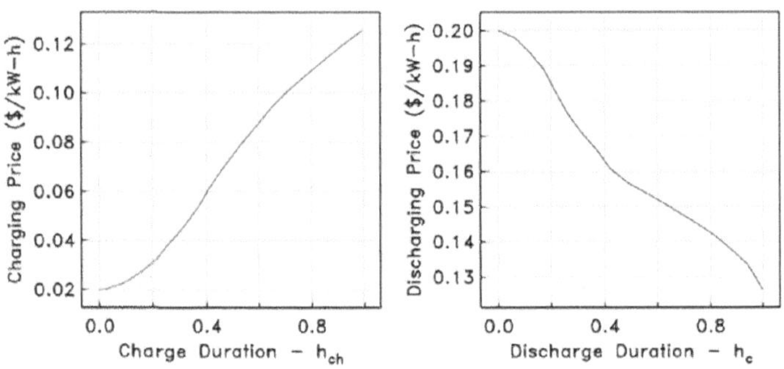

FIGURE 10.16
Charging and discharging price functions that correspond to the price function shown in Figure 10.15 [9].

and **variable costs**. This is defined as the following form.

$$C_{\text{tot}} = C_1 K + S h_d + C_{\text{fom}} \quad [\$/\text{kW-yr}], \tag{10.9}$$

where

$$\begin{aligned} C_1 &= \text{capital cost [\$/kW installed]} \\ K &= \text{capital recovery factor [1/yr]} \\ S &= \text{specific variable cost [\$/kW-h generated]} \\ h_d &= \text{plant service factor [operating hrs/yr]} \\ C_{\text{fom}} &= \text{fixed operating \& maintenance cost [\$/kw-yr]} \end{aligned}$$

The specific variable cost includes the energy cost of charging, C_{ch}, and discharging, C_d, the energy reserve. This specific variable cost is then given as

$$S = C_{ch} + C_d + C_{\text{vom}}, \tag{10.10}$$

where C_{vom} is the cost of Operation & Maintenance (O&M). Now the cost of charging is given as

$$C_{ch} = P_c \frac{w_c}{w_t} = P_c \beta^{-1} \text{ [\$/kW-h generated]}, \tag{10.11}$$

where P_c is the **charging price function** with units of [$/kW-h]. The ideal situation (to generate capital) is the $P_c < P_d$ where P_d is the discharging price function with units of [$/kW-h].

Now, the coefficient for discharging, C_d is defined as

$$C_{ch} = P_f \frac{\dot{m}_f}{\dot{w}_t}, \tag{10.12}$$

where P_f is the **fuel price** with units of [$/kg-fuel], and $\dot{m}_f/\dot{w}_t$ is the **specific fuel consumption** with units of [kg-fuel/kW-h].

The capital costs, C_1, include all of the costs of installation, thus

$$\begin{aligned} C_1 \;=\; & r_w C_c + C_t + r_g C_g + r_w C_{in} + C_{re} + C_{RC} + C_r + C_s \\ & \text{[\$/kW installed]}, \end{aligned} \tag{10.13}$$

where the coefficients C_c, C_t, C_g, and C_{in} are the costs/kW to install the compressor, turbine, generator and intercoolers, respectively. The subscripts re, RC, r, and s, correspond to the reheaters, recuperator, reservoir, and supplemental, respectively. Now $r_w = \dot{w}_c/\dot{w}_t$ is the **compressor-to-turbine capacity ratio**, $r_g = \dot{w}_g/\dot{w}_t$ is the **generator-to-turbine capacity ratio**. Typically

$$\dot{w}_g = \max\left[\dot{w}_c, \dot{w}_t\right], \tag{10.14}$$

so that

$$r_g = \max\left[r_w, 1\right]. \tag{10.15}$$

We now define a **discharge-charge ratio**, r_h given as

$$\begin{aligned} r_h &= \frac{h_d}{h_c} \\ &= \frac{\dot{w}_c}{\dot{w}_t}\frac{w_t}{w_c} \\ &= r_w \beta, \end{aligned} \tag{10.16}$$

where h_d and h_c are the hours per year of discharging and charging, respectively. We also note that $\frac{\dot{w}_c}{\dot{w}_t}$ represents the design condition, and $\frac{w_t}{w_c}$ represents the thermodynamic condition.

10.5.2 Net Benefit

In order to optimize the operation of a CAES power plant it is necessary to define a "net benefit" which represents a metric of merit. The net benefit, B, in this case is given as

$$B = (P_d - S)\, h_d - C_1 K - C_{\text{fom}} \;\; [\$/\text{kw-yr}], \tag{10.17}$$

where $P_d = f(h_d)$ is the discharging price which is a function of the discharge duration.

The object is then to maximize B, where we note that

$$B = f\left(r_{mt}, R, r_b, h_d, \epsilon_{RC}, m, n\right), \tag{10.18}$$

where ϵ_{rc} is the effectiveness of the recuperator, of which $0 \leq \epsilon_{RC} \leq 1$.

The following provide ranges of the independent variable.

$$\begin{aligned}
r_{st} \leq r_{mt} \leq 4.91;&\ r_{mt} = T_4/T_1 = \text{max. temp. ratio}\\
;&\ r_{st} = T_3/T_1 = \text{max. temp. ratio}\\
0 \leq r_h;&\ r_h = \text{discharge-charging duration ratio}\\
0 \leq h_d \leq \gamma;&\ h_d = \text{discharge duration}\\
;&\ \gamma = \text{a constraint that prevents charging}\\
;&\ \text{discharging at the same time}\\
;&\ \text{therefore, } h_d\left(1 + \frac{1}{r_h}\right) \leq 8,760\\
0 \leq \epsilon_{RC} \leq 1;&\ \epsilon_{RC} = \text{recuperator effectiveness}\\
1 \leq m \leq m_{\max}&\\
1 \leq n \leq n_{\max}&\\
0.01 \leq P_{HF} \leq 0.1;&\ \text{heat price [\$/kW-h]}
\end{aligned}$$

As an example of the process of optimization, the heat price, P_{HF}, was varied between 0.01 and 0.1 \$/kW-h. The resulting optimal values of the isentropic temperature ratio, $R^* = (T_3/T_1)_s$, and of the recuperator effectiveness, ϵ are presented in Figure 10.17 together with the corresponding maximum values of the net annual benefit, B [9].

Figure 10.17 indicates that the optimal recuperator effectiveness, ϵ_{RC}, increased with increasing heat price, P_{HF}, reaching a value of 0.51 for the maximum heat price considered. The optimal isentropic temperature ratio, $R^* = (T_3/T_1)_s$, also increased with increasing the heat price, up to a heat price of $P_{HF} = 0.078$. Above that heat price, the optimal isentropic temperature ratio asymptoted to a value of $R^* = 3.5$. The net annual benefit, B^*, decreased smoothly with increasing heat price.

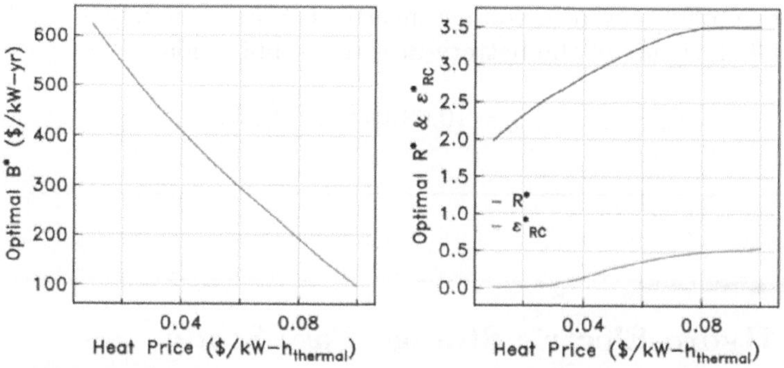

FIGURE 10.17
Result of optimization based on a range of heat price for a CAES power plant [9].

10.6 Battery Case Study

We consider an electro-chemical battery energy storage. For such as system the rated energy stored, E_{rated} in [W-h] is

$$E_{\text{rated}} = C_{\text{rated}} V_{\text{nominal}}, \tag{10.19}$$

where C_{rated} is the amp-hour capacity of the battery, and V_{nominal} is the nominal voltage of the battery. In the use of batteries, there is a general restriction on the "depth of discharge" (DOD) to ensure a long operating life. The standard is a DOD of 50% of capacity.

The **average** battery efficiency is approximately 80% at the start of its useful life. At the end of its useful life, the efficiency drops to approximately 50% at the end of its useful life. Therefore the average efficiency of a battery is approximately 68%.

Example 10.1

Consider a deep-cycle lead acid battery in which $V_{\text{nominal}} = 60\text{V}$, and $C_{\text{rated}} = 1,200$ A-h. The usable energy is then

$$\begin{aligned} E_{\text{usable}} &= E_{\text{rated}} \cdot \text{DOD} \\ &= (1,200)(60)(0.5) \\ &= 36[\text{kw-h}]. \end{aligned} \tag{10.20}$$

We can define the efficiency for the battery "system" to include the battery and the power inverter that converts A.C. to D.C. for charging. Thus

$$\eta_{\text{battery/inverter}} = \eta_{\text{battery}} \eta_{\text{inverter}}. \tag{10.21}$$

The average efficiency of a voltage inverter is approximately 85%. Therefore the overall efficiency of the battery-inverter combination is

$$\eta_{\text{battery/inverter}} = (0.68)(0.85) = 0.578 \ (57.8\%). \tag{10.22}$$

10.7 Hydro-Electric Storage Case Study

This section considers the energy that can be stored and the efficiency of hydro-electric storage. The premise, as shown in Figure 10.18, is that water is pumped from a lower reservoir using wind power when electrical demand is low. When electrical demand is high, the water from the upper reservoir is released to pass through a turbine to generate electricity. The energy generated in this process is

$$E_{\text{hydro}} = \rho g \, h \, \text{VOL} \eta, \tag{10.23}$$

where

$$\begin{aligned}
\text{VOL} &= \text{water volume stored } [\text{m}^3] \\
h &= \text{stored water elevation (pressure head) } [\text{m}] \\
\rho &= \text{water density } [1{,}000 \text{ kg/m}^3] \\
g &= \text{gravitational constant } [9.8 \text{ m/s}^2] \\
\eta &= \eta_t \eta_{pipe} \\
\eta_t &= \text{turbine efficiency } (0.60) \\
\eta_{\text{pipe}} &= \text{pipe flow efficiency } (0.90).
\end{aligned}$$

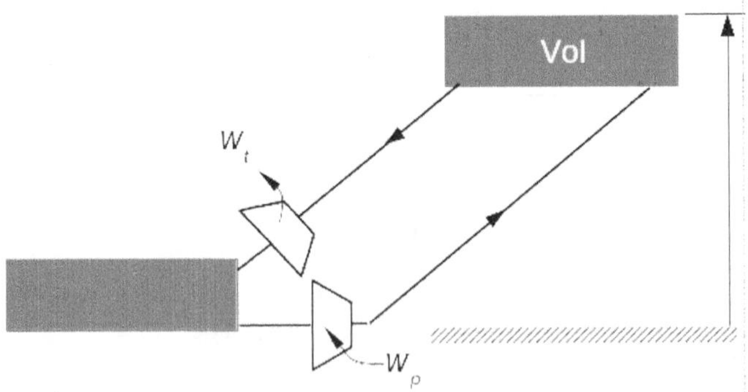

FIGURE 10.18
Schematic of a hydro-electric storage configuration.

Noting that 1 J = 1 W, the stored energy in units of [kW-h] is

$$E = \frac{\rho g \, \text{VOL} \, h\eta}{3,600}, \tag{10.24}$$

or the required volume of water needed to supply a given amount of energy is

$$\text{VOL} = \frac{3,600E}{\rho g h\eta}, \tag{10.25}$$

where in both cases 3,600 s/h appears as a conversion between hours and seconds.

Example 10.2

Determine the volume of water at an elevation of 50 m. that is needed to produce 100 kW-h of electric power.

$$\begin{aligned} \text{VOL} &= \frac{3,600E}{\rho g h\eta} \qquad (10.26) \\ &= \frac{(3,600)(100)}{9.8(50)(0.60)(0.90)} \\ &= 1,359\,\text{m}^3 \\ &= 50\,\text{m by } 20\,\text{m by } 1.4\,\text{m deep.} \end{aligned}$$

10.8 Buoyant Hydraulic Energy Storage Case Study

Wind turbines in deep off-shore locations are supported by floating structures, such as shown in Figure 10.19. This has led to a concept for storing electric energy that is similar to pumped hydro-electric storage but instead uses buoyant hydraulic energy in the floating structures [10].

The buoyant energy is stored through the potential energy of the mass of the floating structure. Figure 10.20 illustrates the concept. The floating structure has an opening at its lowest point that can allow water to enter an internal compartment. When the water enters the compartment, it passes through a turbine to generate electricity. In this case, the floating structure will sink lower in the water. The water can be pumped out of the compartment by reversing the turbine to act as a pump. This requires electric power from the wind turbine. As the water is pumped out of the compartment, it rises higher in the water. This sequence is illustrated in Figure 10.21. When the

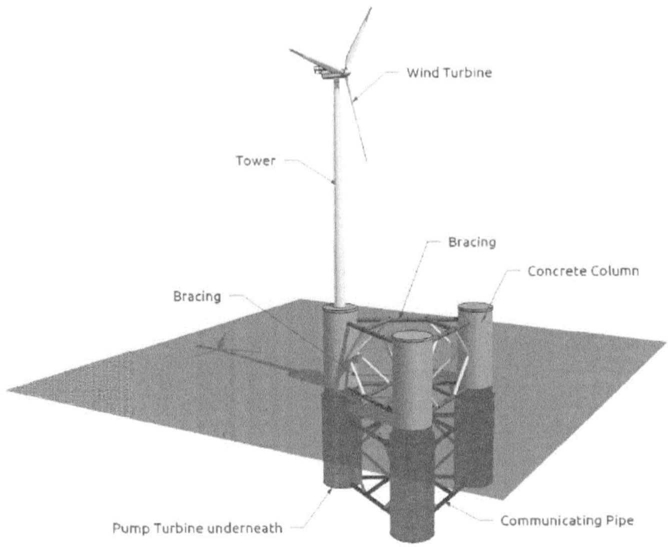

FIGURE 10.19
Example of a floating off-shore platform supporting a wind turbine [10].

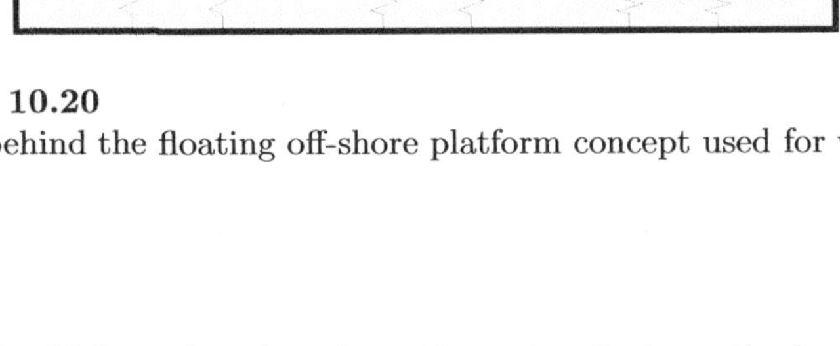

FIGURE 10.20
Principle behind the floating off-shore platform concept used for wind energy storage.

float is at its highest elevation above the water, it stores the largest amount of energy. That energy is converted to electricity when the water is allowed to fill the compartment, entering through the electric turbine.

The schematic shown in Figure 10.22 is used to analyze this buoyant energy storage system. The usable energy depends on buoyant mass and the size of the internal compartment. Considering an idealized system, where the total mass is concentrated, and the reservoir has a cylindrical shape, then the maximum occurs when the compartment is half full, at which point the immersion depth

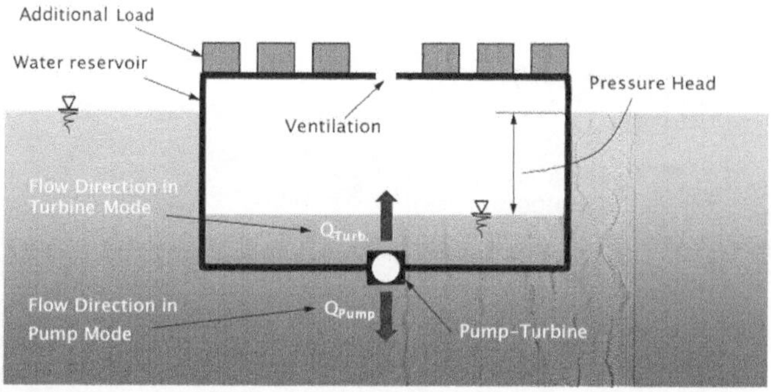

FIGURE 10.21
Example of a sequence of floating position based on the amount of water contained in an internal compartment of the floating structure [10].

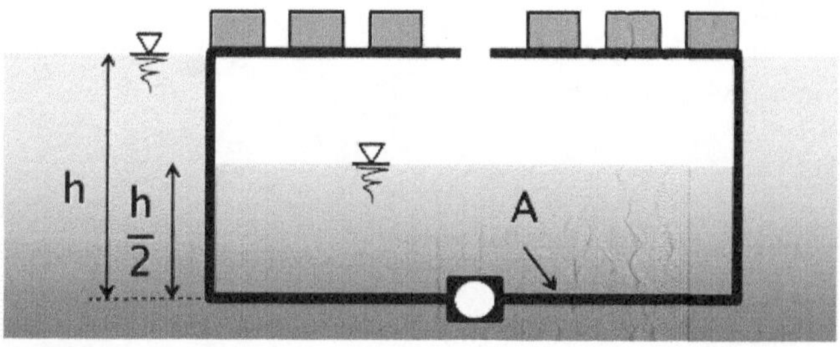

FIGURE 10.22
Schematic representation of the buoyant energy storage [10].

is denoted by h. The maximum amount of stored energy is then

$$\begin{aligned} E &= mg\frac{h}{2} \\ &= \rho A\frac{h}{2}g\frac{h}{2}\eta_t \\ &= \rho A g\frac{h^2}{4}\eta_t, \end{aligned} \tag{10.27}$$

where A is the projected area of the floating structure so that $A(h/2)$ is the volume of displaced water, and η_t is the efficiency of the turbine ($\simeq$60%). Rearranging the previous equation,

$$\begin{aligned} m &= \rho A\frac{h}{2} \\ &= \frac{2E}{gh}. \end{aligned} \tag{10.28}$$

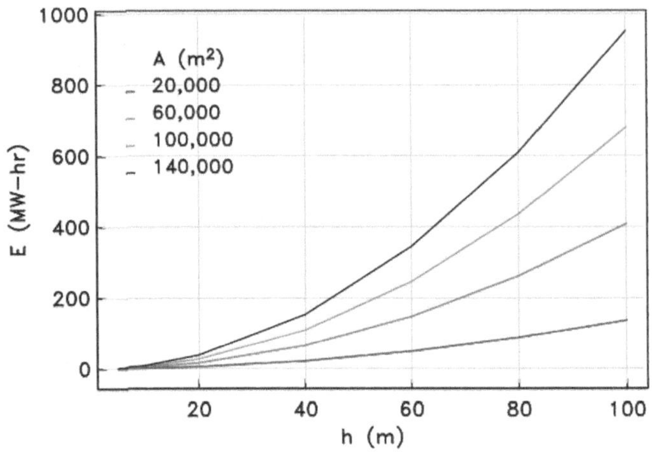

FIGURE 10.23
Relation between the projected area of the floating structure and the immersion depth for a given stored power level.

The gravimetric energy density is

$$\begin{aligned} \rho_{\text{grav}} &= \frac{E}{m} \\ &= g\frac{h}{2}. \end{aligned} \tag{10.29}$$

The volumetric energy density is

$$\begin{aligned} \rho_{\text{vol}} &= \frac{E}{hA} \\ &= \frac{m}{2}\frac{g}{A} \\ &= \rho g\frac{h}{4}. \end{aligned} \tag{10.30}$$

The relation between the projected area of the floating structure and the immersion depth for a given stored power level is shown in Figure 10.23. For example, a floating structure with a projected area of 40,000 m^2 that can change elevation by 20 m can store 10 MW-h of energy. Like the pumped hydro-electric system, the buoyant energy system has a short response time and an unlimited number of charge-discharge cycles.

Problems

1. For A CAES system, the "Energy Storage Effectiveness" is

$$\beta = \frac{w_t}{w_c}.$$

Considering all of the pipe losses to be minimal so that

$$\sigma_t = \sigma_c = 1,$$

and all of the efficiencies

$$\eta_c = \eta_t = \eta_{\text{eff}} = 0.5,$$

and the number of inter-coolers and re-heaters

$$(n-1) = (m-1) = 3,$$

generate a plot of β as a function of the maximum temperature ratio, T_5/T_1, and the thermal isentropic temperature ratio, T_2/T_1 while staying within the following limits

$$\frac{T_3}{T_1} \le \frac{T_5}{T_1} \le 4.91,$$

and

$$\eta_c \le \frac{T_2}{T_1} \le 3.5.$$

2. For the CAES system in Problem 1, the "Net Benefit" , B, can be simply stated as the difference between the revenue generated from the stored air minus the cost of storing the air, or

$$B = \left(P_d - \frac{P_c}{\beta}\right) h_d,$$

where P_d is the sale price of the energy (\$/kW-h), P_c is the purchase price of the energy to compress the air (\$/kW-h), and h_d is the discharge duration (h).

Assuming that $P_d = 0.08\$/\text{kW-h}$, and that $P_c = 0.04\$/\text{kW-h}$, and $h_d = 4\text{h}$, generate a plot of B as a function of the values of β obtained from Problem 1.

3. For a pumped hydraulic storage system like that shown in Figure 10.8,

(a) Generate a plot of the energy (MW-h) as a function of water volume (m^3) and elevation (m). Assume that the turbine efficiency is $\eta_t = 1$.

(b) What is the volume of water at an elevation of 100 m that is needed to generate 1 GW-h of electrical energy?

4. The following relates to a comparison between pumped hydraulic and bouyant hydraulic storage systems.

   (a) Compare the relation for energy in the pumped hydraulic storage system to that of the buoyant hydraulic storage system. How do they differ?

   (b) Generate a plot of the energy (MW-h) as a function of area of the floating platform (m^2) and height of the internal water level (m).

   (c) What is the area of the floating platform for a 100 m height of the internal water level to generate 1GW-h of energy?

5. A farmer proposes to store energy generated by his wind turbine by pumping water from a lower pond to an upper pond. The upper pond is 10 m. above the level of the lower pond.

   (a) If the wind turbine operated 4 hours daily at its 100 kW rated power, what is the daily volume of water that is needed to store the daily total amount of electric energy?

   (b) If batteries were used instead, where the energy stored in the battery is

   $$E = CV,$$

   where
   V is the nominal battery voltage,
   C is the amp-hour capacity, and
   E has units of Watt-hours.
   If each battery has a nominal voltage of $V = 60\,\text{V}$, and an amp-hour capacity of $C = 1,200$ A-h, How many batteries are needed to store the energy from the wind turbine?

   (c) List any advantages the hydro-power storage has over the battery storage.

6. (a) Why is electric storage important to wind energy?

   (b) List **five** methods of electric energy storage.

   (c) For each, rate them in terms of their **discharge time at rated power**, with 1 being the highest and 5 being the lowest.

   (d) Why is discharge time at rated power important for electric storage?

References

1. U.S. Department of Energy, "20% wind energy by 2030", July 2008. http:www.osti.gov/bridge.
2. F. Diaz-Gonzalez, A. Sumper, O. Gomis-Bellmunt, and R. Villafafila-Robles, "A review of energy storage technologies for wind power applications", *Renewable and Sustainable Energy Reviews*, 2012, 16(4):2154–2171.
3. H. Ibrahim, A. LLinca, and J. Perron, "Energy storage systems – Characteristics and comparisons", *Renewable and Sustainable Energy Reviews*, 2008, 12(5):1221–1250.
4. P. Van den Bossche, F. Vergels, J. Van Mierlo, J. Matheys, and W. Van Autenboer, "SUBAT: An assessment of sustainable battery technology", *Journal of Power Sources*, 2006, 162(2):913–919.
5. M. Winter and R. J. Brodd, "What are batteries, fuel cells, and supercapacitors?" *Chemical Reviews*, 2004, 104:4245.
6. J. R. Miller and A. F. Burke, "Electrochemical capacitors: Challenges and opportunities for real-world applications", *Electrochemical Society Interface*, 2008, 17(1):53.
7. Electricity Storage Association, https://energystorageassociation archive.org/.
8. P. Vadasz, "Compressed air storage: Optimal performance and techno-economical indices", *International Journal Applied Thermodynamics*, 1999, 2(2):69–80.
9. R. Klar, M. Aufleger, and M. Thene, "Buoyant energy: Decentralized offshore energy storage in the European power plant park", University of Innsbruck, Technikerstrase 13a, 6020 Innsbruck, Austria, http://www.buoyant-energy.com.

11

Economics

In the process of assessing changes in the design of a wind turbine, it is important to evaluate what the impact such changes would have on the system cost. This includes the initial capital (IC) cost , balance of station (BOS) cost, operation and maintenance (O&M) cost, levelized replacement (LR) cost, and the annual energy production (AEP) revenue which balances these costs. Many of these affect the other. For example, increasing AEP may increase IC cost.

The levelized cost of electricity (COE) has been used as an attempt to evaluate the total system impact of any change in wind turbine designs. The levelized COE attempts to limit the impact of financial factors, such as the cost of capital in wind farm development, so that the true impact of technical changes can be assessed. It is often difficult to determine the total impact of increasing power rating or rotor diameter on the economics of the wind turbine.

The DOE and NREL [1] have compiled statistics on a range of wind turbine rated power levels in order to develop scaling relationships. These have mainly focused on three-bladed, upwind rotor, pitch-regulated, variable-speed designs. The design variables are primarily the rotor diameter, hub height, and rated power (in kW). Other factors are types of gear boxes and electric generators. The results of the developed models that lead to costs are in 2002 dollars. These can be brought to present dollars using the Consumer Price Index that is readily available at online web sites. The ultimate metric of merit in the economic analysis is the COE that is needed to offset the total system cost and be competitive with non-renewable electric energy sources.

11.1 Cost of Energy, *COE*

The cost of energy, COE, is determined using the following formula:

$$\mathrm{COE} = \frac{(\mathrm{FCR})(\mathrm{ICC})}{\mathrm{AEP_{net}}} + \mathrm{AOE}, \tag{11.1}$$

DOI: 10.1201/9781003611486-11

where

$$\begin{aligned}
\text{COE} &= \text{levelized cost of energy [\$/kW-h]} \\
\text{FCR} &= \text{fixed charge rate [1/yr]} \\
\text{ICC} &= \text{initial capital cost [\$]} \\
\text{AEP}_{\text{net}} &= \text{net annual energy production [kW-h/yr]} \\
\text{AOE} &= \text{annual operating expenses} \\
&= \frac{\text{LLC}}{\text{AEP}_{net}} + \frac{O\&M + \text{LRC}}{\text{AEP}_{\text{net}}} \\
\text{LLC} &= \text{land lease cost} \\
\text{O\&M} &= \text{levelized O\&M cost} \\
\text{LRC} &= \text{levelized replacement/overhaul cost.}
\end{aligned}$$

The fixed rate charge, FCR, is the annual amount per dollar, of initial capital cost needed to cover the capital cost, a return on debt and equity, and various other fixed charges. The FCR includes construction financing, financing fees, return on debt and equity, depreciation, income and property taxes, and insurance. The FCR is set as 0.1158 per year.

The initial capital cost, ICC, is the sum of costs of the wind turbine system and the balance of station, BOS, cost. The primary elements of the wind turbine system include

- wind turbine rotor including
  - rotor blades
  - rotor hub
  - pitch mechanism and bearings
  - spinner, nose cone
- drive train, nacelle including
  - low-speed shaft
  - bearings
  - gearbox
  - mechanical break, high-speed coupling, associated components
  - generator
  - variable-speed electronics
  - yaw drive and bearing
  - main frame
  - electrical connections
  - hydraulic and cooling systems

  - nacelle cover

- control, safety system and conditioning monitoring
- tower
- balance of station, including
  - foundation/support structure
  - transportation
  - roads, civil work
  - assembly and installation
  - electrical interface/connections
  - engineering permits

With regard to off-shore wind turbines, the following initial capital costs also need be considered.

- marinization, to handle the marine environment
- port and staging equipment
- personal access equipment
- scour protection
- security bond to cover decommissioning
- offshore warranty premium.

Annual operating expenses (AOE) include land or ocean bottom lease cost, levelized $O\&M$ cost, and levelized replacement/overhaul cost (LRC). Land lease costs (LLC) are the rental or lease fees charged for the wind turbine installation. LLC is expressed in units of [$/kW-h].

O&M costs in [$/kW-h] are the largest portion of the annual operating expenses, AOE. O&M includes

- labor, parts, and supplies for scheduled turbine maintenance;
- labor, parts, and supplies for unscheduled turbine maintenance;
- parts and supplies for equipment and facilities maintenance; and
- labor for administration and support.

The levelized replacement/overhaul cost (LRC) in [$/kW] is the cost of major replacements and overhauls over the life of the wind turbine.

The *net* annual energy production (AEP) represents the projected energy output of the turbine based on a given annual average wind speed. The *gross* AEP is adjusted for factors such as the rotor coefficient of power, mechanical and electrical conversion losses, blade soiling losses, array losses, and machine availability.

11.2 Component Estimate Formulas

11.2.1 Rotor Blade Mass

There exists a direct correlation between the mass (weight) of wind turbine rotor and its radius [1–3]. This is shown in Figure 11.1 which relates total mass/blade (kg) to the rotor radius for different design criteria [2,3]. Considering that mass scales with volume (length cubed), we expect that the mass of the rotor would scale with the rotor radius to a power. These is supported by the relations shown by the different curves in the figure. For the baseline rotor design, the mass per blade is

$$m = 0.1452R^{2.9158}, \tag{11.2}$$

which is close to the length cubed relation we would expect. The weight scaling based on the static load is somewhat different at

$$m = 0.2113R^{2.8833}. \tag{11.3}$$

The combination of these leads to a final design with a mass scaling per rotor given by

$$m = 0.1527R^{2.6921}. \tag{11.4}$$

The use of advanced (fiberglass) materials [2,3] can reduced the weight of the rotor, providing a mass scaling of

$$m = 0.4948R^{2.53}. \tag{11.5}$$

11.2.2 Rotor Blade Cost

The increased mass of the rotor that comes with increasing the rotor radius translates into an increase in the cost of the rotor. These costs include material, tooling, labor, overhead, and profit. Overhead and profit were assumed to be 28% of the material and labor costs. The material costs were taken to scale as R^3, that is, approximately as the volume of material that made up the rotor. The result [4] is shown in Figure 11.2 which relates total cost (in 2002 dollars) of the rotor to the rotor radius for different materials and fabrication methods. For the baseline rotor, the cost/blade is

$$\text{Cost} = 3.1225R^{2.879}, \tag{11.6}$$

which is again close to the length cubed relation we would expect. The trend in the baseline rotor material costs per rotor is

$$\text{Baseline material cost} = 0.4019R^3 - 955.24. \tag{11.7}$$

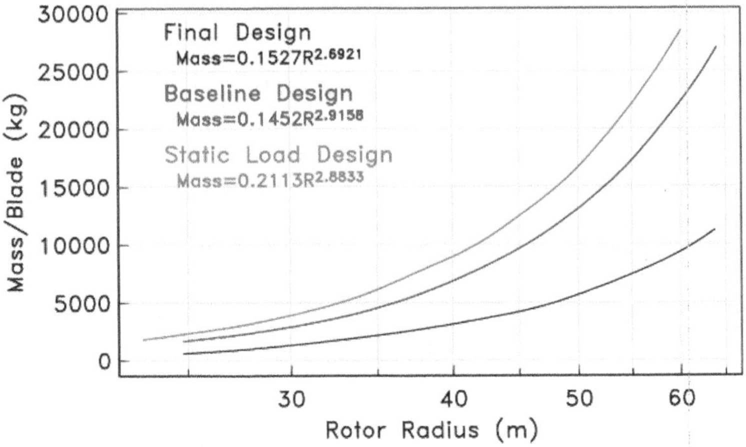

FIGURE 11.1
Wind turbine rotor blade mass correlation with rotor radius.

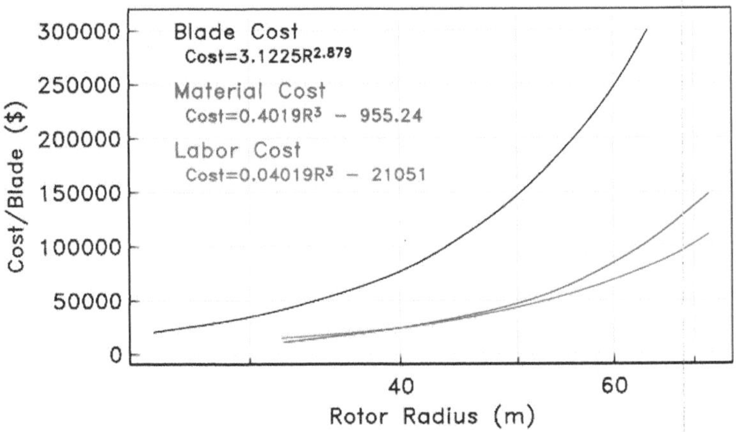

FIGURE 11.2
Wind turbine rotor blade cost, labor cost, and baseline and advanced material cost correlations with rotor radius.

Also shown in the figure is the labor costs per blade as a function of the rotor radius, which is as follows:

$$\text{Labor cost} = 0.04019R^3 - 21{,}051, \tag{11.8}$$

which is assuming the use of advanced (fiberglass) materials that are common to modern wind turbines.

11.2.3 Rotor Hub Cost

The rotor hub is the structure on which the rotor blades mount. Since the rotor hub has to support the weight of the rotor, its mass is expected to scale approximately linearly with the mass of the rotor. This is in fact the case as given by the following relation [5].

$$\text{Hub mass} = 0.954(\text{Single blade mass}) + 5,680.3. \tag{11.9}$$

The historic hub cost (in 2002 dollars) then scales with the hub mass as

$$\text{Hub cost} = \text{Hub mass} + 5,680.3. \tag{11.10}$$

11.2.4 Pitch Mechanism and Bearings Cost

The pitch mechanism rotates the rotor blades while in wind speed Region III, where the wind turbine produces its rated power. Since the torque produced by the pitch mechanism and loads on the rotor bearings depend on the aerodynamic loads on the rotor, their masses are expected to scale with the mass of the rotor. For the pitch bearing, the mass scales linearly with the total (three) blade mass as

$$\text{Total pitch bearing mass} = 0.1295(\text{Total (3) blade mass}) + 491.31. \tag{11.11}$$

The total mass of the pitch mechanism was then found to scale with the total pitch bearing mass as

$$\text{Total pitch mechanism mass} = 1.328(\text{Total pitch bearing mass}) + 555. \tag{11.12}$$

The total pitch system (pitch mechanism plus bearings) cost (in 2002 dollars) was determined as a function of the rotor diameter, D, to be

$$\text{Total pitch system cost} = 0.4801D^{2.6578}. \tag{11.13}$$

11.2.5 Spinner Nose Cone Cost

The spinner nose cone fits over the rotor hub to provide an aerodynamic profile. The mass and cost are given by the following two relations:

$$\text{Nose cone mass} = 18.5D - 520.5, \tag{11.14}$$

and

$$\text{Nose cone cost} = 5.57(\text{Nose cone mass}), \tag{11.15}$$

where the nose mass is scaled with the rotor diameter, and the cost is in 2002 dollars.

11.2.6 Low-Speed Shaft Cost

The rotor hub attaches to the low-speed shaft. This shaft then transmits the rotor torque to the gear box. The mass and cost are given by the following two relations,

$$\text{Low-speed shaft mass} = 0.0142D^{2.888}, \tag{11.16}$$

and

$$\text{Low-speed shaft cost} = 0.0100D^{2.887}, \tag{11.17}$$

where the mass is again scaled with the rotor diameter, and the cost is in 2002 dollars.

11.2.7 Main Bearings Cost

The low-speed shaft rotates on a set of main bearings. The forces on these bearings are directly related to the weight and aerodynamic loading of the rotor, which should scale with the rotor disk diameter. The mass and cost of the main bearings were found to be given by the following two relations,

$$\text{Main bearing mass} = (0.000123D - 0.000123)D^{2.5}, \tag{11.18}$$

and

$$\text{Main bearing cost} = 35.2(\text{Main bearing mass}), \tag{11.19}$$

where the mass is again scaled with the rotor diameter, and the cost is in 2002 dollars.

11.2.8 Gearbox Cost

The gear box steps up the rotation speed of the rotor to a speed that is necessary for the generator to produce the rated power. The input to the gearbox comes from the torque (kN-m) transmitted through the low-speed shaft. As mentioned, the torque on the low-speed shaft scales with the aerodynamic torque produced by the rotor disk. There are three standard gearbox configurations of which each have a mass and cost. The following lists the three configurations[5,6].

1. Three-stage Planetary/Helical Gearbox

$$\text{Mass} = 70.94(\text{Low-speed shaft torque})^{0.759}, \tag{11.20}$$

$$\text{Cost} = 16.45(\text{Rated power})^{1.249}. \tag{11.21}$$

2. Medium-speed Single-stage Drive

$$\text{Mass} = 88.29(\text{Low-speed shaft torque})^{0.774}, \tag{11.22}$$

$$\text{Cost} = 74.10(\text{Rated power}). \tag{11.23}$$

3. Multi-path Drive

$$\text{Mass} = 139.69(\text{Low-speed shaft torque})^{0.774}, \tag{11.24}$$

$$\text{Cost} = 15.26(\text{Rated power})^{1.249}. \tag{11.25}$$

In all of these relations, torque has units of kN-m, and rated power is in kW.

11.2.9 Mechanical Brake/High-Speed Coupling Cost

The mechanical break is intended to prevent rotor rotation when the wind speed exceeds the cut-out velocity. The brake needs to overcome the aerodynamic torque produced by the rotor disk, and therefore its mass and cost should scale appropriately with the torque or power as given by the following relations [5].

$$\text{Brake/coupling cost} = 1.9894(\text{Rated power}) - 0.1141, \tag{11.26}$$

$$\text{Brake/coupling mass} = 0.1(\text{Brake/coupling cost}). \tag{11.27}$$

11.2.10 Electric Generator Cost

The generator and gearbox are a coupled arrangement. Therefore like the gearbox, there are three configurations. One additional configuration not included in the list of gearbox options, is direct drive. The mass and cost of these four arrangements are given in the following, where again rated power is in kW.

1. High-speed Generator with Three-stage Planetary/Helical Gearbox

$$\text{Mass} = 6.47(\text{Rated power})^{0.9223}, \tag{11.28}$$

$$\text{Cost} = 65.00(\text{Rated power}). \tag{11.29}$$

2. Medium-speed Permanent Magnet Generator with Single-stage Drive

$$\text{Mass} = 10.51(\text{Rated power})^{0.9223}, \tag{11.30}$$

$$\text{Cost} = 54.73(\text{Rated power}). \tag{11.31}$$

3. Permanent Magnet Generators with Multi-path Drive

$$\text{Mass} = 5.34(\text{Rated power})^{0.9223}, \tag{11.32}$$

$$\text{Cost} = 48.03(\text{Rated power}). \tag{11.33}$$

4. Permanent Magnet Generator with Direct Drive

$$\text{Mass} = 661.25(\text{Low-speed shaft torque})^{0.6060}, \tag{11.34}$$

$$\text{Cost} = 219.33(\text{Rated power}). \tag{11.35}$$

11.2.11 Variable-Speed Electronics Cost

The variable speed electronics consists of a power converter that can manage the power level under variable speed operation. The converters are designed based on the rated power (kW). As such, the same is the case with respect to cost as shown in the following relation [7].

$$\text{Cost} = 79.0(\text{Rated power}). \tag{11.36}$$

11.2.12 Yaw Drive and Bearing Cost

The yaw drive rotates the rotor disk plane to be perpendicular to the wind direction. The yaw bearing supports the full weight of the rotor and all of the components in the nacelle. The following scales the yaw drive and bearing mass and cost on the rotor diameter, D.

$$\text{Mass} = 0.00144D^{3.314}, \tag{11.37}$$

$$\text{Cost} = 0.0678D^{2.964}. \tag{11.38}$$

11.2.13 Mainframe Cost

The mainframe is the internal structure inside of the nacelle that supports the main bearings, gearbox and generator. The mass and cost are then broken down into four arrangements presented with the electric generator. These were all found to scale with the rotor diameter.

1. High-speed Generator with Three-stage Planetary/Helical Gearbox

$$\text{Mass} = 2.233D^{1.953}, \tag{11.39}$$

$$\text{Cost} = 9.489D^{1.953}. \tag{11.40}$$

2. Medium-speed Permanent Magnet Generator with Single-stage Drive

$$\text{Mass} = 1.295D^{1.953}, \tag{11.41}$$

$$\text{Cost} = 303.96D^{1.067}. \tag{11.42}$$

3. Permanent Magnet Generators with Multi-path Drive

$$\text{Mass} = 1.721D^{1.953}, \tag{11.43}$$

$$\text{Cost} = 17.92D^{1.672}. \tag{11.44}$$

4. Permanent Magnet Generator with Direct Drive

$$\text{Mass} = 1.228D^{1.953}, \tag{11.45}$$

$$\text{Cost} = 627.28D^{0.850}. \tag{11.46}$$

In addition to the internal support structure, allowance is made for platforms and railings to enable safe inspections and maintenance. The mass and cost of these are based on the respective mainframe mass.

$$\text{Platform mass} = 0.125(\text{Mainframe mass}), \tag{11.47}$$

$$\text{Platform cost} = 8.7(\text{Platform mass}). \tag{11.48}$$

11.2.14 Electrical Connections Cost

The electrical connections include electronic switching gear, and any tower wiring. The cost estimate is \$40/kW of rated power (in 2002 dollars). Thus

$$\text{Cost} = 40(\text{Rated power}). \tag{11.49}$$

11.2.15 Hydraulic and Cooling Systems Cost

The hydraulic and cooling systems mass and cost are estimated to be a fixed percentage of the wind turbine rated power. Thus

$$\text{Mass} = 0.08(\text{Rated power}), \tag{11.50}$$

and

$$\text{Cost} = 12(\text{Rated Power}). \tag{11.51}$$

11.2.16 Nacelle Cover Cost

The Nacelle cover shields the internal components of the nacelle from the weather. The cost and mass are

$$\text{Cost} = 11.537(\text{Rated power}) + 3849.7, \tag{11.52}$$

and

$$\text{Mass} = 0.1(\text{Nacelle cost}). \tag{11.53}$$

11.2.17 Control, Safety System, Condition Monitoring Cost

The control, safety, and monitoring system is taken to be a fixed cost of \$35,000 (in 2002 dollars) for land-based wind turbines. The estimated cost is \$55,000 for off-shore wind turbines because of their more extensive requirements [1].

11.2.18 Tower Cost

The tower is a steel tubular structure that supports the mass of the rotor and all of the internal components of the nacelle. It needs to withstand the compression loads of the combined mass of these components, as well as the

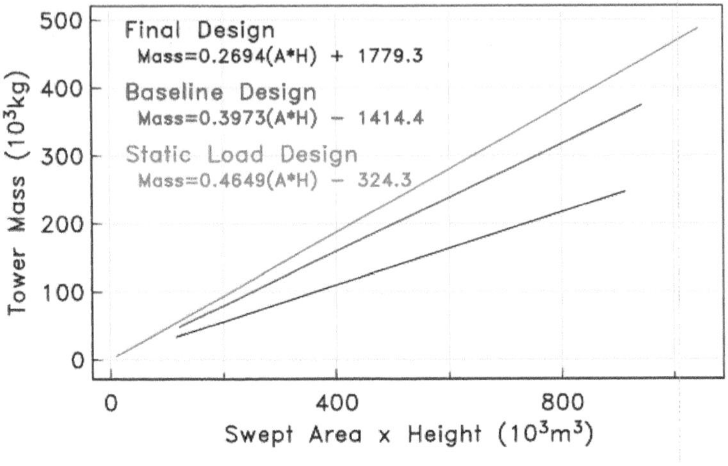

FIGURE 11.3
Wind turbine tower mass correlation with the product of the rotor swept area and hub height.

bending loads produced by the axial forces on the rotor, which scale with the rotor disk area. The maximum bending stress scales with the hub height of the rotor and therefore that is a factor in the tower mass and cost [3,8]. Historic data of the mass of the tower as a function of the product of the rotor area and hub height is presented in Figure 11.3.

Based on the historic data, the mass of a baseline tower design is

$$\text{Baseline design mass} = 0.3973(\text{Rotor area})(\text{Hub height}) - 1414.4, \quad (11.54)$$

Based on the static loads the tower mass is

$$\text{Static load design mass} = 0.4649(\text{Rotor area})(\text{Hub height}) - 324.3. \quad (11.55)$$

For a final *advanced* design, the mass of the tower is given by

$$\text{Final design mass} = 0.2694(\text{Rotor area})(\text{Hub height}) + 1779.3. \quad (11.56)$$

Assuming a 2002 cost of steel of \$1.50/kg, the cost of the tower is

$$\text{Cost} = 1.50(\text{Mass}). \quad (11.57)$$

11.2.19 Transportation Cost

The transportation of the wind turbine large rotors is a considerable factor in the cost of a new wind turbine. Since the rated power scales with the rotor diameter, the cost of transportation is estimate based on the rated power with units of \$/kW. Thus starting with a transportation cost factor [8] given as

$$\text{Transportation cost factor} = 1.581 \times 10^{-5}(\text{Rated power})^2 - 0.0375(\text{Rated Power}) + 54.7, \quad (11.58)$$

the transportation cost is

$$\text{Transportation cost} = (\text{Transportation cost factor})(\text{Rated power}). \quad (11.59)$$

11.2.20 Roads, Civil Work Cost

Most often, new roads or other civil improvements such as increasing the width of existing roads or bridges are needed to gain access to a wind turbine location. Estimates for this involve a cost factor and the rated power of the wind turbine on which the size and mass of the components scale [8]. The cost factor has units of \$/kW of rated power given by

$$\text{Roads, civil work cost factor} = 2.17 \times 10^{-6}(\text{Rated power})^2 - 0.0145(\text{Rated power}) + 69.54, \quad (11.60)$$

and then the roads and civil work cost is

$$\text{Roads, civil work cost} = (\text{Roads, civil work cost factor})(\text{Rated power}). \quad (11.61)$$

11.2.21 Assembly and Installation Cost

In correlating historic factors related to the cost of assembly, the two most important wind turbine design parameters were found to be the hub height and rotor diameter. This is not too surprising of an observation considering that one would expect that the degree of difficulty of assembly would increase with elevation and rotor blade size. The cost in 2002 dollars was then estimated [8] to be

$$\text{Cost} = 1.965[(\text{Hub height})(\text{Rotor diameter})]^{1.1736}. \quad (11.62)$$

11.2.22 Electrical Interface/Connections Cost

The electrical interface covers the turbine transformer and the individual share of cables from the wind turbine to the substation. Based on historic data [9], the cost estimate in 2002 dollars, is

$$\text{Cost} = (\text{Electrical interface/connections cost factor})(\text{rated power}), \quad (11.63)$$

where the cost factor is given as

$$\text{Electrical interface/connections cost factor} = 3.49 \times 10^{-6}(\text{rated power})^2 - 0.0221(\text{rated power}) + 109.7, \quad (11.64)$$

and in which the cost factor has units of [\$/kW].

11.2.23 Engineering and Permit Cost

The cost of engineering and permits involves the design of the entire wind energy facility and the procurement of permits needed to erect the facility. In the case of a wind farm, this cost is based on a turbine-by-turbine basis. The costs depend highly on the location, environmental conditions, availability of electrical grid access, and local permitting conditions. The cost estimate in 2002 dollars is [9]

$$\text{Cost} = (\text{Engineering and permit cost factor})(\text{Rated power}), \tag{11.65}$$

where the cost factor is given as

$$\text{Engineering and permit cost factor} = 9.94 \times 10^{-4}(\text{rated power}) + 20.31, \tag{11.66}$$

and in which the cost factor has units of [\$/kW].

11.2.24 Levelized Replacement Cost

Levelized replacement cost is a sinking fund factor that is intended to cover long-term replacements and overhaul of major turbine components, such as blades, gearboxes, and generators. The cost estimate in 2002 dollars is

$$\text{Cost} = (\text{Levelized replacement cost factor})(\text{rated power}), \tag{11.67}$$

where the cost factor is given as

$$\text{Levelized replacement cost factor} = 10.7(\text{rated power}), \tag{11.68}$$

and in which the cost factor has units of [\$/kW].

11.2.25 Operations and Maintenance Cost

Operations and Maintenance ($O\&M$) cost covers the day-to-day operations costs that include scheduled and unscheduled maintenance of the wind turbine(s). Based on historical operations of land-based wind farms, the recommended $O\&M$ costs are \$0.007/kW-h. Thus

$$\text{Cost} = 0.007(\text{AEP}), \tag{11.69}$$

where AEP has units of [kW-h] and costs are in 2002 dollars.

11.2.26 Land Lease Cost

Wind turbines normally pay lease fees for land used for wind farm development. This cost is principally based upon the land used by the turbine. The factors applied in different wind farm developments vary widely depending on the wind class of the particular site, the nature and value of the land, and the potential market price for the wind. An estimate [5] of the lease costs is

$$\text{Cost} = 0.00108(\text{AEP}), \tag{11.70}$$

where AEP has units of [kW-h] and costs are in 2002 dollars.

11.3 Example Cost Breakdown

An example [1] of the component cost breakdown for a land-based 1,500 kW (rated) wind turbine with a rotor diameter of 70 m and a hub height of 65 m is shown in Table 11.1. In the example, the gear box is a "medium-speed, single stage" and the electric generator is a "medium-speed permanent magnet with single stage drive". These costs are in 2002 dollars.

Considering the lumped components that make up the rotor, the most expensive component corresponds to the rotor blades. For the components that make up the drive train and the nacelle, the most expensive component is the gear box, followed next by the variable speed electronics and the main frame. The tower is comparable in cost to the gear box. These components then make up the largest portion of the turbine capital cost (TCC). The balance of station cost (BOS) is approximately one-third of the initial capital cost (ICC), which is the sum of the TCC and BOS costs.

The annual cost of O&M, replacement and land lease totals \$107,452, which is approximately 7.6% of the ICC made up of the sum of the Turbine Capital and BOS Costs. Based on the wind conditions (Weibull coefficients) and the power curve ($V_{\text{cut-in}}$, V_{rated} and $V_{\text{cut-out}}$), wind turbine capacity factor was determined to be 86.1%. The net annual production (AEP) was then

$$\text{AEP} = (0.861)(24)(365)(1,500) = 11312\,\text{MW-h}. \tag{11.71}$$

The annual operating expenses (AOE) are then

$$\begin{aligned}
\text{AOE} &= \frac{\text{LLC}}{\text{AEP}_{\text{net}}} + \frac{\text{O\&M} + \text{LRC}}{\text{AEP}_{\text{net}}} && (11.72)\\
&= \frac{\$17,217}{11,312,000}_{\text{kW-h}} + \frac{\$79,185 + \$16,050}{11,312,000}_{\text{kW-h}} && (11.73)\\
&= 0.00950, && (11.74)
\end{aligned}$$

where all of the operating expenses have been normalized by the AEP in units of [kW-h].

TABLE 11.1
Component Cost Breakdown for a Land-Based 1,500 kW (rated) Wind Turbine with a Rotor Diameter of 70 m and a Hub Height of 65 m [1]

Component	Mass (kg)	Cost ($1,000)
Rotor System		
Blades	13,845	387.5
Hub	18,878	24.6
Pitch mechanism and bearings	2,282	38.5
Spinner, nose cone	775	4.3
Drive train, Nacelle		
Low-speed shaft	3,026	2.03
Bearings	347	12.3
Gearbox	18,438	11.1
Mech. brake, HS-coupling etc.	298	3.0
Generator	8,931	82.1
Variable spd. electronics	-	118.5
Yaw drive and bearing	1,875	20.0
Main frame	5,196	28.3
Internal platform	650	6.3
Electrical connections	-	60.0
Hydraulic, cooling system	120	18.0
Nacelle cover	2,116	21.2
Control, safety system, condition monitoring	-	35
Tower	118,993	178.5
Turbine Capital Cost (TCC)	**197,068**	**1,151**
Balance of Station (BOS)	-	260.6
Transportation	-	51.0
Roads, civil Work	-	79.0
Assembly & Installation	-	48.3
Electrical interface/connections	-	0.0
Engineering & Permits	-	32.7
Foundations	-	49.5
Initial Capital Cost (ICC)	**1,403**	**169,804**
Levelized replacement cost/yr (LRC)	-	**16.1**
(O&M) per turbine per year	-	**79.2**
Land lease cost (LLC)	-	**12.2**
Capacity factor	-	**86.1%**
Net Annual Energy Production (AEP MW-h)	-	**11,312**
Fixed rate charge (FCR)	-	**11.85%**
COE ($/kW-h)	-	**0.02428**

The COE is then

$$\begin{aligned} \text{COE} &= \frac{(\text{FCR})(\text{ICC})}{\text{AEP}} + \text{AOE} \qquad (11.75) \\ &= \frac{(0.1185)(\$1,411,641)}{4,312,000_{\text{kW-h}}} + 0.00950 \\ &= 0.0243, \end{aligned}$$

where the fixed rate charge (FRC) is taken to be 11.85%. This represents the cost of capital to fund the project.

This is the minimum cost that the owner of the wind turbine could sell the electricity to cover their initial and maintenance costs. This is normally expressed as dollars per kilowatt-hour, which in the example is then \$0.0243/kW-h. This COE is competitive during high electricity demand conditions, but may not competitive during low demand conditions. Thus, it motivates the need for electric energy storage that was discussed in Chapter 10.

11.4 Optimized Economics

This section considers if there exists an optimum horizontal wind turbine design that *minimizes the initial cost* and subsequently the investment in establishing a wind turbine site. These costs are ultimately reflected in the COE, which is the breakpoint above which the wind turbine site is financially profitable.

As presented in this chapter, the economic factors involved in establishing a wind turbine site simply involve the wind turbine rotor radius, hub height, rated power, and wind conditions at the site. The rated power is primarily determined by the rotor radius, which removes it as one of the independent variables. In a levelized comparison of the economic effect of the rotor radius and hub height, the wind conditions at the site would remain unchanged.

In examining the question of an economically optimum wind turbine and site design, we will employ the evolutionary genetic algorithm approach introduced in Chapter 8 on Wind Farms.

The economic cost model is that presented in the previous section of this chapter. The object function is simply to minimize the COE with independent variables being the rotor radius and the hub height. This is expressed in the following equation:

$$\text{Obj} = \min[\text{COE}(R, H)]. \tag{11.76}$$

Included as a spatial constraint on the hub height is that

$$H \geq 2.5 \cdot R. \tag{11.77}$$

The wind farm terrain at the site is assumed to be homogeneous with a uniform terrain elevation and surface roughness. The wind turbines are considered to be identical with the same power curves, $P = f(U)$, where U is the wind speed at rotor hub height. The probability of wind speeds is based on Weibull coefficients determined at the site as $c = 5.42$ m/s, $k = 2.12$ and $z_{\text{ref}} = 10$ m.

The rated power is simply estimated as

$$P_{\text{rated}} = \eta \frac{Cp\rho V_{\text{rated}}^3 \pi R^2}{2}, \tag{11.78}$$

TABLE 11.2
Functional Evaluation of Initial Population

	Initial Random Population		Feval	Remarks
Solution 1	R^1	H^1	COE^1	-
Solution 2	R^2	H^2	COE^2	-
Solution 3	R^3	H^3	COE^3	$\min[\text{COE}(R, H)]$
Solution 4	R^4	H^4	COE^4	-

where $Cp = 0.45$ is the coefficient of power, and $\eta = 0.9$ is the product of the efficiencies of the gear box and generator.

The wind turbine is assumed to have three blades. The cut-in wind speed is 4 m/s. The cut-out wind speed is 20 m/s, and the rated wind speed is 11 m/s. The air density is the standard, 1.225 kg/m^2.

A solution generated by an evolutionary algorithm is referred to as a "chromosome/individual" which is made up of genes. In this application, the genes are the two independent variables, R and H. In the selection of decision variables and problem dimension, the values were assigned as integers in which the resulting R and H were rounded to the nearest meter.

Following the genetic algorithm stages described in Chapter 8, the Stage 1 Population Generation consisted of four chromosomes that each comprised two genes (R and H). As before, chromosome refers to the solution/individual of the formulated optimization problem, while the collection of such chromosomes is referred to as a population. The chosen system size was 4 by 2. Table 11.2 shows the $4x^2$ array that was initially randomly populated.

In Stage 2, all the chromosomes of random population go through a "fitness evaluation" to measure the quality of the Solution created by EA. The fitness evaluation of chromosome/solution is carried out by calculating the objective function value as Feval $= \text{COE}^i, i = 1, 2, \ldots, 4$. The Solution with the minimum Feval is then identified. In this example, this was arbitrarily selected to be that of Solution 3.

In Stage 3, the chromosomes are selected from the initial population to act as "parent" for reproduction, based on the fitness of the solution/individual. This tells how to choose individuals in the population that will create offspring for the next generation. The fittest solution will have a higher probability to be selected as a parent. The details of this stage are presented in Chapter 8.

Stage 4 involves a "cross-over" involving selecting mating chromosomes, determining cut point for crossover, and updating the population. This is followed by Stage 5 in which a mutation operation is performed to generate a random sharp change in a chromosome that is crucial to the convergence of the algorithm. Finally Stage 6 involves the "survival selection and replacement" in which the chromosomes exhibiting the lowest feval in the current and previous populations are merged through a predetermined replacement mechanism to form a new population for the next generation.

TABLE 11.3
Results from Consecutive Runs of the Economic Genetic Algorithm in Which Wind Turbine Radii and Hub Heights Are Sought That Minimize the Cost of Electricity

R	*H*	COE ($/kW-h)
36	125	0.02598
28	125	0.02365
36	125	0.02598
10	125	0.03323
28	172	0.02365
28	125	0.02365

TABLE 11.4
Ranges of COEs for Land-Based and Off-shore wind Turbine Installations [1]

	Land-based	Offshore
Installed capital cost (ICC)	$1,400–$2,900/kW	$4,500–$6,500/kW
Annual operating expenses (AOE)	$9–$18/MW-h	$15–$55/MW-h
Capacity factor	18%–53%	30%–55%
Fixed rate charge (FRC)	6%–13%	8%–15%
Operational life	20–30 years	20–30 years
COE	$60–$100/MW-h	$168–$292/MW-h

The optimization process is repeated until the objective function from one iteration to the next stops changing within a given tolerance. The result of a series of consecutive runs of the economic genetic algorithm is shown in Table 11.3. Note that as a result of the random variability in the genetic algorithm, there is some small variability in the results from one run to the next. However, the general result is that it is more financially economical to select *smaller diameter wind turbines* over larger wind turbines as well as hub heights that are on the order of $3.5 \cdot R$. This results suggests a evaluation of the current trend of fielding ever larger diameter wind turbines.

11.5 Summary

The previous formulas and examples are designed to provide a reasonable estimate of the cost of a new wind turbine project at a land-based site. Table 11.4 provides an historic range of the COEs for new wind turbine installations up to the year 2011. Included in the table are land-based and offshore installations. It is readily apparent that the COE is significantly higher for offshore wind turbines that primarily stems from the higher installed capital costs and operating costs. The operation lifetimes of land-based and offshore are comparable.

Problems

1. We want to examine the effect of the wind turbine rotor diameter on the total cost of installation broken down by

   (a) the cost of the rotor,
   (b) the cost of the drive train,
   (c) the cost of the tower,
   (d) and the balance of station (BOS)

   of which the total of the above is the initial capital cost (ICC).

   For this we will assume that this is a land-based HAWT of baseline design characteristics with a fixed rated power of 1,500 kW. Assume the wind turbine uses a high-speed generator with a three-stage planetary/helical gear box.

   (a) Investigate varying the rotor diameter from 60 to 80 m, where the hub height will be

   $$H_{\text{hub}} = D - 5,$$

   where the diameter has units of meters.

   (b) Plot the costs of the rotor, drive train, tower and BOS as a function of rotor diameter, D.

2. Based on Problem 1, we want to examine how the rotor diameter affects the COE.

   (a) For the wind conditions at the proposed wind turbine site, the turbine capacity factor is estimated to be 32.82%. Based on this calculate the net annual production (AEP).
   (b) Plot the COE as a function of the rotor diameter. Assume a fixed rate lending charge (FCR) of 11.85%.
   (c) Comment on the impact of the rotor diameter on the COE. Note that we assumed that the wind turbine could generate the same rated power even if the rotor diameter were reduced.

3. The gear box on a HAWT has the highest failure rate of any of the other components.

   (a) Plot the cost for the three types of gearboxes for rated power from 1,000 kW to 8 MW.

   (b) Which is the most cost effective at the lower power rating? Which is most cost effective at the higher power rating? Should that have a bearing on which to select?

4. In evaluating the result from the Genetic Algorithm economic analysis, discuss what factors might have had the largest impact on lowering the COE.

   (a) Rank the elements in order from largest to smallest impact on the COE.

5. The genetic algorithm economic analysis indicated that a site consisting of smaller radii wind turbines is more economical that one with larger radii turbines.

   (a) Discuss the impact this might have on a wind farm design based on having a specific total energy capture.
   (b) Would the arrangement of the wind turbines in the wind farm change?
   (c) Would the wind farm energy density be different? Explain.

References

1. L. Fingersh, M. Hand, and A. Laxon, "Wind turbine design cost and scaling model", Technical Report NREL/TP-500-40566, National Renewable Energy Laboratory, December 2006.
2. D. Griffin, "WindPACT turbine design scaling studies technical area 1 – Composite blades for 80 to 120 meter rotor", NREL/SR-500-29492, National Renewable Energy Laboratory, April, 2001.
3. D. Malcom and A. Hansen, "WinPACT turbine rotor design study", NREL/SR-500-32495, National Renewable Energy Laboratory, April 2006.
4. "Offshore wind energy: Ready to power a sustainable Europe, concerted action on offshore wind energy in Europe", NNE5-1999-562, Final Report, Delft University of Technology, The Netherlands, December 2001.
5. G. Bywaters, V. John, J. Lynch, P. Mattila, G. Norton, J. Stowell, M. Salata, O. Labath, A. Chertok, and D. Hablanian, "Northern power systems WindPACT drive train alternative design study report", NREL/SR-500-35524, National Renewable Energy Laboratory, October 2004.
6. R. Poore and T. Lettemmaier, "Alternative design study report: WindPACT advanced wind turbine drive train design study",

NREL/SR-500-33196, National Renewable Energy Laboratory, August 2003.

7. K. Smith, "WindPACT turbine design scaling studies technical area 2: Turbine, rotor and blade logistics", NREL/SR-500-29439, National Renewable Energy Laboratory, June 2001.
8. D. Shafer, K. Strawmyer, R. Conley, J. Guidinger, D. Wilkie, T. Zellman, and D. Beradett, "WindPACT turbine design scaling studies: Technical area 4 – Balance-of-station-cost", NREL/SR-500-29950, National Renewable Energy Laboratory, July 2001.

12

Design Summary and Trade Study

In the process of assessing changes in the design of a wind turbine, it is important to evaluate the impact that design decisions have on such performance and operation aspects as power generated, potential operation lifetime, purchase and operation costs, and the cost of electricity (COE) it produces.

This chapter is intended to examine some of those aspects in the form of a Trade Study. The study utilizes formulas that were presented in Chapters 2, 4, 7, and 11 which represent the atmospheric boundary layer and wind characteristics, aerodynamic performance of a wind turbine rotor, structural design, and wind energy economics. The codes used to generate the plots in this chapter are given in the Appendix Sections A.2 and A.3.

In this example, we consider a wind turbine with the following characteristics.

Number of blades, $B = 3$

Tip speed ratio, $\lambda = 7$

Blade radius (m), $33 \leq R \leq 60$

Wind speed (m/s), $5 \leq V_\infty \leq 14$

Constant rotor section shape, NACA 4415 airfoil

$C_l = 0.368 + 0.0942\alpha$

$C_d = 0.00994 + 0.000259\alpha + 0.0001055\alpha^2$

The angle of attack, α, has units of degrees

Rotor $\theta_{cp} = -2°$

r/R	c/R	θ_T
0.1	0.0830	45
0.2	0.0919	25.6
0.3	0.0775	15.7
0.4	0.0628	10.4
0.5	0.0522	7.4
0.6	0.0450	4.5
0.7	0.0376	2.7
0.8	0.0337	1.4
0.9	0.0277	0.4
1.0	0.0216	0.00

DOI: 10.1201/9781003611486-12

Here we note that the chord length scaled linearly with the rotor radius.

With regard to the structural properties of the rotor. For the NACA 4415 airfoil, the thickness-to-chord, t/c is 0.15. If we consider a box beam main spar that is internal to the rotor, such as shown in Figure 12.1, with a width, $w = 0.35c$ and a height, $h = t = 0.15c$, the bending moment of inertia in the M_y (upward) direction is $I = wh^3/12$ or $I = (0.35c)(0.15c)^3/12$. Similarly, the bending moment of inertia in the M_z direction is $I = hw^3/12$ or $I = (0.15c)(0.35c)^3/12$. The coordinate frame is similarly shown in Figure 12.2. The modulus of elasticity for a thin glass/epoxy structure is estimated to be $E = 9$ GPa.

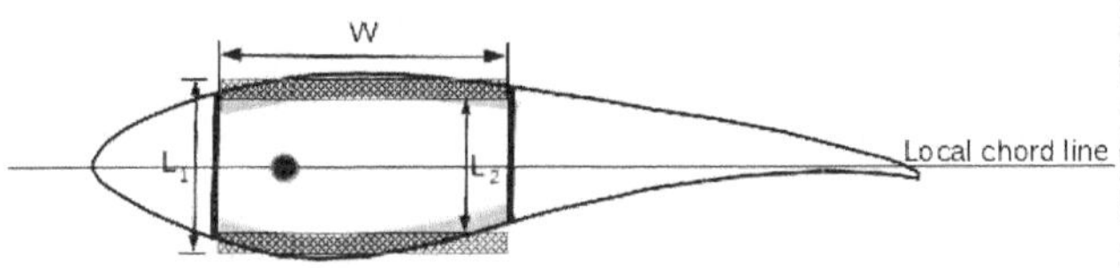

FIGURE 12.1
Rotor blade cross-section illustrating internal structure to resist bending.

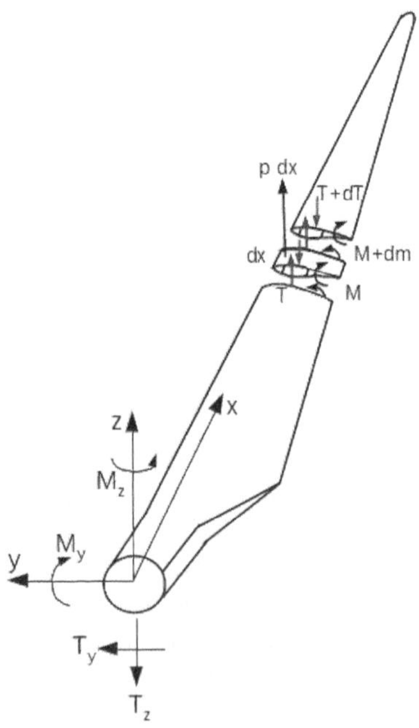

FIGURE 12.2
Rotor blade load and bending coordinate system.

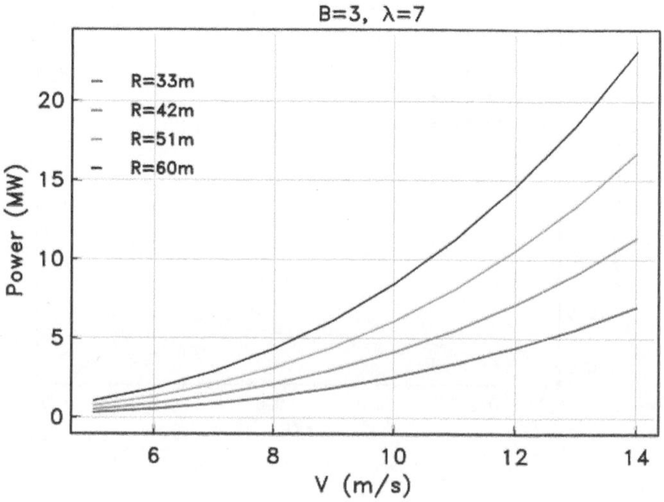

FIGURE 12.3
Effects of wind speed on the maximum total power for different rotor radii.

12.1 Design Power

The first step in designing a wind turbine is to size the rotor to match the power requirement. This corresponds to the "design" power based on a design wind speed. Later, consideration of the wind conditions at the site is needed to determine the annual expected power (AEP). Based on the conditions listed above, Figure 12.3 shows the effect of wind speed on the maximum power for four different rotor radii from the smallest examined (33 m) to the largest examined (60 m). This has the expected trend with P proportional to V_∞^3 and R^2. Based on the choice of a design wind speed, the necessary radius to produce the design power can then be selected.

12.2 Design Structure

Since wind turbines are designed for an operational life of at least 20 years, an important consideration is the structural design. One aspect of this is the turbine blade deflection which if excessive, can reduce the operational life of the rotor, and also impart unsteady loads on other components such as bearings and the gear box. Therefore, Figures 12.4–12.7 examine the effect of

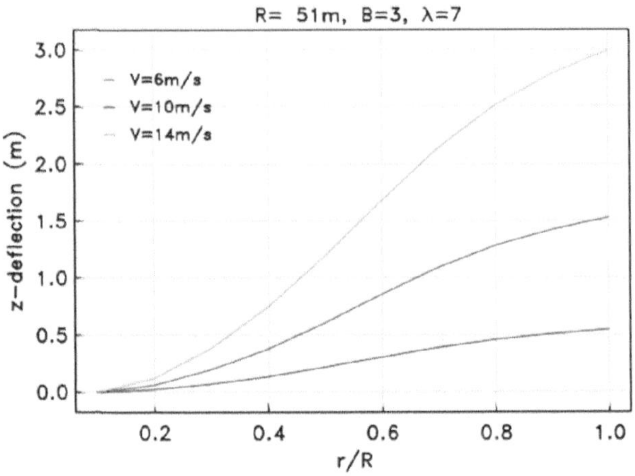

FIGURE 12.4
Rotor z-component deflection at different radial locations for $R = 51$ m at three wind speeds.

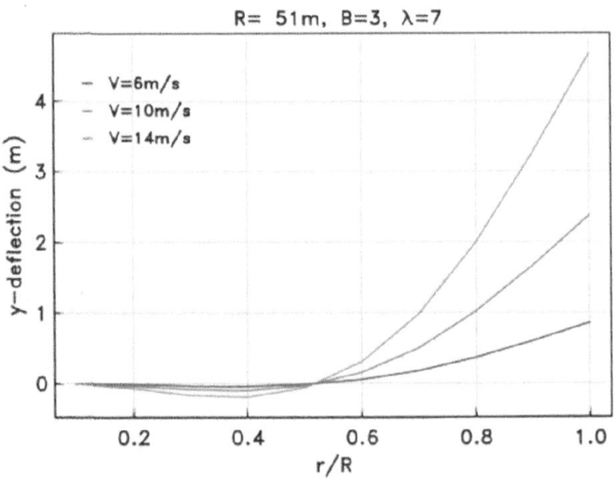

FIGURE 12.5
Rotor y-component deflection at different radial locations for $R = 51$ m at three wind speeds.

rotor diameter and wind speed on the rotor deflection. Figure 12.4 shows the z-component of the bending deflection in the radial direction along a 51 m radius rotor for three wind speeds of 6, 10, and 14 m/s. The corresponding y-component of the bending deflection in the radial direction is shown in Figure 12.5. These illustrate a nonlinear increase in the rotor bending at every

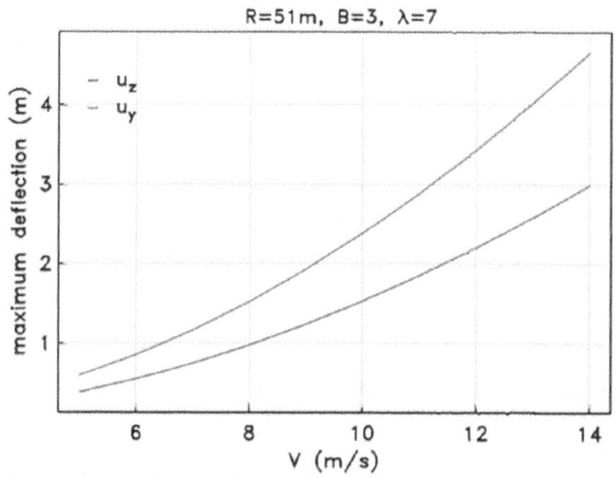

FIGURE 12.6
Rotor maximum deflection as a function of wind speed for $R = 51$ m.

radial location with increasing wind speed. The maximum deflection obviously occurs at the rotor tip. Figure 12.6 documents maximum deflection for the two bending components. This clearly indicates the nonlinear relation in rotor deflection with wind speed. It also shows that for this rotor design, the y-component (in the direction of M_y) deflection is noteably larger than that of the z-component (in the direction of M_z) in magnitude. The reason for this is largely due to the lower bending moment of inertia in the M_y direction due to the box structure height being smaller than its width. This is made worse towards the tip of the blade where, as the chord length is reduced, the fixed thickness-to-chord ratio causes a commensurate decrease in the height of the box beam.

Considering the highest wind speed investigated (14 m/s), Figure 12.7 shows that the maximum deflection components increase linearly with rotor radius. Close examination of the change in maximum deflection with radius indicates that the y-component deflection increases slightly faster than the z-component. This again is specific to the design, particularly on the rotor constant pitch angle, θ_{cp}.

12.3 Design Economics

In order to investigate the economic impact of the design decisions, there needs to be an understanding of the conditions at the site. This includes wind speed occurrence data, the hub height for the wind turbine, and the cut-in, rated

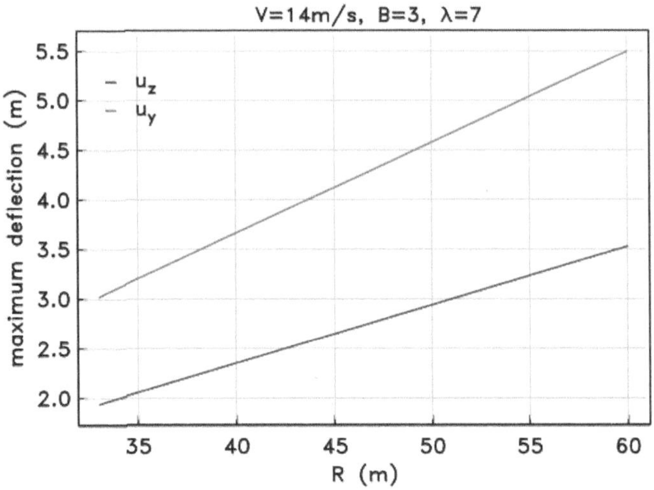

FIGURE 12.7
Rotor maximum deflection as a function of rotor radius for a wind speed of 14 m/s.

and cut-out wind speeds for the design. In this design study, the following conditions were used:

Rotor radius, $33 \leq R \leq 60$ m

Hub height, $H = 2.25R$ m

Air density, $\rho = 1.225$ kg/m^3

Cut-in wind speed, $V_{\text{cut-in}} = 4 - 7$ m/s

Rated wind speed, $V_{\text{rated}} = 11$ m/s

Cut-out wind speed, $V_{\text{cut-out}} = 20$ m/s

Weibull coefficient, $k_{\text{ref}} = 2.12$

Weibull coefficient, $c_{\text{ref}} = 5.42$ m/s

Meteorological data, $z_{\text{ref}} = 10$ m

The hub height was assumed to scale linearly with the rotor radius. Weibull coefficients were used to determine the probable wind speeds at the hub height based on a standard reference height of 10 m and a wind speed of 11 m/s. Based on the cut-in, rated and cut-out wind speeds, the capacity factor was

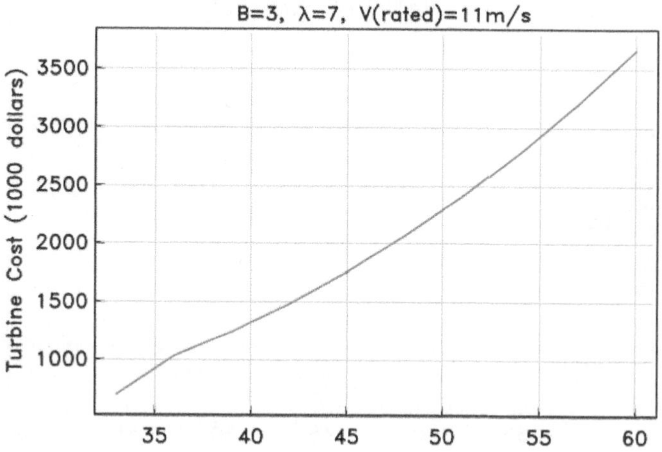

FIGURE 12.8
Wind turbine cost as a function of the rotor radius.

found to be a relatively high 90.7%. Rotor radii from 33 to 60 m were investigated. Table 12.1 provides a summary of the mass and cost of wind turbine components, and the balance of station costs for a 51 m rotor with $V_{\text{cut-in}} = 4$ m/s. The cut-in velocity does not affect these costs. It does however affect the net annual production and COE. These are presented in the economic summary for this case in Table 12.2. This case yielded a net annual production AEP of 9839.92 MW-h, an operation and maintenance cost of \$68879.44, an annual operating expenses AOE of 0.0094 \$/kw-h, and a COE of 0.0422\$/kw-h.

The trends for the range of rotor radii and cut-in wind speed examined in this trade study are presented in Figures 12.8–12.12. Figures 12.8 and 12.9 show the trends in the purchase and installation costs of a wind turbine as a function of the rotor diameter. These are not a function of the cut-in wind speed.

As observed in Figure 12.8, the purchase cost of the wind turbine increased nonlinearly with the rotor diameter. Here the purchase cost of the wind turbine increased by a factor of three when the rotor radius was increased by only a factor of two. The balance of station cost is observed to vary linearly with rotor radius, however a factor of two increase in the rotor radius produced an almost four-times increase in the installation cost of the wind turbine.

The annual energy production and COE depend on the capacity factor which is a function of the hub height and probability of wind speeds being between the cut-in and cut-out wind speeds. This is illustrated in Figure 12.10 which shows the capacity factor as a function of the hub height for two cut-in wind speeds of 4 and 7 m/s. As indicated in Table 12.1, the hub height used in the economic analysis scaled with the rotor diameter. For either of the cut-in

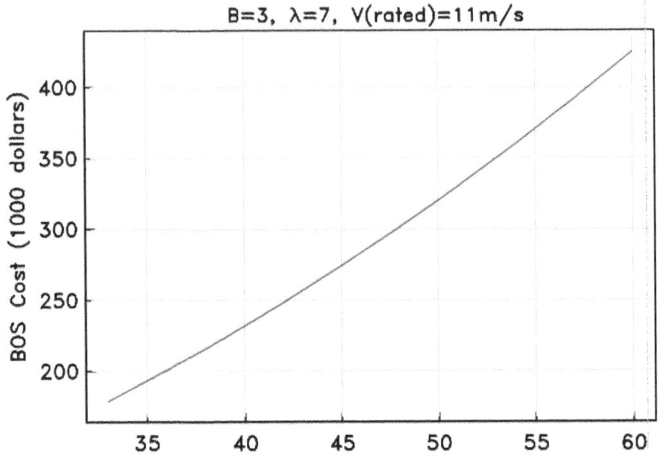

FIGURE 12.9
Balance of station cost as a function of the rotor radius.

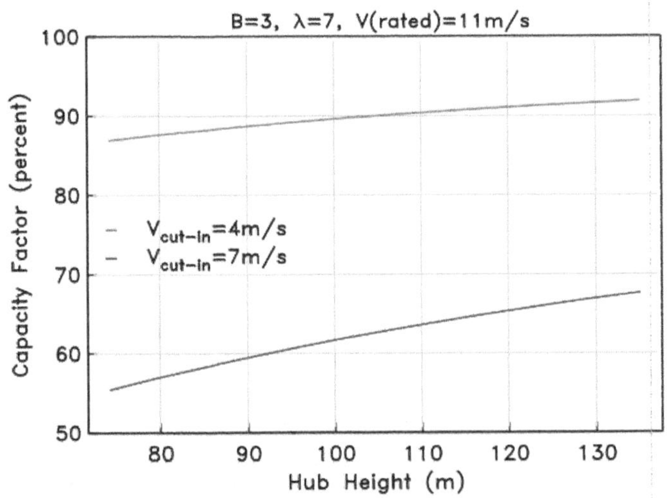

FIGURE 12.10
Capacity factor as a function of the rotor hub height.

wind speeds, the capacity factor varies nonlinearly with the hub height. In general however, the capacity factor increases with hub height, which reflects the higher wind speeds that occur at higher elevations in the atmospheric boundary layer. The degree to which this occurs depends on the cut-in wind speed relative to the Weibull parameters. As Figure 12.10 indicates, for the

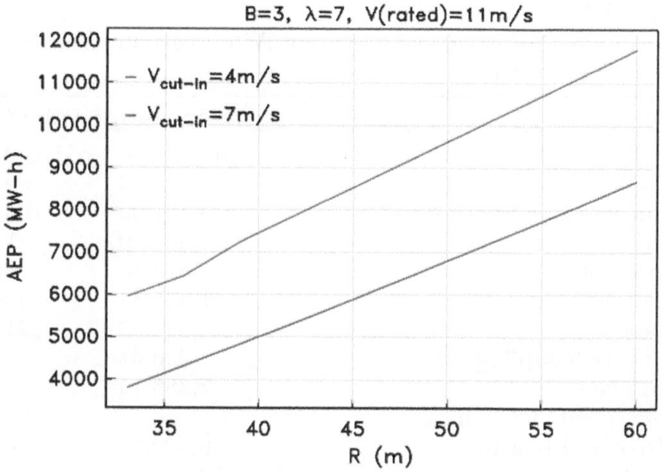

FIGURE 12.11
Annual energy production as a function of the rotor radius.

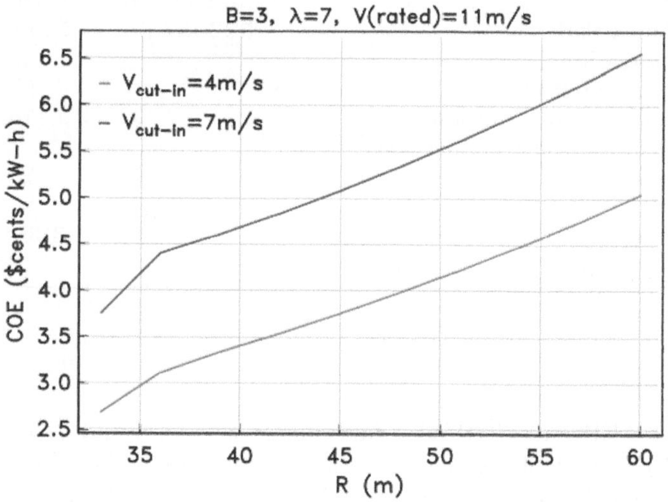

FIGURE 12.12
Cost of electricity as a function of the rotor radius.

Weibull parameters used in this case study, there was a substantial drop-off in the capacity factor between cut-in wind speeds of 4–7 m/s. This will have a significant impact on the annual power generation (AEP) and COE.

TABLE 12.1
Mass and Cost (2002$) Breakdown for Wind Turbine Case Study with $R = 51$ m and $V_{\text{cut-in}} = 4$ m/s

Component	Mass (kg)	Cost ($1,000)
Rotor blade	41,464.80	1,140,327.75
Rotor hub	45,237.72	50,918.02
Pitch mechanism	5,861.00	104,662.03
Spinner nose	1,366.50	7,611.40
Low speed shaft	8,976.88	6,008.17
Main bearing	1,305.34	45,948.32
Gear box	15,894.53	91,750.78
Mech. brake/coupling	246.31	2,463.16
Electric generator	7,483.14	67,766.81
Variable speed electronics	-	97,817.97
Yaw drive and bearing	6,529.24	60,914.48
Mainframe	10,840.91	42,266.16
Internal platform	1,355.11	13,144.60
Electrical connection	-	49,528.08
Hydraulic and cooling systems	99.05	14,858.42
Nacelle cover	1,813.48	18,134.83
Control and safety system	-	35,000.00
Tower	371,115.90	556,673.87
Turbine total	**522067.38**	**2405794.75**
Transportation	-	40,249.53
Roads and civil work	-	67,993.39
Assembly and installation	-	116,944.90
Electrical interface/connections	-	0.00
Engineering and permit	-	26,671.83
Foundation	-	78,108.81
Balance of Station (BOS)	-	**329,968.50**
Levelized replacement	-	12,382.00

Plots of AEP and COP as a function of the rotor radius are shown in Figures 12.11 and 12.12 for the two cut-in wind speeds. For each of the cut-in wind speeds, the annual energy production closely varies linearly with the rotor radius. In each case, doubling the rotor radius approximately doubled the AEP.

The COE shown in Figure 12.12 increases somewhat nonlinearly with increasing rotor diameter. The COE is probably the metric of merit for the wind turbine system, since it needs to be competitive with other forms of electric power generation. Therefore even though the annual electric power production increases with increasing rotor radius, the COE also increases. The effect of the cut-in wind speed has a particularly strong impact on the COE. Therefore, sizing of the wind turbine to match the wind conditions at

TABLE 12.2
Economic Summary for Wind Turbine Case Study with $R = 51$ m and $V_{\text{cut-in}} = 4$ m/s

Fixed rate charge	11.85%
Capacity factor (%)	90.71
Net annual production AEP(MW-h)	9,839.92
Operation and maintenance cost ($)	68,879.44
Land lease cost ($)	10,627
Annual operating expenses AOE $/kw-h	0.0094
Cost of electricity COE $/kw-h:	0.0422

the site are particularly important. Trade studies like this are extremely useful to investigate the impact of design decisions like this that include as many aspects as possible of the wind energy system.

Problems

1. For the conditions of the case study, plot the effect of the cut-out wind speed on the annual energy production and cost of electricity.
2. Based on the results from the case study, if a 1,000 MW-h AEP is required at the site, which would be the most cost-effective, (a) one wind turbine with an approximate 50 m rotor radius or (b) multiple wind turbines with smaller radii whose combined capacity equals a 1,000 MW-h AEP?
3. The maximum rotor deflection depends on the modulus of elasticity of the material used in making the rotor. Determine the sensitivity this has on the maximum deflection of the rotor. What other factors of the rotor design and/or operation can reduce the deflection?
4. Discuss how electric energy storage could be included in the economic trade study. How does this impact the COE requirement?

13

New Concepts

Traditional horizontal wind turbines continue to evolve and become more efficient through a combination of improved rotor aerodynamic designs, introduction of active feedback aerodynamic control, and the use of better materials. Even with these improvements, such wind turbine designs are still constrained by the Betz limit, which specifies the maximum amount of energy that can be extracted from the wind to be 59.3% of the available energy. Thus there is an interest in developing new, less traditional approaches that might overcome the Betz limit or otherwise offer other benefits. This section discusses some of these possible concepts listing their potential benefits as well as possible limitations.

13.1 Vertical Axis Wind Turbine

Vertical axis wind turbines (VAWTs) are receiving a second look as an alternative to HAWTs. Their chief advantages are that individual VAWTs utilize less area, do not depend on the wind direction, and can be more closely packed in arrays in wind farms to provide a potentially higher energy density than wind farms made up of HAWTs. Because of their slow rotor spinning speed, VAWTs are also indicated to be more environmentally friendly, with virtually no aerodynamic noise, and with a much lower impact on flying species such as bats and birds.

An example of a modern VAWT is shown in Figure 13.1. This is a type that are based on a helical rotor shape. Such a system could be suitable for single homes as a supplemental power source.

An interesting adaptation of the helical VAWT is shown in Figure 13.2. In this case, the VAWT is oriented in the horizontal direction and suspended over a highway road. In this concept [2], the natural wind energy could be supplimented by the"wind" induced by the passage of vehicles that pass underneath the wind turbines. The electric energy produced by this concept could be used for road lighting or signage. Another example is shown in Figure 13.3 in which VAWTs line the median next to a roadway. Here again the natural wind is supplimented by the flowfield induced by the passing vehicles. Other examples of roadway applications of wind energy are emerging.

 DOI: 10.1201/9781003611486-13

FIGURE 13.1
Example of a modern vertical axis wind turbine design [1].

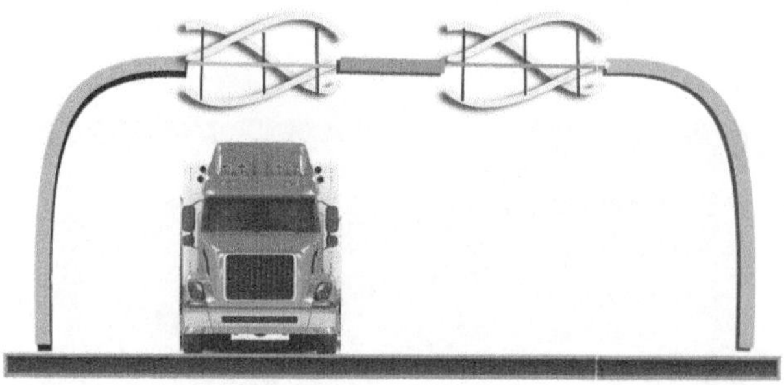

FIGURE 13.2
Concept for highway electric power generation using overhead horizontally oriented helical VAWTs.

A pilot concept for wind farms made up of groups of small VAWTs is shown in Figure 13.4 [3]. This pilot wind farm consists of 10 m. tall wind vertical wind turbines that each generate 3–5 kW of power. They are grouped in pairs where the two wind turbines in the pair rotate in opposite directions. The designers indicate that this minimizes the amount of drag on each wind turbine in the pair, enabling them to spin faster, and maximizing the power efficiency of the farm as a whole. A criticism of the vertical wind farm approach is that because of the use of smaller wind turbines, the number of wind turbines and the land area required would significantly exceed that if larger conventional HAWTs were used.

An alternative to a wind farm of smaller VAWTs is the concept for a Gigawatt rated vertical wind turbine [2]. This is a magnetically levitated

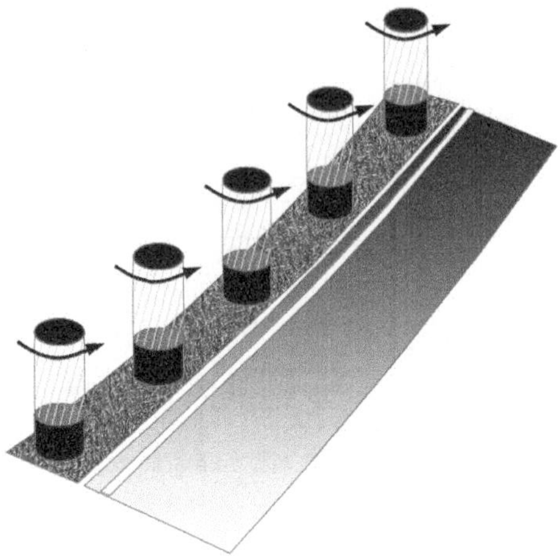

FIGURE 13.3
Concept for highway electric power generation using a series of small VAWTs lining the roadway median strip.

FIGURE 13.4
Photograph of a pilot test of a concept for wind farms made up of small VAWTs [3].

(MagLev) wind turbine concept that would be scaled to be capable of providing power to 750,000 homes. The magnetic levitation would eliminate the friction on the bearing support at the base of the wind turbine. A criticism of this concept is that the electro-magnetic bearing requires a continuous amount of energy. Most likely this would utilize cryogenic cooling to minimize electric losses in the bearing. The concept was invented in 1981, and there are reported to be several of the MagLev wind turbines operating in China. The power rating of these is not however published.

13.2 Wind Focusing Concepts

The Betz limit results from having an open rotor disk about which the air can be deflected as a result of the blockage it presents. A number of concepts have emerged that are designed to incorporate shrouds or ducts that encircle the rotor. Before presenting these concepts, it is useful to provide some analysis.

13.2.1 Shrouded Rotors

A schematic of a shrouded wind turbine rotor is shown in Figure 13.5. The shroud that is placed around the rotor disk is designed to constrain the stream tube in a way in which the velocity is accelerated from V_∞ to V_d. From momentum theory presented in Chapter 4, the coefficient of power, C_p, for an *unshrouded* wind turbine is

$$C_{P_{us}} = \frac{P}{\frac{1}{2}\rho A_d V_\infty^3}, \tag{13.1}$$

where the subscript "us" signifies an unshrouded wind turbine. In Equation 13.1, P is the power extracted from the wind which corresponds to

$$P = TV_{d_{us}} = TV_\infty[1 - a], \tag{13.2}$$

where T is the thrust acting on the rotor disk.

For the shrouded rotor, the wind velocity at the rotor disk, V_{d_s} is determined by the change in the cross-section area of the duct ahead of the rotor disk. For a contracting cross-section ahead of the rotor disk that is shown in Figure 13.5, $V_{d_s} > V_\infty$, whereas for the unshrouded rotor, $V_{d_{us}} = V_\infty(1 - a)$, where $a > 0$ so that in general $V_{d_{us}} < V_\infty$.

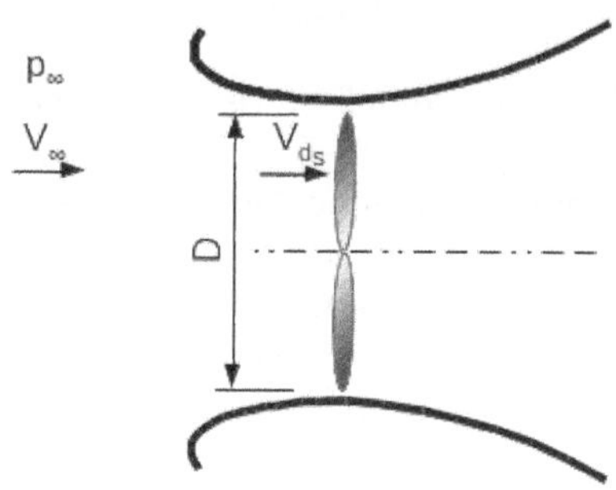

FIGURE 13.5
Schematic drawing of a shrouded horizontal wind turbine.

For the *shrouded* rotor, the coefficient of power is

$$C_{P_s} = \frac{P}{\frac{1}{2}\rho A_d V_\infty^3} = \frac{T V_{d_s}}{\frac{1}{2}\rho A_d V_\infty^2 \frac{V_\infty}{V_{d_s}} V_{d_s}}. \tag{13.3}$$

From Chapter 4, the thrust coefficient for the unshrouded rotor is

$$C_{T_{us}} = \frac{T}{\frac{1}{2}\rho A_d V_\infty^3}. \tag{13.4}$$

Therefore Equation 13.3 becomes

$$C_{P_s} = C_{T_{us}} \frac{V_{d_s}}{V_\infty} = C_{T_{us}} \epsilon, \tag{13.5}$$

where $\epsilon = V_{d_s}/V_\infty$.

Again from Chapter 4 for an unshrouded rotor, the power coefficient is

$$C_{P_{us}} = C_{T_{us}}(1 - a). \tag{13.6}$$

Therefore combining Equations 13.5 and 13.6 to eliminate $C_{T_{us}}$ one obtains

$$C_{P_s} = C_{P_{us}} \frac{\epsilon}{1 - a}. \tag{13.7}$$

Considering the Betz optimum for which $a = 1/3$, even a straight duct without area contraction upstream of the rotor, $\epsilon = 1$, will produce a larger power coefficient than an unshrouded rotor. Any amount of area contraction that accelerates the air velocity, $\epsilon > 1$, will produce an increase in the power coefficient above that of an unshrouded rotor.

It is also easy to show that ratio of the shrouded and unshrouded power coefficients scales with the ratio of the mass flow through the rotor disk, namely

$$C_{P_s} = C_{P_{us}} \frac{\dot{m}_s}{\dot{m}_{us}}. \tag{13.8}$$

Thus there is an advantage to funneling the air stream in a ducted rotor arrangement.

As a caution, the previous analysis neglects viscous losses in the boundary layers on the walls of the shroud. In addition, the area diffusion portion of the shroud needs to be carefully designed to avoid strong adverse pressure gradients that could result in flow separation on the shroud walls. Flow separation would result in large pressure losses in the duct that would reduce the mass flow through the shroud and therefore lower the rotor power coefficient. Finally the structural requirements of large shrouds on multi-megawatt wind turbines provide a significant challenge.

One approach of a shrouded rotor that is marketed under the name "Wind Lens" was developed by a group at the Kyushu University Research Institute

FIGURE 13.6
Ducted horizontal wind tunnel undergoing wind tunnel tests at Clarkson University.
Photograph courtesy of K.D. Visser, Clarkson University.

for Applied Mechanics (RIAM) in Japan [4,5]. The wind lens consists of a circular contraction duct that fits around the rotor. Another example is shown in Figure 13.6. This shows a ducted wind turbine concept developed at Clarkson University undergoing wind tunnel tests.

Another concept aimed at directing the wind around a horizontal axis wind turbine is illustrated in Figure 13.7. It consists of a passive concave mound that is located at the base of a horizontal wind turbine that is intended to accelerate the surface wind approaching the rotor disk. A commercial version of this concept is marketed as the "Wind Donut" [4,6]. The designers of this claim that it increases the turbine power output by 15%–30%.

A concept that is a combination of wind orienting and rotor ducting is shown in Figure 13.8. This consists of a tapered inlet that collects the wind and then passes it through a duct in which a wind turbine (or multiple wind turbines) is located. The system shown in Figure 13.8 was designed and built by SheerWind Inc. [7–9]. The system is designed to accelerate the air delivered to the turbine by as much as four times, which, in combination with the

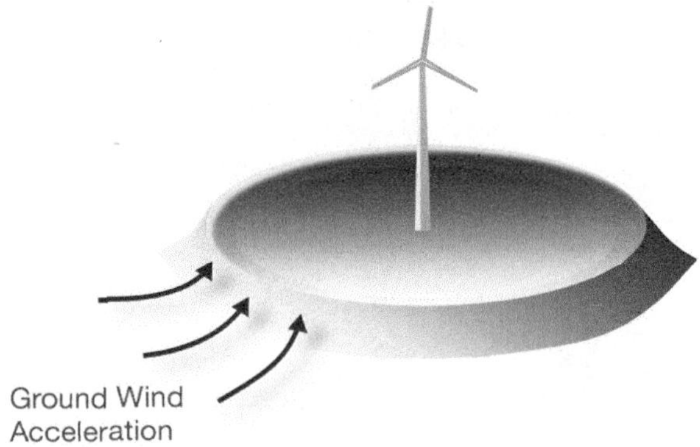

FIGURE 13.7
Artificial hill concept to accelerate ground wind around wind turbines.

FIGURE 13.8
Photograph of SheerWind 2.2 MW INVELOX wind capture system under construction in China.
Source: SheerWind, Chaska, MN.

ducted fan effect, is reported to increase the energy capture by as much as 600% compared to conventional designs. They document that the system can generate electricity in wind speeds as low as 1 m.p.h.

The concept of focusing the wind for energy harvesting has also entered into building architecture. An example is shown in Figure 13.9. This is an 11 story parking garage that incorporates a dozen vertical axis wind turbines. The wind turbines are stacked in two double-helical columns along a corner of

FIGURE 13.9
Building design that incorporates wind energy.
Image courtesy of HOK — Steve Hall © Hedrich Blessing.

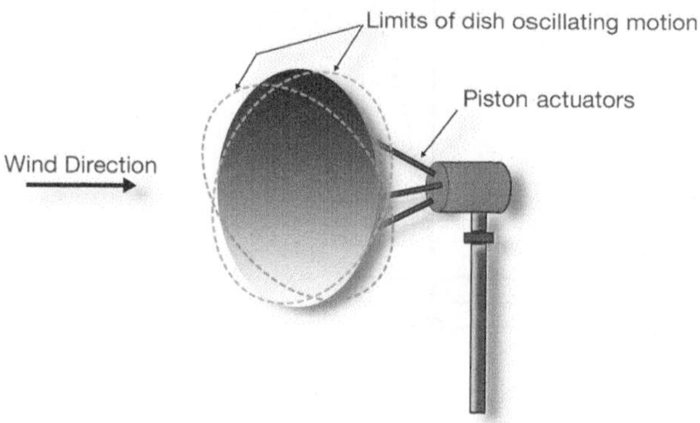

FIGURE 13.10
Illustration of an oscillating wind disk blade-less wind turbine concept.

the building. The wind turbines generate enough power to power the exterior lighing of the building.

13.3 Bladeless Wind Turbine Concepts

Both horizontal and vertical aerodynamic wind turbines rely on converting aerodynamic lift on rotating wing sections into electrical work. The following are complete departures from these concepts that are categorized as "blade-less wind turbines". One of these developed by Saphon Energy[10] in Tunisia involves a flexible disk that oscillates and deflects in a wind stream such as illustrated in Figure 13.10 The motion of the disk drives hydraulic pistons that turn an impeller pump that drives an electric generator. The designers claim that the design overcomes the Betz limit.

Another bladeless concept is referred to as the "Wind Stalk" [11]. This consists of a flexible pole that is attached at its base to a stack of photoelectrically active disks. The flexible poles are designed to deflect and oscillate in the wind through a combination of their aerodynamic drag and wake instability. Their motion is converted into electric energy by the piezoelectric generators. Figure 13.11 shows a concept of hundreds of wind stalks in a wind farm that is intended to resemble a field of wheat.

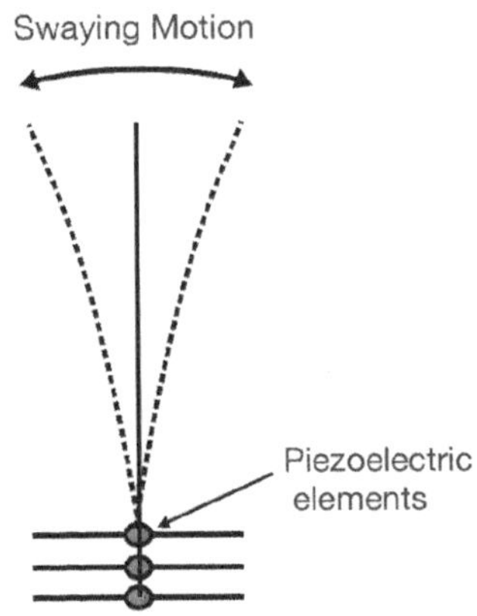

FIGURE 13.11
Flexible wind stalks blade-less wind turbine concept.

FIGURE 13.12
Altaeros energies 35 ft lighter-than-air flying wind turbine.
Source: Altaeros Energies.

FIGURE 13.13
"Sky Serpent" tethered flying wind turbines [16].

13.3.1 Airborne Wind Turbine Concepts

There are a number of airborne wind turbine concepts. The motivation for these is to place the wind turbines at high altitudes that are at the edge of the atmospheric boundary layer where the highest wind speeds occur. The concept in Figure 13.12 is an example of a helium-filled lighter-than-air flying wind turbines. The wind turbine is tethered to the ground by a electric transmission line. The concept incorporates a duct that accelerates the air past horizontal rotor disk [12]. In another concept [13], a lighter-than-air wind turbine rotates around a horizontal axis to generate electrical energy. Both concepts can orient themselves with respect to the wind direction.

A rigid tethered flying wind turbine system, referred to as an "energy kite" is under development by the Makani designers [14,15]. The design has a 30 ft wing span and is intended to generate 30 kW of power. It will use a strong flexible tether that will allow it to reach altitudes of 80–350 m. When airborne, it is designed to fly in a vertical oval that subtends these two altitudes.

Another type of tethered wind turbine known as the "Sky Serpent" [16] is shown in Figure 13.13. This consists of an array of small rotors on a single flexible shaft that is attached to a generator. One end of the shaft is held aloft by helium balloons. The objective of the concept was to increase efficiency by insuring that each rotor catches undisturbed air. This requires achieving an optimal angle for the shaft in relation to the wind direction and having an ideal spacing between the rotors.

13.4 Other Concepts

There are a number of other wind energy concepts that have also emerged. The following summarizes some of these concepts.

13.4.1 Lateral Axis Wind Turbine

The lateral axis wind turbine design shown in Figure 13.14 [17] rotates on a horizontal axis similar to a Ferris wheel. The rotor blades rotate in an epicyclical path around the central shaft. The advantages of this design are unclear.

13.4.2 Tree-Shaped Wind Turbine

A new concept uses tree-shaped wind turbines [18,19] for an aesthetic approach to wind energy. These use numerous small multi-colored vertical wind turbines to represent the "leaves" of the trees. These trees can be placed in an urban

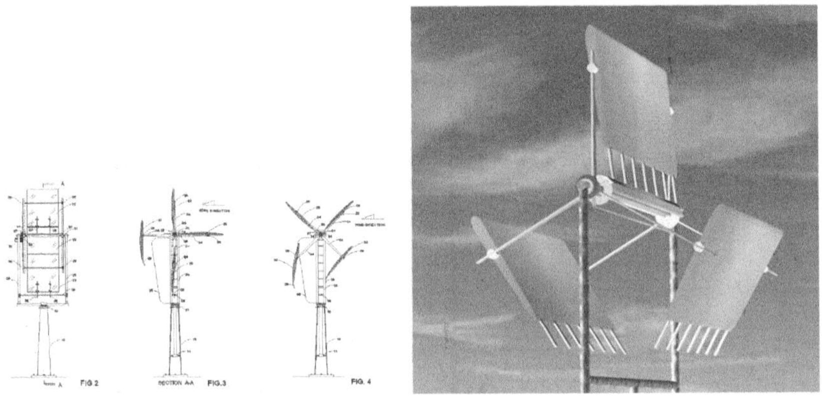

FIGURE 13.14
Lateral axis wind turbine design [17].

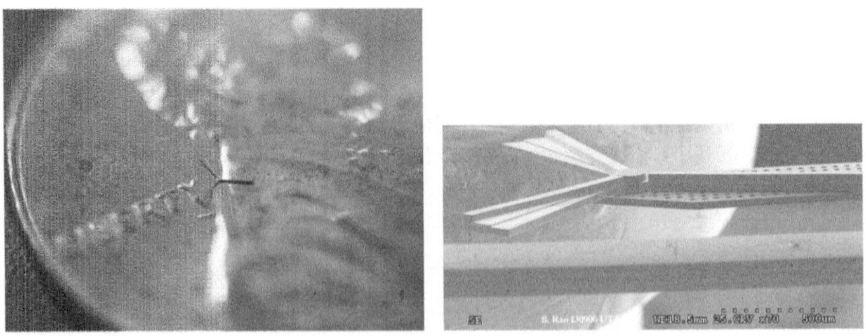

FIGURE 13.15
Ultra-tiny micro wind turbine design developed at the University of Texas Arlington [19,21].

environment where they can exploit small air currents flowing along buildings and streets. They could also eventually be installed in backyards and urban centers. A 26 ft high prototype located in Paris, France can generate electricity in wind speeds as low as 4.5 m.p.h.

13.4.3 Wind Turbine Phone Charger

Another concept is a small utility wind portable wind turbine that consists of a 12 in. tall, three-bladed VAWT [20]. It has a built-in 15,000 mA-h battery, a 15 W generator, and a USB port. It can charge battery operated devices having standard USB ports, such as a cell phone. Other examples of small vertical and horizontal wind turbines for powering small appliances are emerging in the market place.

13.4.4 Miniature Wind Turbine

The researchers at the University of Texas at Arlington have designed an ultra-tiny micro-windmill shown in Figure 13.15 that they claim is capable of generating enough wind energy to recharge cell phone batteries [19]. The scale of these tiny wind turbines is such that ten of them can be mounted on a single grain of rice.

13.4.5 Wind Powered Street Border Lights

A new concept for a wind-generated road border lighting system [22] is schematically shown in Figure 13.16 [20]. These are VAWTs that rotate due

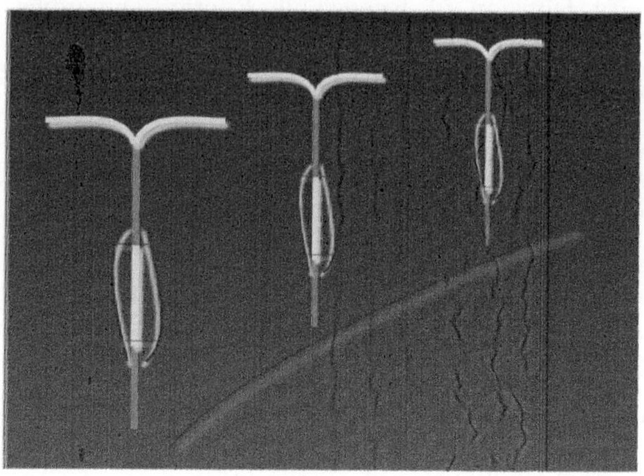

FIGURE 13.16
Road lighting concept using helical VAWTs on light poles.

to the wind generated by passing vehicles. The energy is captured and stored during the day time and used to illuminate the core of the turbines at night, marking the edge of the roadway.

These are just a few examples of concepts for wind turbines that aim to surpass the Betz limit, reduce the environmental impact of wind turbines, or aesthetically integrate wind power into architecture or designs. The ideas appear to be limitless.

References

1. National Renewable Energy Laboratory (NREL) image gallery, image 39786.jpg
2. C. Houghton, "Wind turbine design: The most amazing windmills in the world", Wind Science, December 26, 2013, https://www.mwps.world/news/2021/04/26/the-most-amazing-wind-turbinesdesigns/.
3. J. O. Dabiri, "Potential order of magnitude enhancement of wind farm density via counter-rotating vertical axis wind turbines", *Renewable and Sustainable Energy*, 2011, 3:043104, .
4. R. Das, "Future of wind energy", Greeniacs.com, April 2, 2013, http://www.greeniacs.com/GreeniacsArticles/Energy/Future-of-Wind-Energy.html.
5. "Wind energy concentration system, wind lens", RIAM, Kyushu University Division of Renewable Energy Dynamics, November 2014.
6. A. Liszewski "Wind energy donut for more efficient turbines", Alternative Energy News, May 2017
7. D. Allaei, D. Tarnowski, and Y. Andreaopoulos, "INVOLEX with multiple turbine generator systems", *Energy*, 2015, 93(1):1030–1040.
8. D. Allaei and Y. Andreaopoulos, "INVOLEX: Description of a new conceptin wind power and its performance evaluation", *Energy*, 2014, 69:336–344.
9. B. Yirka, "ShearWind claims its Invelox wind turbine produces 600% more power", Phys.Org., May 2013, http://phys.org/news/2013-05-sheerwind-invelox-turbine-power.html.
10. Seeker, "Bladeless wind turbine inspired by sails", November 12, 2012, http://www.seeker.com/bladeless-wind-turbine-inspired-by-sails-1766042479.html.

11. D. Quick, "Windstalk concept is a wind farm without turbines", *Environment*, October 13, 2010, http://www.gizmag.com/windstalk-concept/16647/.
12. Alteros Energies, https://youtu.be/ykIVutc0QP8l.
13. L. Blain, "Magenn floating wind generators take advantage of high altitude winds", *Environment*, February 2009, http://newatlas.com/magenn-mars-floating-wind-generator/11109/.
14. Energy Without Carbon, "New wind turbine designs", http://www.energy-without-carbon.org/NewWindDesigns.
15. http://www.makanipower.com/concept.
16. K. Thompson, "Ten times the turbine", Popular Science, May 13, 2008, http://www.popsci.com/scitech/article/2008-05/ten-times-turbine.
17. P. Dvorak, "The lateral axis wind turbine", Alternative Energy News, March 2005, http://www.alternative-energy-news.info/lateral-axis-wind-turbine/.
18. M. Williams, "Harnessing the Wind: The Wind Tree", https://www.herox.com/blog/246-harnessing-the-wind-the-wind-tree.
19. C. Phillips, "New tree-shaped wind turbines to be installed on the streets of Paris", Newsweek World, January, 2015, http://www.newsweek.com/new-tree-shaped-wind-turbine-be-installed-streets-paris-296591.
20. J. O'Callaghan, "Forget plugs, charge your mobile with a wind turbine: Portable propeller harnesses breezes to power up a phone", Daily Mail, Science and Technology, April, 2014, http://www.dailymail.co.uk/sciencetech/article-2604207/Trinity-portable-wind-turbine-charge-mobile-phone.htm.
21. B. Dodson, "World's smallest windmills to power cell phones", New Atlas Science, January 13, 2014, http://newatlas.com/worlds-smallest-windmill-energy-harvesting/30425/.
22. A. Schwartz, "Turbine light illuminates highway with wind", Inhabit Innovation, February 5, 2010, http://inhabitat.com/turbine-light-powers-highway-lights-with-wind/.

14

Capstone Design - Deployable Wind Turbines for Disaster Response Missions

This chapter presents a capstone design project that incorporates many aspects of wind energy design. It involves the design of a mobile horizontal axis wind turbine that can be erected singly or in a group to provide electric power to a remote site where no electric infrastructure exists. To be transportable, the design must adhere to transport packaging space constraints. Compared to conventional wind turbine applications, this project places a premium on rapid transport, installation, and quick teardown upon mission completion. Assembly and anchoring of the wind turbine are important considerations. The system must also include energy storage to account for variations in wind conditions. The following summarizes key objectives for the mobile wind energy system design.

1. A portable wind energy system that can be transported in a standard 20 ft shipping container.
2. Electric power generation to support a portion of the loads of a company-sized element or disaster response team (50–100 personnel) with deployments from weeks to months.
3. A highly mobile setup and takedown in hours.
4. A lightweight and packable system that can be erected by a single person.
5. A wind turbine with a rated power between 10–20 kW.
6. Energy storage of 100 kW-h per wind turbine.

The motivation for the project comes from a Department of Defense Domestic and International Disaster Response Review that includes both defense and disaster response applications for a deployable wind turbine. Disaster response is part of a broader set of activities including disaster recovery, disaster preparation, and disaster mitigation. Examples include the aftermath of hurricanes, floods and fires, and events that do not necessarily involve widespread destruction of infrastructure such as disease outbreak, famine, or care of displaced persons. A major focus of response activities is assisting the population affected by the disaster by providing temporary electric power to key facilities.

DOI: 10.1201/9781003611486-14

TABLE 14.1
Dimensions of a Standard 20 ft Shipping Container

Exterior (m)			Interior (m)			Door (m)		Weight (kg)
6.1	2.4	2.6	5.9	2.3	2.4	2.3	2.3	2,290

14.1 Wind Turbine Design

As discussed in previous chapters, the available wind at a site represents the upper bound of energy production for a wind turbine. The design and operation of the wind turbine determine how much of the available wind energy can be converted into usable electricity. As wind conditions will vary from site to site, a deployable wind turbine design needs to take into account for such variations.

For portability, a constraint in the design is that all of the components that make up the wind turbine must fit within a standard 20 ft. shipping container. These containers are designed to be carried on tractor-trailer trucks and on military transports. For reference, Table 14.1 lists the exterior and interior dimensions and weight of a standard shipping container.

In meeting the requirement of being a highly mobile setup and takedown that can be completed in hours:

1. the rotors blades need to be a one piece design,
2. the tower can be a maximum of three pieces, and
3. a simple base support system is needed.

Other considerations for the design include:

1. having a constant rotor section shape that is the same along the rotor span.
2. employing a direct drive (without gear box) electric generator,
3. utilizing lead acid batteries (nominal voltage of 60 V, and a capacity of 1,200 A-h) for electric storage, and
4. utilizing the storage container for the base of the wind turbine.

The constant rotor section shape will make the rotor fabrication easier and lighter, which reduces cost so that a spare rotor might be included in the container. Employing a direct drive electric generator eliminates the gear box, which statistically has the highest failure rate in the wind turbine. A direct drive generator is feasible for the proposed power range (10–20 kW) with the armature not being excessively large. Lead-acid batteries are the most economical method of electric energy storage. Their storage capacity, rate of

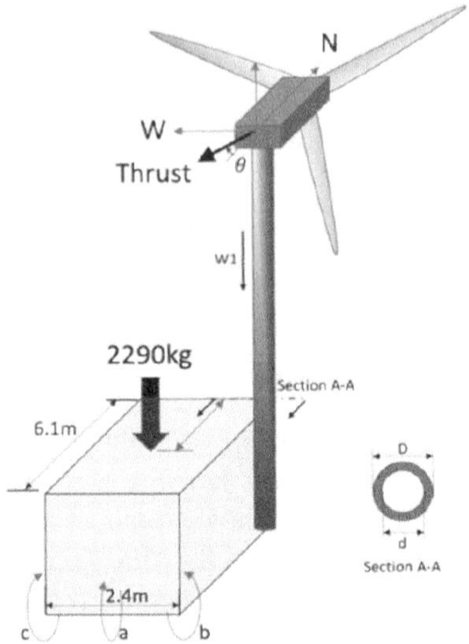

FIGURE 14.1
Example of deployable wind energy system concept that makes use of the storage container for the wind turbine base.

discharge, and number of charge-discharge cycle capabilities are suitable for the proposed limited duration missions.

It is preferred and possibly required that the deployable turbine foundation not significantly disturb the local ground surface. Therefore, designs need to consider approaches that rely on ballast weights and/or deployable outriggers. One concept that makes use of the storage container for the wind turbine base is shown in Figure 14.1.

14.1.1 Site Wind Conditions

The wind conditions at the deployable site are important in achieving the best wind turbine performance. The wind conditions will certainly vary from one site to another. Therefore the design should be one that can best accommodate variability in wind conditions. Factors that will have an impact on performance include the cut-in, rated and cut-out wind speeds. Ultimately, clear documentation on the wind turbine physical and performance characteristics is necessary to predict the number of wind turbines and storage capability, to meet the electric power requirements at the site.

General wind conditions will be assumed in which

1. the local wind speed has a Weibull probability distribution with coefficients, $k = 2.17$ and $c = 12.12$ m/s measured at a 10 m elevation, and
2. there is a range of predominant wind directions that span $\pm 45°$.

14.1.2 Rotor Design

The application suggests a rotor design that is low cost, light weight, and whose performance is reasonable insensitive to wind conditions. Based in these objectives, the following are assumed:

1. a conservative turbine coefficient of power of $C_p = 0.45$, and an overall electric energy conversion efficiency is $\eta = 0.9$, and
2. a nominal tip-speed-ratio of $\lambda = 7.5$.
3. a simple rotor section shape with lift and drag characteristics of $C_l(\alpha) = 0.368 + 0.0942\alpha$ and $C_d(\alpha) = 0.00994 + 0.000259\alpha + 0.0001055\alpha^2$.

The rotor design will include the following steps:

1. Choose a hub height.
2. Choose the number of rotor blades.
3. Determine the the rotor radius based on $C_p = 0.45$, and rotor chord, $c(r)$, and twist, $\theta_t(r)$ distributions. For a design rated wind speed, determine the probability of occurrence based on the Weibull coefficients corrected for the hub height.
4. Using the procedure illustrated in Figure 5.6 of Chapter 5, trim off the inboard portion of the rotor, to obtain the final $c(r)$ distribution that will be used to determine the power.

Assuming the tower can be made up three sections. Based on Table 14.1, the largest segment to fit inside the container is 5.9 m. Assuming the tower is made up of three sections with a 10 cm overlap at two of the section joints, the total length of the tower will be 17.5 m, which corresponds to the turbine hub height. H. The rotor blades are limited to being one piece. In order to fit within the container, the largest span is 5.9 m. Assuming the center hub radius is 10 cm, the rotor radius will be 6 m, or a rotor diameter of 12 m. Table 14.2 summarizes the calculations used to estimate the wind turbine design power.

The design and cut-out wind speeds for the design were selected to provide a design power that was within the specified range (10–20 kW) while also occurring over a *reasonable* number of hours with a 24 hours period. In this case, based on the Weibull probability, rated power would be produced for 15.7 hours per day. This is more than ample to charge the batteries in a 24 hours period.

The next step in the rotor design is to determine the rotor blade chord and twist distributions. The wind turbine will bed designed to have three rotor blades, that is $B = 3$. This follows the steps detailed in Table 5.2 of Chapter 5. In determining the rotor blade twist angle distribution, $\theta(r)$, the angle of attack that is used is the one corresponding to $(C_l/C_d)_{\max}$. Figure 14.2 is a plot of (C_l/C_d) versus angle of attack, α based on the prescribed rotor section shape with lift and drag characteristics. This shows that $(C_l/C_d)_{\max}$

TABLE 14.2
Calculations for Wind Turbine Design Power

z_{ref}(m)	10
k_{ref}	2.17
c_{ref} (m/s)	12.12
n	0.23
H (m)	17.5
V_{rated} (m/s)	8
V_{cutout} (m/s)	20
D (m)	12
ρ (kg/m^3)	1.225
k_{new}	2.282
c_{new} (m/s)	13.785
$\mathcal{P}(V > V_{\text{rated}})$	0.749
$\mathcal{P}(P_{\text{rated}} < V < V_{\text{cutout}})$	0.653
No. of hours in day	15.664
λ	7.5
C_p	0.45
η	0.9
Design power (kW)	1.44E+01

occurs at $\alpha = 6°$. This therefore corresponds to $\alpha(r)_{(C_l/C_d)_{\max}}$ is used in the expression to obtain the radial distribution of the rotor twist, $\theta(r)$.

The resulting rotor radial distributions $\phi(r)$ and $\theta(r)$ are shown in Figure 14.3. The rotor twist angle goes to a small negative angle of attack at the tip. The radial chord distribution, chord distribution is shown in Figure 14.4. The solid curve and data points corresponds to the distribution based on Equation 5.17 in Chapter 5. The straight line drawn through two points in the original distribution at $r/R = 0.5$ and $r/R = 0.9$ and used to extrapolate to the inner radius positions to provide the final chord distribution. As pointed out in Chapter 5 this reduces the weight and cost of the rotor without significant loss of energy capture performance.

Having determined the rotor shape, perform a blade element momentum (BEM) analysis to determine the following wind turbine characteristics:

1. the induction factors, $a(r)$ and $a'(r)$ distributions,
2. the differential thrust, $dT_n(r)$ (N) distribution,
3. the differential Power, $dP(r)$ (W) distribution,
4. the total wind turbine power,
5. the total normal thrust.

The total power and probability of the design wind speed is used to determine the number of batteries needed to proves 100 kW-h storage and the number of hours needed to charge the batteries. The differential thrust normal to the

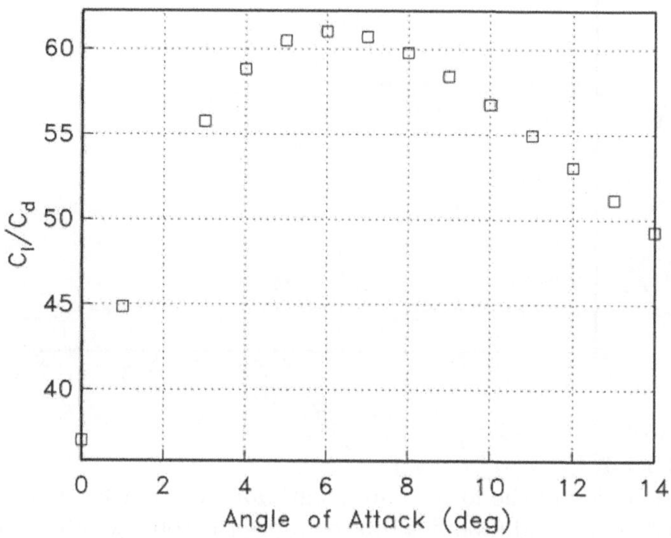

FIGURE 14.2
Airfoil section lift-to-drag ratio as a function of angle of attack used to find the angle of attack of $(C_l/C_d)_{\max}$.

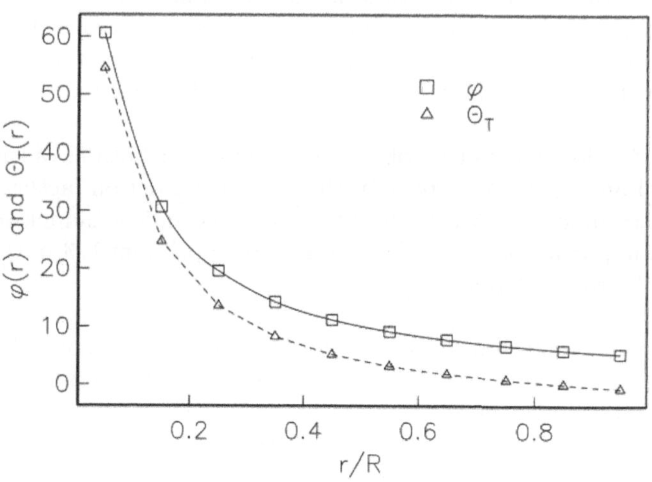

FIGURE 14.3
Relative wind angle, $\phi(r)$ and blade twist angle, $\theta(r)$ along the rotor radius for a Betz design.

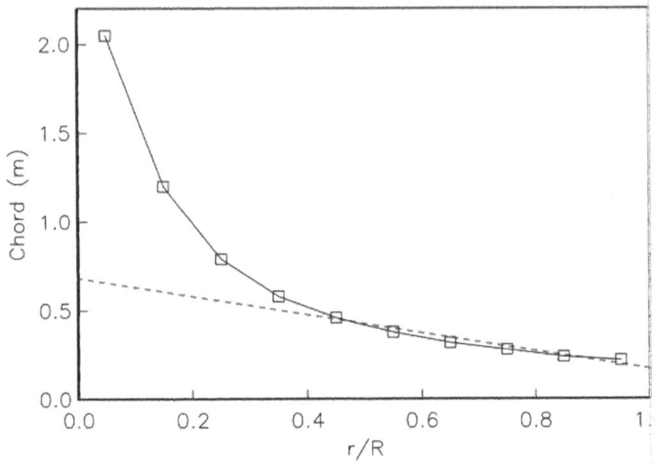

FIGURE 14.4
Radial distribution of the local chord length for the rotor for a Betz optimum design (solid curve) and modification to reduce the rotor weight (dashed line).

rotor plane is summed to get the total thrust force which is used in the design of the tower and base.

The rotor blade radius and the radial distribution of the chord and twist were incorporated into a BEM code. The rotor constant pitch angle , θ_{cp}, was varied to determine the value that would maximize the total wind turbine power. The result is shown in Figure 14.5. This shows that the maximum power occurred with a $\theta_{cp} = 0°$. This value was used in the determination of the characteristics of the wind turbine design documented in the following figures.

Figure 14.6 documents the rotor blade radial distributions of the induction factors, $a(r)$ and $a(r)'$. As expected, the angular induction factor, a', is small and decreases quickly towards the rotor tip. The inflow induction factor, a, being more important is fairly close to the Betz optimum 1/3 in the outboard portion of the rotor blade.

The radial distribution of the differential thrust that is normal to the rotor plane, $dT_n(r)$, is shown in Figure 14.7. This shows an almost linear increase of the normal differential thrust with increasing radial position. The aspect that is important in designing the tower structure and tower base is the total thrust, which is 3,663 N.

Finally, the radial distribution of the differential power, $dP(r)$, is shown in Figure 14.8. This shows an almost linear increase in the differential power with increasing radial position until nearing the rotor tip, where the Prandtl tip-loss model takes effect. The total power is 13.421 kW. As expected, this is less than the original 14.4 kW that was estimate in the rotor design (Table 14.2). However, the rated power still meets the requirement of being between 10 and 20 kW.

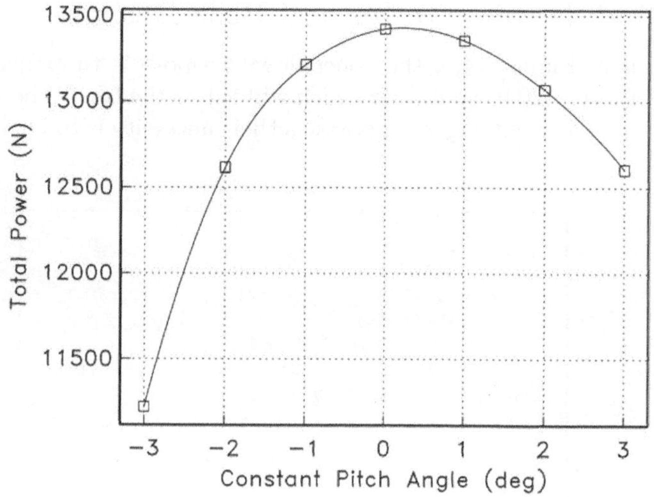

FIGURE 14.5
Effect of the constant rotor pitch, θ_{CP} on the total wind turbine power.

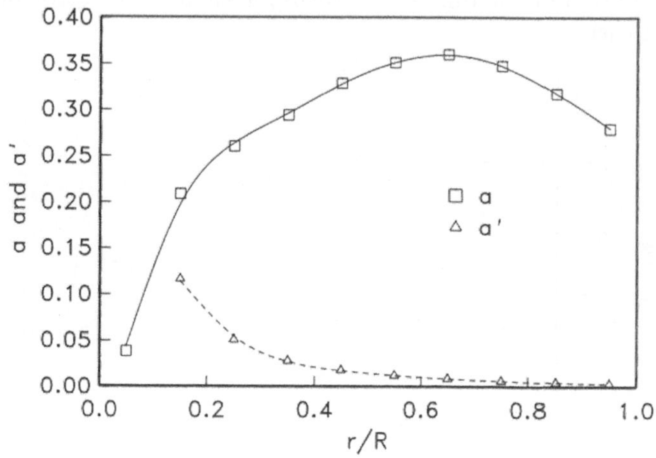

FIGURE 14.6
Radial distribution of the induction factors, $a(r)$ and $a'(r)$ for the wind turbine.

14.2 Wind Turbine Tower and Base Designs

As illustrated in Figure 14.1, the concept we propose is to attach the wind turbine tower to the 20 ft. container which will act as the base. The illustration does not indicate any outrigger supports as their necessity is to be determined.

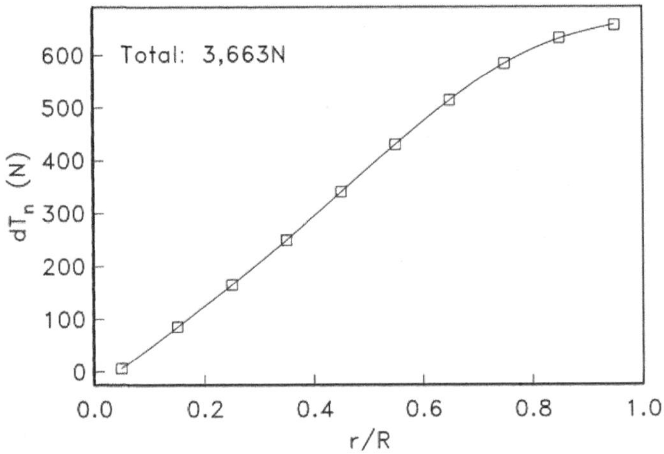

FIGURE 14.7
Radial distribution of the differential thrust normal to the rotor plane, $T_n(r)$ for the wind turbine.

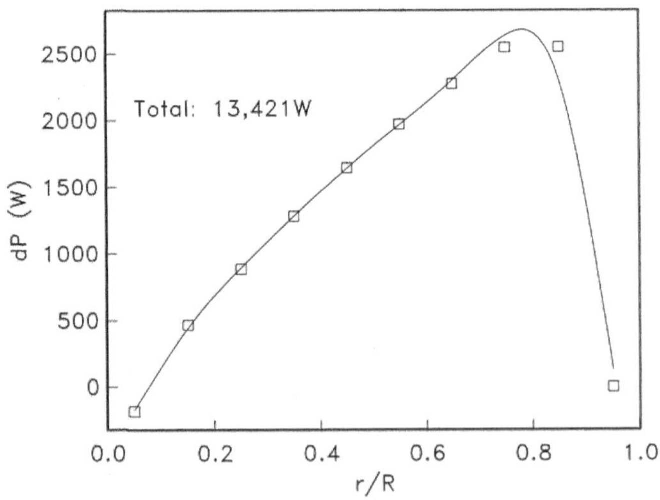

FIGURE 14.8
Radial distribution of the differential power, $dP(r)$, for the wind turbine.

TABLE 14.3
Computed Weights of Key Wind Turbine Components

Component	Formulas	Mass (kg)
Single rotor blade (kg):	$0.4948(\text{Rotor radius})^{2.53}$	46.3
Rotor hub (kg) :	$0.954(\text{Single blade mass}) + 5,680.3$	5,724.5
Electric generator (kg):	$10.51(\text{Rated power})^{0.9223}$	115.3
Nacelle (kg):	$1.228(\text{Rotor diameter})^{1.953}$	157.3
Tower (kg):	$0.2694(\text{Rotor area})(\text{Hub height}) + 1,770$	2,303.2
Total (three blades)		**8,439**

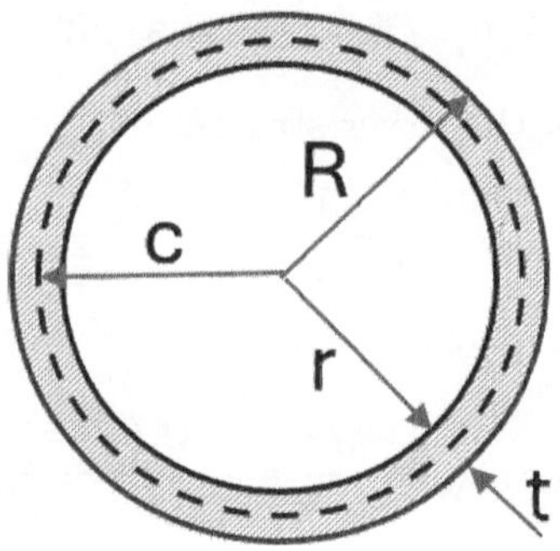

FIGURE 14.9
Tower hollow tube general cross-section.

In addition to the thrust load that acts on the rotor hub, Table 14.3 lists formulas from which to derive the mass of various components of the wind turbine. These formulas are the same ones used in the economic analysis in Chapter 11. The total reflects the mass with three rotor blades. The total weight (mass times the gravitation constant) from the wind turbine components that acts along the tower axis is denoted as w_1 in Figure 14.1

14.2.1 Tower

The tower needs to be designed to withstand the maximum bending stress and shear stress that primarily results from the rotor-normal thrust, T_n, on the rotor. For minimum weight, the tower is considered to be a constant diameter hollow tube with an outside diameter, D and and inside diameter, d. A cross-section schematic is shown in Figure 14.9. The material is assumed to be steel with a maximum yield stress of $\sigma_u = 2.9 \times 10^5$kPa. A factor of safety of 2 is assumed.

Considering the tower tube hollow structure, the wall thickness is

$$t = R - r = \frac{D}{2} - \frac{d}{2} = \frac{D - d}{2}. \tag{14.1}$$

For this cross-section, the bending moment of inertial is

$$I = \frac{\pi}{64}(D^4 - d^4). \tag{14.2}$$

The BEM analysis indicated that the rotor-normal thrust force is 3,663 N. This is assumed to be concentrated at the hub height. As illustrated in Figure 14.1, the tower is attached to the side of the container. We will assume that the container is rigid. Therefore the unsupported length of the tower is the hub height (17.5 m) minus the supported length corresponding to the height of the container (2.6 m) equal to 14.9 m. Based on this, the maximum bending moment on the tower is

$$M_{\text{max}} = T_n \cdot (H - 2.6) = 54{,}578.7\,\text{N-m}. \tag{14.3}$$

The compressive stress in the tower structure is

$$\sigma_{\text{compression}} = \frac{W_1}{A_{\text{cross-section}}} = \frac{W_1}{\frac{\pi}{4}(D^2 - d^2)}. \tag{14.4}$$

The bending stress in the tower structure is

$$\sigma_{\text{bending}} = \frac{Mc}{I} = \frac{M\left(\frac{D-d}{4}\right)}{\frac{\pi}{64}(D^4 - d^4)}, \tag{14.5}$$

where c is approximated as the radius half way between r and R, or

$$c = r + \frac{t}{2} = \frac{d}{2} + \frac{D-d}{4} = \frac{D-d}{4}. \tag{14.6}$$

The total stress on the tower structure is the stress due to the bending moment that is opposed by the compression stress due to the weight of the wind turbine components, or

$$\sigma_{\text{tot}} = \sigma_{\text{bending}} - \sigma_{\text{compression}}, \tag{14.7}$$

where

$$\sigma_{\text{tot}} \leq 2\sigma_u. \tag{14.8}$$

The approach used in designing the wind turbine tower as a hollow tube was to assume a tube wall thickness and then to determine the outer diameter of the tube that would meet the requirement that the maximum shear stress due to bending would be less than one-half the ultimate shear stress of the material. In keeping with the objective of minimizing the weight, the wall thickness chosen was $t = 2$ mm. With that, the effect of the outer diameter on the shear stress was examined. The result is shown in Figure 14.9 which shows $\sigma_{\text{tot}}/\sigma_u$ as a function of the outer diameter, D for a fixed wall thickness, $t = 2$ mm. The horizontal line marks where $\sigma_{\text{tot}}/\sigma_u = 0.5$, which sets the minimum outer diameter that satisfies that the shear stress in the wall of the tower is less than the ultimate shear stress of the material by a factor of safety of 2. As illustrated in Figure 14.10, the *minimum* outer diameter that meets that requirement is $D = 7$mm.

14.2.2 Base

The proposed design will use the container as the mobile wind turbine base. It is then necessary to determine if this will prevent tipping of the wind turbine under the proposed wind conditions. In evaluating this, three moments, a, b and c have been denoted in Figure 14.1. Moment "a" is the restoring moment to prevent tipping along the long axis of the container which as illustrated, is aligned with the North direction. The sum of the moments about "a" result from thrust on the rotor, the weight of the wind turbine components whose total mass is given in Table 14.3, and and the weight of the container, all multiplied by their respective moment arm lengths. This is shown in Equation 14.9. Here, the container mass is treated as a point source at its centroid location.

$$\begin{aligned}\sum M_a &= -T \cdot H + W_{\text{carrier}} \cdot L_{\text{carrier}}/2 + W_1 \cdot L_{\text{carrier}} \\ &= -(3,663\,\text{N})(17.5\,\text{m}) + (2,292\,\text{kg})(9.8\,\text{N/kg})(3.05\text{m}) \\ &\quad + (8,439\,\text{kg})(9.8\,\text{N/kg})(6.1\,\text{m}) \\ &= 508,829\,\text{N-m}.\end{aligned} \tag{14.9}$$

Based on this, there is no possibility of the mobile wind turbine tipping due to a wind from the North.

Moment "c" is the restoring moment for a wind from the North-East (45° from North). With the long axis of the container aligned with the North

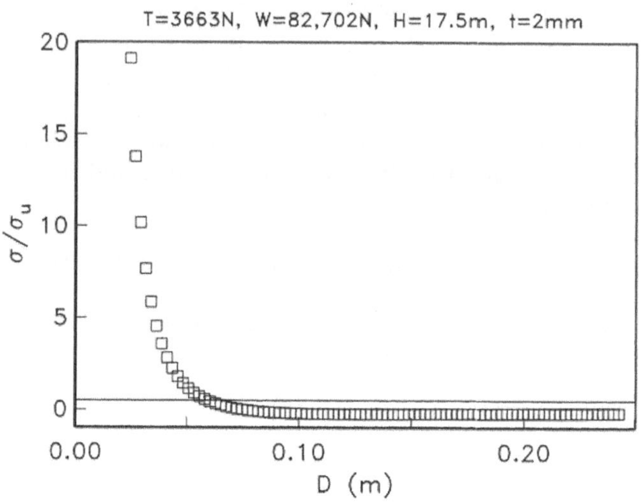

FIGURE 14.10
Maximum stress in tower structure normalized by the ultimate material stress. Horizontal line marks $\sigma_{\text{tot}}/\sigma_u = 0.5$.

direction, the component of the rotor thrust is $T_n \sin(45°)$. Following this, Moment "c" is given by Equation 14.11.

$$\sum M_c = -T\sin(45°) \cdot H + W_{\text{carrier}} \cdot W_{\text{carrier}}/2 + W_1 \cdot W_{\text{carrier}} \quad (14.10)$$

$$= -((0.707)(3,663)\,\text{N})(17.5\,\text{m}) + (2,292\,\text{kg})(9.8\,\text{N/kg})(1.2\,\text{m}) + (8,439\,\text{kg})(9.8\,\text{N/kg})(2.4\,\text{m})$$

$$= 180,095\,\text{N-m}. \quad (14.11)$$

Based on this, there is no possibility of the mobile wind turbine tipping due to a wind from the North-East.

Finally, Moment "b" is the restoring moment for a wind from the North-West. We note that with the placement of the wind turbine tower on the East edge of the container, it does not contribute to Moment "b". Moment "b" is therefore given by Equation 14.12.

$$\begin{aligned} \sum M_b &= -T\sin(45°) \cdot H + W_{\text{carrier}} \cdot W_{\text{carrier}}/2 + W_1 \cdot W_{\text{carrier}} \\ &= -((0.707)(3,663)\,\text{N})(17.5\,\text{m}) + (2,292\,\text{kg})(9.8\,\text{N/kg})(1.2\,\text{m}) \\ &= -18,390\,\text{N-m}. \end{aligned} \quad (14.12)$$

Based on the negative value of the restoring moment, "b", the container base alone cannot prevent the wind turbine from tipping with a wind from the North-West. The solution is to add an outrigger on the East side of the container. This is illustrated in Figure 14.11.

For the outrigger, it is assumed to be without weight and rigid, and have some width, W, and length, L. The moment restoring force is assumed to be concentrated at the geometric center of the outrigger at location $L/2$. The force exerted by the outrigger is assumed to be the weight density, WD (N/m^2), corresponding to the sum of all the component weights divided by the area of ground contact that includes the container and the outrigger. The weight density is therefore

$$\text{Weight density} = (W_{\text{carrier}} + W_1)/(L_{\text{carrier}} \cdot W_{\text{carrier}} + L_{\text{outrigger}} \cdot W_{\text{outrigger}}). \quad (14.13)$$

The weight density is a function of the contact area, $W \cdot L$, of the outrigger. The moment produced by the outrigger is

$$\begin{aligned} M_{\text{outrigger}} &= F_{\text{outrigger}} \cdot L_{\text{outrigger}}/2 \\ &= L_{\text{outrigger}} \cdot W_{\text{outrigger}} \cdot WD \cdot L_{\text{outrigger}}/2 \\ &= L^2_{\text{outrigger}} \cdot W_{\text{outrigger}} \cdot WD/2. \end{aligned} \quad (14.14)$$

As a design, $W = 0.25$ m was chosen. Then the only unknown is the length of the outrigger, which needs to be sufficient to offset the negative pitching moment "b".

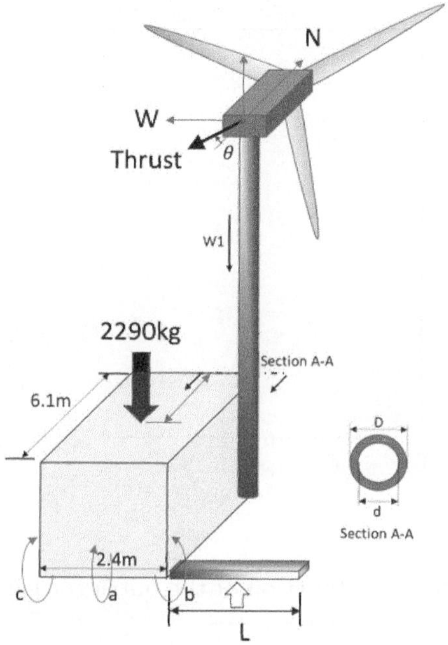

FIGURE 14.11
Deployable wind energy system concept including an outrigger of width W and length, L, on the storage container that is used to balance moment "b".

Figure 14.12 shows the value of Moment "b" as a function the length, L of an outrigger with a width of 0.25 m. Without the outrigger, Moment "b" is $-18,390$ N-m. The object is to have a positive Moment "b" which will prevent the wind turbine from tipping in a wind from the North-West. The horizontal dotted line corresponds to Moment "b" equal to zero. The outrigger needs to be of sufficient length for Moment "b" to be positive. Based on Figure 14.12, the minimum is approximately 4.71 m. As this is a fairly long outrigger, the equivalent moment could be obtained by *two* outriggers of half the length that are located on the east side of the container.

14.3 Wind Turbine Arrangement

14.3.1 Total Power

This part of the design is to calculate the total power generated by an arrangement of the mobile wind turbines that make up a wind farm. In doing this,

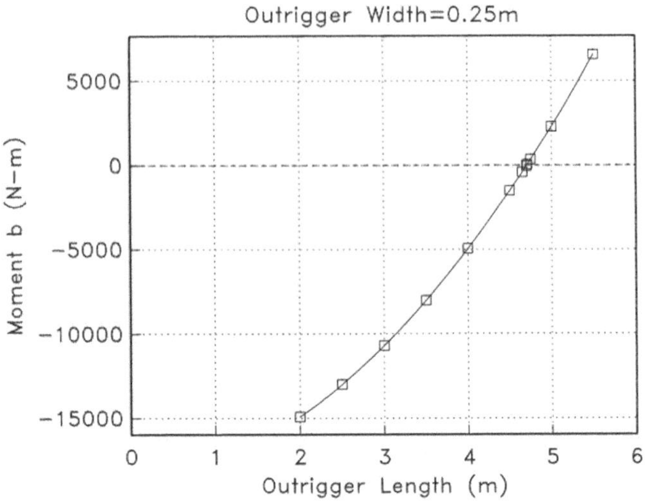

FIGURE 14.12
Restoring moment "b" as a function of the outrigger length, L.

it is necessary to take into account the effect of wakes of upwind turbines on the power generated by downwind turbines.

In the current design, the rated power of each of the wind turbines based on the BEM analysis, was 13.421 kW. Ideally, six wind turbines would produce 80.5 kW A hypothetical arrangement of the six wind turbines in a 50 m by 40 m (x by y) compound is shown in Figure 14.13. In determining power losses due to wake effects, it is assumed that

1. the surrounding roughness is open inland country with hedges and buildings having an equivalent roughness height of $z_0 = 0.01$ m,
2. that the long axis of the wind farm is aligned with the North direction,
3. the predominant wind direction is again from the North, North-West and North-East.

The approach to determining wind turbine wake effects follows the methods that were described in Chapter 8. All of the mobile wind turbines in the proposed arrangement are identical and therefore have the same properties. The properties that are relevant to the wake effects are listed in Table 14.4. We note that the thrust coefficient, C_T, was based on the normal thrust, T_n, that was obtained in the BEM analysis. The inflow induction factor, a, was then calculated from this thrust coefficient, namely

$$a = 0.5\left(1 - \sqrt{1 - C_T}\right). \qquad (14.15)$$

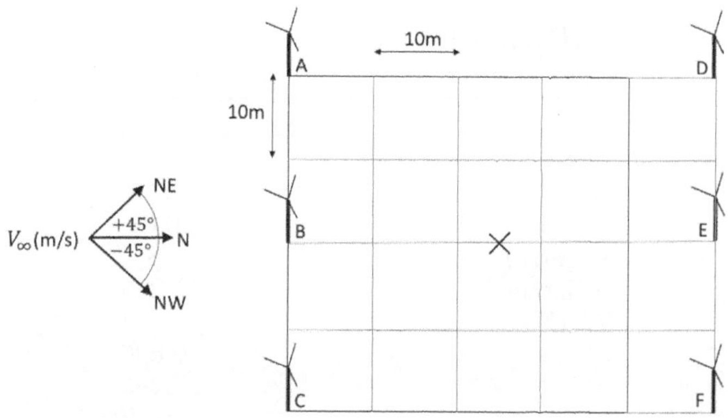

FIGURE 14.13
Hypothetical mobile wind farm for generating power at a disaster response site.

This approach provides an average inflow induction factor over the whole rotor. The value of $a = 0.292$ in Table 14.4 appears to be a reasonable average of the radial distribution on the rotor as shown in Figure 14.5. The effective radius of the wind turbines, r_r, is based on the inflow induction factor and as expected is larger than the physical rotor radius. The wake spreading boundaries is tangent to the effective rotor with radius, r_r. The wake entrainment constant, α, is a function of the hub height, z_{hub}, and the roughness height

$$\alpha = \frac{0.5}{\ln\left(\frac{z_{\text{hub}}}{z_0}\right)}. \tag{14.16}$$

This in turn determines the wake spreading and radius, r_1, at a downstream location, x, namely

$$r_1 = \alpha x + r_r. \tag{14.17}$$

From Chapter 8, the wake deficit velocity is

$$u_{\text{def}_{ij}} = \frac{2a}{1+\alpha\left(\frac{\vec{x}_{ij}}{r_r}\right)^2}, \tag{14.18}$$

and the velocity to the wake affected wind turbine is

$$u_j = V_\infty(1 - u_{\text{def}_{ij}}). \tag{14.19}$$

The power produced by the wake affected wind turbine is then

$$P_j = P_{\text{rated}}\left(\frac{u_j}{V_\infty}\right)^3. \tag{14.20}$$

TABLE 14.4
Wind Farm Power Calculation Input

Input Values	
r_0 (m)	6
z_{hub} (m)	17.5
z_0 (m)	0.01
V (m/s)	8
ρ (kg/m^3)	1.225
T_n (N)	3,663
C_T	0.826
a	0.292
α	0.0670
r_r (m)	7.823
P_{rated} (kW)	13.421

The coordinate locations of the six turbines are listed in Table 14.5. Figures 14.14 and 14.15 show the wind turbine locations and their rotor radii to the same scale. Figure 14.14 shows the wake pattern between Turbines A and D for the wind emanating from the North. Also drawn to scale is the effective radius, r_r of the upstream wind turbine (A), and the radius of the wake at the distance A-D, r_1, which in Table 14.6 is 11.170 m for the North wind direction.

Figure 14.14 illustrates that for the North wind direction, the wake of Turbine A only affects Turbine D. This is similarly the case for Turbines B and E, and Turbines C and F. In these cases for the North wind direction, Table 14.6 indicates that the power generated by Turbines D, E, and F will be 8.067 kW. Thus the total power produced by the six wind turbines under a North wind is (3)(13.421) kW + (3)(8.067) kW or 64.464 kW.

Figure 14.15 shows the wake pattern for the case of the wind out of the North-East. As a result of the symmetry of the turbine layout, this is also representative of the wind out of the North-West. Shown to scale is the wake development from Turbine C which is seen to only affect Turbine D. It is evident that there are no other wind turbines in the arrangement that will be in the wakes of any of the other upwind wind turbines.

The wake development distance between Turbines C and D follows the dashed line, which marks the center of the wake. The distance is $\vec{x}_{C-D} = \sqrt{(x_D - x_C)^2 + (y_D - y_C)^2}$, which in Table 14.6 is 64.031 m. That distance is projected normal to the dashed line at which point the wake radius, r_1, is drawn. The wake radius is observed to fully encompass the Turbine D rotor radius.

Table 14.6 indicates that the power generated by Turbine D with a wind from the North-East is 9.581 kW. All the other wind turbines are not affected and will therefore deliver the rated power. Thus the total power produced by the six wind turbines under a North-East wind is (5)(13.421) kW + (1)(9.581)

TABLE 14.5
Wind Farm Turbine Coordinates

Turbine	x (m)	y (m)
A	0	40
B	0	20
C	0	0
D	50	40
E	50	20
F	50	0

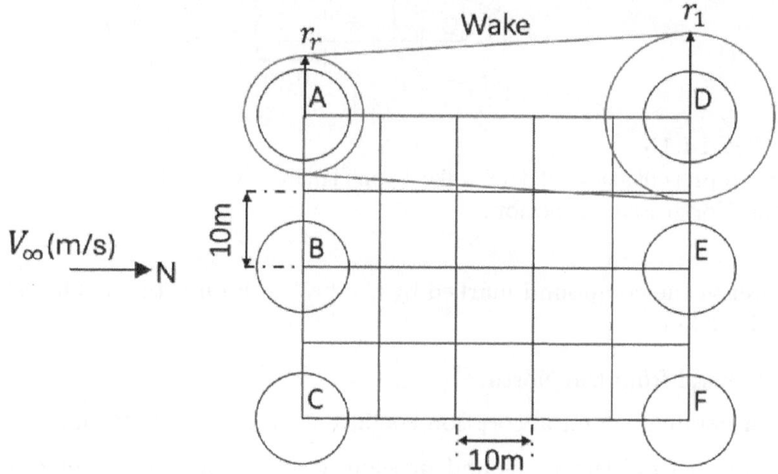

FIGURE 14.14
To-scale representation of the wake from Turbine A on Turbine D for wind from the North direction.

TABLE 14.6
Wind Farm Power

Wind	Wake Case	$\vec{x}$ (m)	u_{def} (m/s)	r_1 (m)	V_D (m/s)	P (kW)
N	A-D	50.000	0.156	11.170	6.751	8.067
N-E	C-D	64.031	0.106	12.109	7.150	9.581

kW or 76.681 kW. As a result of symmetry, the same total power level will be generated for the winds out of the North-West.

14.3.2 Sound Levels

Another point of interest with the mobile wind farm is the acoustic levels generated by the wind turbines. The objective is to estimate the sound levels in

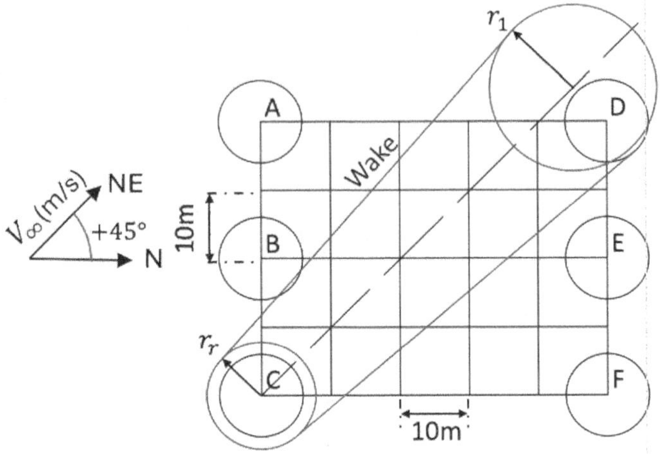

FIGURE 14.15
To-scale representation of the wake from Turbine C on Turbine D for wind from the North-East direction.

the center of the compound marked by the "X" in Figure 14.13. The estimate will be based on

1. a wind from the North,
2. a nominal sound absorption coefficient of $\alpha = 0.005$ dB/m,
3. and a wind turbine sound pressure level at the hub height that is based on data shown in Figure 9.10 of Chapter 9.

As was presented in Chapter 9, we will employ a simple model based on a conservative assumption of hemispherical sound propagation over a reflective surface that includes air absorption, namely

$$L_{W_p} = L_{W_s} - 10\log_{10}\left(2\pi R^2\right) - \alpha R, \qquad (14.21)$$

where L_{W_s} is the sound power level (dB) measured at the sound source, L_{W_p} is the sound power level (dB) measured at a radial distance R from a sound source. For multiple wind turbines, the total sound power level is the sum of the sound power levels due to each turbine. The model predicts a -6dB decrease in the sound power level for every doubling of the distance between the source and receptor. However, one caveat of the sound propagation model is that the sound power level decay rate with each doubling of the distance depends on if the receptor location is upwind or downwind of the sound source. In the latter case, data indicates that the decrease in the sound power level with distance is half that of the upwind case, or -3dB for every doubling of the distance from the source. Therefore, for the wind turbine arrangement shown in Figure 14.13 in a wind from the North, a receptor at the center of

TABLE 14.7
Wind Turbine Generated Sound Levels at the Compound Center

Turbine	x (m)	y (m)	z (m)	R (m)	L_{W_p} (dB)	P (N/m^2)
A	0	40	17.5	18.39	78.27	1,343.12
B	0	20	17.5	18.20	78.32	1,357.57
C	0	0	17.5	18.39	78.27	1,343.12
D	50	40	17.5	18.39	61.63	29.13
E	50	20	17.5	18.20	61.73	29.76
F	50	0	17.5	18.39	61.63	29.13
Center	25	20	0	-	-	-
Total P (N/m^2)		4,131.84				
Total L_{W_p} (dB)		83.15				

the compound ("X" location) is downwind of Turbines A,B and C and upwind of Turbines D, E, and F.

The sound power level of wind turbines is shown to scale with the rotor diameter. In the present design with the 12 m diameter rotor, based on Figure 9.10 in Chapter 9, the sound power level at the hub height is estimated to be 95 dB. As was listed in Table 14.4, for each of the wind turbines, $z_{\text{hub}} = 17.5$ m. The sound propagation distance is measured along a diagonal from the hub height to the ground. The respective distances, R, from each of the wind turbine hub heights to the center of the compound ("X" location) are listed in Table 14.7. The sound power levels produced by each of the wind turbines at the compound center, L_{W_p}, are subsequently listed. These are converted to individual power levels, P using

$$L_p = 10 \log_{10}(P/P_0), \tag{14.22}$$

where $P_0 = 2 \times 10^{-5}\,\text{N/m}^2$.

The individual power levels generated by the wind turbines are summed and then converted to the total sound power level. These are listed at the bottom of Table 14.7, with the total sound power level being 83.15dB. This sound level is about equivalent to street traffic.

14.4 Battery Storage

One of the requirements was that each mobile wind turbine design include lead acid batteries of a nominal voltage of 60 V, and a capacity of 1,200 A-h. The rated wind speed was selected to be one in which based on the prescribed Weibull coefficients and hub height, would result in a reasonable numbers of hours in a 24 hour period for the generated rated power to charge the batteries. As perviously presented, with the chosen rated wind speed of 8 m/s, the rated

TABLE 14.8
Component Cost Breakdown for a Land-Based Mobile Wind Turbine with a Rotor Diameter of 12 m and a Hub Height of 17.5 m

Component	Cost (2002 $)
Rotor System	
Blades	2,411.7
Hub	11,437.8
Pitch mechanism and bearings	354.5
Spinner, nose cone	0
Drive train, Nacelle	
Low-speed shaft	12.7
Bearings	23.8
Gearbox	0
Mech. brake, HS-coupling etc.	26.6
Generator	2,943.6
Variable spd. electronics	1,060.3
Yaw drive and bearing	107.1
Main frame	4,308.3
Internal platform	201.2
Electrical connections	536.8
Hydraulic, cooling system	161.1
Nacelle cover	4,004.5
Control, safety system, condition monitoring	35,000
Tower	0
Turbine Capital Cost (TCC)	**62,589**

TABLE 14.9
Dimensions of a Standard 40 ft Shipping Container

Exterior (m)			Interior (m)			Door (m)		Weight (kg)
12.2	2.4	2.6	12.0	2.3	2.4	2.3	2.3	3,629

power, 13.421 kW, is expected to be produced on a daily basis for 15.7 hours of the 24 hours period.

As discussed in Chapter 10, because of the "depth of discharge" requirement to ensure long operating life, the useable energy in a deep-cycle lead acid battery is nominally only 50% of the voltage-current product. In addition, there are losses in the "battery system" that includes ability of the battery to store a charge (≈68%) and the inefficiency in the power inverter (≈85%) that converts the AC voltage produced by the wind turbine generator to a DC voltage necessary to charge the battery. Therefore the useable energy in a single 60V, 1,200 A-h lead-acid battery is

$$E_{\text{usable}} = (60)(1,200)(0.5)(0.68)(0.85) = 20.8\,\text{kW-h}. \tag{14.23}$$

The requirement is that each wind turbine have an energy storage capacity of 100 kW-h. Therefore this will require each wind turbine to have five batteries.

For six wind turbines, the total storage capacity would be 600 kW-h. In the worst case of a wind from the North, accounting for wake losses, the power generated by the six wind turbines was 64.464 kW. Assuming all of the generated power were directed towards charging the batteries, they would be charged in 9.3 hours. This is well within the predicted 15.7 hours of rated wind speed. In another scenario, approximately 41% of the power generated from the wind could be used to power the compound, with the remaining wind power fully charging the batteries over the 15.7 hours of rated wind speed. The values are more favorable with the wind from the North-East and North-West in which the power from the six wind turbines is larger due to smaller wake loss.

14.5 Estimated Cost

The following is an estimate of the cost for the mobile wind turbine . This is based on the rated power of 13.42 kW, a rotor diameter of 12 m and a hub height of 17.5 m. The electric generator is a permanent magnet direct drive that does not require a gear box. The cost estimate in 2002 dollars is given in Table 14.8. Applying the CPI, the cost of the mobile wind turbine in 2024 dollars is $111,538.

14.6 Summary

The objective was to design a mobile horizontal axis wind turbine for defense and disaster response applications to provide electric power to a remote site where no electric infrastructure exists. To be transportable, the design needed to adhere to transport packaging space constraints. The project placed a premium on rapid transport and installation and quick teardown upon mission completion. Assembly and anchoring of the wind turbine was an important consideration. The system also needed to include energy storage to account for variations in the wind conditions. The specific design requirements included the following.

1. A portable wind energy system that can be transported in a standard 20 ft shipping container.

2. Electric power generation to support a portion of the loads of a company-sized element or disaster response team (50–100 personnel) with deployments from weeks to months.
3. A highly mobile setup and takedown in hours.
4. A lightweight and packable system that can be erected by a single person.
5. A wind turbine with a rated power between 10 and 20 kW.
6. Energy storage of 100 kW-h per wind turbine.

Overall, the design met all of these requirements.

Problems

1. The previous design was dictated to fit within a standard 20' shipping container. Shipping containers also come in 40' lengths. Details are presented in Table 14.9.

   The longer container would allow larger rotors and towers to be transported. Considering the same wind conditions (Weibull coefficients), perform the following design:

   (a) Determine the longest three-piece tower that will fit in the container that when erected determines the hub height.

   (b) Based on the hub height, determine the rated wind speed that gives the same probability of occurrence (hours in a 24 hours day) as in the chapter design.

   (c) Design the largest single-piece rotor that will fit within the container, determining the chord and twist distributions.

   (d) Perform the BEM analysis for a three-bladed design to determine the actual power. Utilize the same section shape that specifies $C_L(\alpha)$ and $C_D(\alpha)$ as in the chapter design.

   (e) Determine the number of these wind turbines it would take to produce the same total power as the six turbines in the chapter design (neglecting wakes), namely (6)(13.421) kW or 80.526 kW

   (f) Discuss the pros and cons to this design over the chapter design.

2. As a continuation of Problem 1, compare the total cost of the wind turbines designed in that problem, to that which produced 80.526 kW in the chapter design.

(a) Assuming that the larger rotor diameters that are possible to be carried in the larger container will allow fewer turbines to produce the same power as the six turbines in the chapter design, is that approach more cost effective?

(b) What parameters in the design are driving the cost?

(c) Is there a combination of the components in the Problem 1 design and the chapter design that might be incorporated to lower the cost?

3. Arrange in a manner that you think could minimize the wake losses, the necessary number of wind turbines designed in Problem 1 to produce the 80.526 kW in the chapter design. These must all fir. within the 50 by 40 m compound.

(a) In this, the wind turbines cannot be spaced any closer than 1.5 rotor diameters apart in the rotor plane direction.

(b) For your arrangement, calculate the wake loss and resulting total (turbine summed) power for wind directions from the North, North-East, and North-West.

(c) Compare this result to that for the chapter design.

(d) Discuss the pros and cons to this design over the chapter design.

Further Reading

1. B. Naughton, B. Houchens, B. Summerville, T. Jimenez, P. Preus, D. Reen, J. Gentle, and E. Lang, *Design Guidelines for Deployable Wind Turbines for Defense and Disaster Response Missions*, Sandia National Laboratories, Albuquerque, NM, 2021. https://docs.nrel.gov/docs/fy22osti/81889.pdf

2. B. Naughton, P. Preus, T. Jimenez, B. Whipple, and J. Gentle, "Market opportunities for deployable wind systems for defense and disaster response", Sandia Report 2020-1517, Sandia National Laboratories, Albuquerque, NM, 2020. https://www.energy.gov/sites/default/files/2020/02/f71/deployable-turbine-market-assessment.pdf

3. D. Lindner, "Air force develops new", Deployable Energy Systems. U.S. Department of Defense, April 6, 2017. https://www.defense.gov/Explore/News/Article/Article/1143863/air-force-develops-new-deployable-energy-systems/

4. EST-Floattech, "World?s first military 100% electric deployable power unit", 2020.

5. Solar Stik, "Solar Stik® , Inc. Ships 100th Hybrid Power System 7000 to the U.S. Army", 2018. https://www.solarstik.com/solar-stik-inc-ships-100th-hybrid-power-system-7000-to-the-u-s-army/
6. International Organization for Standardization, "ISO 668:2020 series 1 freight containers - Classification, dimensions and ratings", ISO, Geneva, Switzerland, 2018.
7. U.S. Department of the Army, *Soldier's Manual and Trainer?s Guide Tactical Power Generation Specialist MOS 91D*, U.S. Department of the Army, Arlington, VA, 2018.

Appendix

A.1 Size Specifications of Common Industrial Wind Turbines

The following are specifications for some of the most popular industrial wind turbines. Note that the blade length refers to the half of the rotor diameter (rotor radius). The rotor blade length is generally about a meter shorter. The hub heights can vary for the same model of wind turbine based on the site location. The values given are the nominal hub heights.

Model	Rated Power (MW)	Blade Length (m)	Hub Ht. (m)	Total Ht. (m)
GE 1.5s	1.5	35.25	64.7	99.95
GE 1.5sle	1.5	38.5	80	118.5
Vestas V82	1.65	41	70	111
Vestas V90	1.8	45	80	125
Vestas V100	2.75	50	80	130
Vestas V90	3.0	45	80	125
Vestas V112	3.0	56	84	136
Gamesa G87	2.0	43.5	78	121.5
Siemens	2.3	46.5	80	126.5
Siemens bonus	1.3	31	68	99
Siemens bonus	2.0	38	60	98
Siemens bonun	2.3	41.2	80	121.2
Suzlon 950	0.95	32	65	97
Suzlon S64	1.25	32	73	105
Suzlon S88	2.1	44	80	124
Repower MM92	2.0	46.25	100	146.25
Enercon E-126	7.6	63.5	135	198.5
Clipper liberty	2.5	44.5	80	124.5
Mitsubishi MWT95	2.4	47.5	80	127.5

Model	Rotor Area (m^2)	RPM Range	Max. Tip Speed (m/s)	Rated Wind Speed (m/s)
GE 1.5s	3,904	11.1–22.2	81.8	12
GE 1.5sle	4,657	?	?	14
Vestas V82	5,281	?–14.4	61.7	13
Vestas V90	6,362	8.8–14.9	70.1	11
Vestas V100	7,854	7.2–15.3	80.0	15
Vestas V90	6,362	9.0–19.0	89.4	15
Vestas V112	9,852	6.2–17.7	103.7	12
Gamesa G87	5,945	9.0–19.0	86.7	13.5
Siemens	6,793	6.0–16.0	75.5	13–14
Siemens bonus	3,019	13.0–19.0	61.7	14
Siemens bonus	4,536	11.0–17.0	67.5	15
Siemens bonun	5,333	11.0–17.0	73.3	15
Suzlon 950	3,217	13.9–20.8	69.7	11
Suzlon S64	3,217	13.9-20.8	69.7	12
Suzlon S88	6,082	?	?	14
Repower MM92	6,720	7.8–15.0	72.9	11.2
Enercon E-126	12,668	5.0–11.7	77.8	?
Clipper liberty	6,221	9.7–15.5	72.9	11.5
Mitsubishi MWT95	7,088	9.0–16.9	84.0	12.5

A.2 Design Trade Code 1: Performance and Structure

```
close all; clear all; clc;

PLOTTING=1; % Turn plotting on/off

% Effect of free-stream speed

% Chord lengths at rR locations (m)
chord=[0.411,0.455,0.384,0.311,0.259,0.223,0.186,0.167,0.137,0.107];
% Twist angles at rR locations (deg)
ThetaT=[45,25.6,15.7,10.4,7.4,4.5,2.7,1.4,0.4,0.00];
% Pitch angle
ThetaCP=-2.0;
% Modulus of Elasticity (GPa)
E=9e9;

% Input parameters

% no. of blades
B = 3;
% Wind speed (m/s)
V = 11.62;
% Air density (kg/m^3)
rho = 1.225;
% Rotor radius (m)
Rscale = 4.953;
% tip-speed-ratio
lambda = 7;

prompt='Input Rs (m)[9]:';
NR=input(prompt);
if isempty(NR)
  NR = 9;
end
prompt='Input R_min (m)[33]:';
Rmin=input(prompt);
if isempty(Rmin)
  Rmin=33;
end
prompt='Input R_max (m)[60]:';
Rmax=input(prompt);
if isempty(Rmax)
  Rmax=60;
end

Rinc=(Rmax-Rmin)/NR;

prompt='Input Vs (m/s)[9]:';
NV=input(prompt);
if isempty(NV)
  NV=9;
end
prompt='Input V_min (m/s)[5]:';
Vmin=input(prompt);
if isempty(Vmin)
  Vmin=5;
end
prompt='Input V_max (m/s)[14]:';
Vmax=input(prompt);
if isempty(Vmax)
  Vmax=14;
end

Vinc=(Vmax-Vmin)/NV;

fid=fopen('O.Project_1b_m','w');

% Outer loop on R

for IR=1:NR+1
   R=Rmin+(IR-1)*Rinc;
% For plotting
```

```
   Rs(IR)=R;

   rscale=R/Rscale;
% At sections (r/R)
   for i=1:10
       rr(i)=R*i/10;
       scalechord(i)=chord(i)*rscale; %scale chord based on R=4.953m radius rotor
       FIz(i)=0.35*scalechord(i)*((0.15*scalechord(i))^3)/12.; %scaled modulus of elasticity; I (m^4)
       FIy(i)=0.15*scalechord(i)*((0.35*scalechord(i))^3)/12.; %scaled modulus of elasticity; I (m^4)
   end

   dr=R/10;

% Inner loop on V

   for IV=1:NV+1

         V=Vmin+(IV-1)*Vinc;
         omega = lambda*V/R;
         % For plotting
         Vs(IV)=V;

% Calculate a and a'

         flag=0;
         i=1;
         while(i <= 10)
           if(flag==0)
              a(i)=0.;
              ap(i)=0.;
              ii=0;
           end
           phi(i) = atand((V*(1-a(i)))/(omega*rr(i)*(1+ap(i)))); % in degrees
           alpha(i) = phi(i) - (ThetaT(i) + ThetaCP);

           Cl(i) = 0.368 + 0.0942*alpha(i);
           Cd(i) = 9.94e-3 + 2.59e-4*alpha(i) + (1.055e-4)*alpha(i)^2;
           Cn(i) = Cl(i)*cosd(phi(i)) + Cd(i)*sind(phi(i));
           Ct(i) = Cl(i)*sind(phi(i)) - Cd(i)*cosd(phi(i));
           sig(i) = (B*scalechord(i))/(2*pi*rr(i));
           ff(i) = (B*(R-rr(i)))/(2.*rr(i)*sind(phi(i)));
           F(i) = (2./pi)*acos(exp(-ff(i)));
           anew = 1./(((4.*(sind(phi(i)))^2)/(sig(i)*Cn(i))) + 1);
           apnew = 1./(((4.*sind(phi(i))*cosd(phi(i)))/(sig(i)*Ct(i))) - 1);
           adiff = abs(anew - a(i));
           apdiff = abs(apnew - ap(i));
           %printf('%d %d %f %f %f %f %f %10.2f %10.2f\n',i,ii,Cl(i),Cd(i),Cn(i),Ct(i),sig(i),ff(i),F(i));
           %printf('%d %d %f %f %10.2f %10.2f\n', i,ii,a(i),ap(i),anew,apnew);
           if (adiff > 0.00001) || (apdiff > 0.00001)
             flag=1;
             a(i)=anew;
             ap(i)=apnew;
             ii=ii+1;
             if (ii > 100)
               printf('Exceeded max count in loop\n')
               break
             end
           else
             flag=0;
             i=i+1;
           end
         end

% Calculate segment tangential and normal forces, power and torque

       Ptot=0.;
       for i=1:10
         Vr(i) = sqrt((V*(1-a(i)))^2 + (omega*rr(i)*(1+ap(i)))^2);
         Ft(i) = B*0.5*rho*Vr(i)^2*Ct(i)*scalechord(i)*dr;
         Fn(i) = B*0.5*rho*Vr(i)^2*Cn(i)*scalechord(i)*dr;
         T(i) = 2*F(i)*rho*V^2*a(i)*(1-a(i))*2*pi*rr(i)*dr;
```

```
        Q(i) = 2*F(i)*ap(i)*(1-a(i))*rho*V*omega*rr(i)^2*2*pi*rr(i)*dr;
        P(i) = omega*Q(i);
        Ptot = Ptot + P(i);
      end

      fprintf(fid,'Input: B,V,R,lambda: %10.5f      %10.5f      %10.5f          %d\n',B,V,R,lambda);
      fprintf(fid,'Output: rr,a,ap,phi,alpha,Ft,Fn,P\n');
      for i=1:10
        fprintf(fid,'%13.8f  %8.5e   %8.5e   %13.8f   %13.8f   %13.8f  %13.8f
        %13.8f\n',rr(i),a(i),ap(i),phi(i),alpha(i),Ft(i),Fn(i),P(i));
      end
      fprintf(fid,'total power/blade=%f\n',Ptot);
      Prated=B*Ptot;
      fprintf(fid,'wind turbine total power: Prated(W)=%f\n',Prated);
      % For plotting
      Power(IR,IV)=Prated;

    %fclose(fid);
% Blade load and deflection calculations section

% Convert element force on a single blade to pressure
      pz(1)=(Fn(1)/B)/rr(1);
      py(1)=(Ft(1)/B)/rr(1);
      for i=2:10
        pz(i)=(Fn(i)/B)/(rr(i)-rr(i-1));
        py(i)=(Ft(i)/B)/(rr(i)-rr(i-1));
      end

% Calculate the shear forces at each blade element starting at the tip

      Tz(10)=0.;
      Ty(10)=0.;
      for i=10:-1:2
          Tz(i-1)=Tz(i) + 0.5*(pz(i-1)+pz(i))*(rr(i)-rr(i-1));
          Ty(i-1)=Ty(i) + 0.5*(py(i-1)+py(i))*(rr(i)-rr(i-1));
      end

% Calculate the moments at each blade element starting at the tip

      Mz(10)=0.;
      My(10)=0.;
      for i=10:-1:2
          Mz(i-1)=Mz(i)-Tz(i)*(rr(i)-rr(i-1))-((1/6)*pz(i-1)+(1/3)*pz(i))*(rr(i)-rr(i-1))^2;
          My(i-1)=My(i)-Ty(i)*(rr(i)-rr(i-1))-((1/6)*py(i-1)+(1/3)*py(i))*(rr(i)-rr(i-1))^2;
      end

% Calculate ky and kz where we assume FIy=FIz=FI,
% and the first principle lies along the chord line, that is nu=0

      for i=1:10
          M1=My(i)*cosd(ThetaT(i))-Mz(i)*sind(ThetaT(i));
          M2=My(i)*sind(ThetaT(i))-Mz(i)*cosd(ThetaT(i));
          k1=M1/(E*FIz(i));
          k2=M2/(E*FIy(i));
          kz(i)=-k1*sind(ThetaT(i)) + k2*cosd(ThetaT(i));
          ky(i)=k1*cosd(ThetaT(i)) + k2*sind(ThetaT(i));
      end

% Calculate Thetaz and Thetay. Thetaz(1)=Thetay(1)=0

      Thetaz(1)=0.;
      Thetay(1)=0.;
      for i=1:9
        Thetaz(i+1)=Thetaz(i)+0.5*(kz(i+1)+kz(i))*(rr(i+1)-rr(i));
        Thetay(i+1)=Thetay(i)+0.5*(ky(i+1)+ky(i))*(rr(i+1)-rr(i));
      end

% Find displacements, uz and uy. uz(1)=uy(1)=0

      uz(1)=0.;
      uy(1)=0.;
```

```
      for i=1:9
        uz(i+1)=uz(i)+Thetay(i)*(rr(i+1)-rr(i))+((1./6.)*ky(i+1)+(1./3.)*ky(i))*(rr(i+1)-rr(i))^2;
        uy(i+1)=uy(i)+Thetaz(i)*(rr(i+1)-rr(i))+((1./6.)*kz(i+1)+(1./3.)*kz(i))*(rr(i+1)-rr(i))^2;
      end

      fprintf(fid,'i,Thetaz,Thetay,uz,uy=\n');
      for i=1:10
         fprintf(fid,'%f %f %f %f %f\n',rr(i),Thetaz(i),Thetay(i),uz(i),uy(i));
      end

      fprintf(fid,'\n');
      fprintf(fid,'\n');

      % For Plotting
      zdef(:,IR,IV)=uz;
      ydef(:,IR,IV)=uy;

   end
% End inner V loop

end
% End outer R loop

fclose(fid);

A=[Rs;Vs];
%printf('The Rs are %f5.2',A(1))
disp('   R (m) V (m/s)'); % Print to screen R's and V's
disp(A')

% Plotting
if (PLOTTING==1)

colorstring='brgk'; % Choose colors for each plot symbol
R_index=[1,4,7,10]; % Select which R to plot
figure(1)
for i=1:length(R_index)
  plot(Vs,Power(R_index(i),:).*1e-6,'Color',colorstring(i),'marker','s','linestyle','none'); hold on;
end
hold off
xlabel('V (m/s)')
ylabel('Power (MW)')
xlim([4.8 14.2])
ylim([-.4 24])
title('\beta=3, \lambda=7')
legend({'R=33m','R=42m','R=51m','R=60m'},'location','northwest')
legend boxoff

r_R=1/(NR+1):1/(NR+1):1;

R_index=[7]; % Select which radius to plot
V_index=[2,6,10]; % Select which velocities to plot
figure(2)
for i=1:length(V_index)
  plot(r_R,zdef(:,R_index,V_index(i)),'Color',colorstring(i),'marker','s','linestyle','none'); hold on;
end
hold off
xlabel('r/R')
ylabel('z-deflection (m)')
xlim([0.08 1.02])
ylim([-.04 3.04])
title('R=51m, \beta=3, \lambda=7')
legend({'V=6m/s','V=10m/s','V=14m/s'},'location','northwest')
legend boxoff

figure(3)
for i=1:length(V_index)
  plot(r_R,ydef(:,R_index,V_index(i)),'Color',colorstring(i),'marker','s','linestyle','none'); hold on;
end
hold off;
xlabel('r/R')
```

```
ylabel('y-deflection (m)')
xlim([0.08 1.02])
ylim([-.25 5.])
title('R=51m, \beta=3, \lambda=7')
legend({'V=6m/s','V=10m/s','V=14m/s'},'location','northwest')
legend boxoff

R_index=[7];
figure(4)
plot(Vs,zdef(end,R_index,:),'bs'); hold on;
plot(Vs,ydef(end,R_index,:),'rs'); hold off;
xlabel('V (m/s)')
ylabel('maximum deflection (m)')
xlim([4.8 14.2])
title('R=51m, \beta=3, \lambda=7')
legend({'u\_z','u\_y'},'location','northwest')
legend boxoff

V_index=[10];
figure(5)
plot(Rs,zdef(end,:,V_index),'sb'); hold on;
plot(Rs,ydef(end,:,V_index),'sr'); hold off;
xlabel('R (m)')
ylabel('maximum deflection (m)')
%xlim([4.8 14.2])
%ylim([-.4 24])
title('V=14m/s, \beta=3, \lambda=7')
legend({'u\_z','u\_y'},'location','northwest')
legend boxoff
end
```

A.3 Design Trade Code 2: Economics

```
close all; clear all; clc;

PLOTTING=1; % Turn plotting on/off

% Economics

% Chord lengths at rR locations (m)
chord=[0.411,0.455,0.384,0.311,0.259,0.223,0.186,0.167,0.137,0.107];
% Twist angles at rR locations (deg)
ThetaT=[45,25.6,15.7,10.4,7.4,4.5,2.7,1.4,0.4,0.00];
% Pitch angle
ThetaCP=-2.0;
% Hub height (m)
H=65;
% Weibull coeffs (m/s,-,m)
cref=5.42;kref=2.12;zref=10;
% Modulus of Elasticity (GPa)
E=9e9;
% I(m^4)
for i=1:10
   FI(i)=0.35*chord(i)*((0.15*chord(i))^3)/12.;
   %fprintf('%d %f\n',i,FI(i))
end

% Input parameters

% no. of blades
B = 3;
% Wind speed (m/s)
V = 11.62;
% Air density (kg/m^3)
rho = 1.225;
% tip-speed-ratio
lambda = 7;

prompt='Input Rs (m)[9]:';
NR=input(prompt);
if isempty(NR)
  NR = 9;
end
prompt='Input R min (m)[33]:';
Rmin=input(prompt);
if isempty(Rmin)
  Rmin=33;
end
prompt='Input R max (m)[60]:';
Rmax=input(prompt);
if isempty(Rmax)
  Rmax=60;
end

Rinc=(Rmax-Rmin)/NR;

prompt='Input V cutin (m/s)[4]:';
Vcutin=input(prompt);
if isempty(Vcutin)
  Vcutin=4;
end
prompt='Input V rated (m/s)[11]:';
Vrated=input(prompt);
if isempty(Vrated)
  Vrated=11;
end
prompt='Input V cutout (m/s)[15]:';
Vcutout=input(prompt);
if isempty(Vcutout)
  Vcutout=15;
end
%Vcutin=5;
%Vrated=11;
%Vcutout=15;
```

```
fid=fopen('0.Project_1c_m','w');

% Outer loop on R

for IR=1:NR+1
   R=Rmin+(IR-1)*Rinc;
% For plotting
   Rs(IR)=R;

% At sections (r/R)
   for i=1:10
       rr(i)=R*i/10;
       %fprintf('%d %f\n',i,rr(i))
   end
   dr=R/10;

% Find power at Vcutin = Prated

   V=Vrated;
   omega=lambda*V/R;

% Calculate a and a'

   flag=0;
   i=1;
   while(i <= 10)
      if(flag==0)
         a(i)=0;
         ap(i)=0;
         ii=0;
      end
      phi(i) = atand((V*(1-a(i)))/(omega*rr(i)*(1+ap(i)))); % in degrees
      alpha(i) = phi(i) - (ThetaT(i) + ThetaCP);

      Cl(i) = 0.368 + 0.0942*alpha(i);
      Cd(i) = 9.94e-3 + 2.59e-4*alpha(i) + (1.055e-4)*alpha(i)^2;
      Cn(i) = Cl(i)*cosd(phi(i)) + Cd(i)*sind(phi(i));
      Ct(i) = Cl(i)*sind(phi(i)) - Cd(i)*cosd(phi(i));
      sig(i) = (B*chord(i))/(2*pi*rr(i));
      ff(i) = (B*(R-rr(i)))/(2.*rr(i)*sind(phi(i)));
      F(i) = (2./pi)*acos(exp(-ff(i)));
      anew = 1./(((4.*(sind(phi(i)))^2)/(sig(i)*Cn(i))) + 1);
      apnew = 1./(((4.*sind(phi(i))*cosd(phi(i)))/(sig(i)*Ct(i))) - 1);
      adiff = abs(anew - a(i));
      apdiff = abs(apnew - ap(i));
      %fprintf('%d %d %f %f %f %f %f %10.2f %10.2f\n',i,ii,Cl(i),Cd(i),Cn(i),Ct(i),sig(i),ff(i),F(i));
      %fprintf('%d %d %f %f %10.2f %10.2f\n', i,ii,a(i),ap(i),anew,apnew);
      if (adiff > 0.00001) || (apdiff > 0.00001)
         flag=1;
         a(i)=anew;
         ap(i)=apnew;
         ii=ii+1;
         if (ii > 100)
           printf('Exceeded max count in loop\n')
           break
         end
      else
         flag=0;
         i=i+1;
      end
   end

% Calculate segment tangential and normal forces, power and torque

   Ptot=0.;
   for i=1:10
      Vr(i) = sqrt((V*(1-a(i)))^2 + (omega*rr(i)*(1+ap(i)))^2);
      Ft(i) = B*0.5*rho*Vr(i)^2*Ct(i)*chord(i)*dr;
      Fn(i) = B*0.5*rho*Vr(i)^2*Cn(i)*chord(i)*dr;
      T(i) = 2*F(i)*rho*V^2*a(i)*(1-a(i))*2*pi*rr(i)*dr;
```

```
      Q(i) = 2*F(i)*ap(i)*(1-a(i))*rho*V*omega*rr(i)^2*2*pi*rr(i)*dr;
      P(i) = omega*Q(i);
      Ptot = Ptot + P(i);
   end

   fprintf(fid,'Input: B,V,R,lambda: %10.5f      %10.5f      %10.5f          %d\n',B,V,R,lambda);
   fprintf(fid,'Output: rr,a,ap,phi,alpha,Ft,Fn,P\n');
   for i=1:10
       fprintf(fid,'%13.8f  %8.5e   %8.5e   %13.8f   %13.8f   %13.8f  %13.8f
       %13.8f\n',rr(i),a(i),ap(i),phi(i),alpha(i),Ft(i),Fn(i),P(i));
   end
   fprintf(fid,'total power/blade=%10.2f\n',Ptot);
   Prated=B*Ptot;
   fprintf(fid,'wind turbine total power: Prated(W)=%10.2f\n',Prated);
   Prated=Prated/1000.; % kW

   %fclose(fid);

   fprintf(fid,'\n');
   fprintf(fid,'\n');

% Economics Section

% Turbine Capital Cost (TCC) and Turbine Total Mass (TTM)

   TCC=0;
   TTM=0;
   H=R*2.25; % account for rotor radius
% For plotting
   Hub(IR)=H;

% ROTOR COSTS

% Rotor mass
   Rmass=0.1452*(R^(2.9156)); % mass per blade

   % Rotor blade cost
   Rotorcost=3.1225*(R^(2.897)); % 2002 dollars
   if(R > 13.1)
      RMcost=(0.4019*(R^3))-955.24; % 2002 dollars
   else
      RMcost=0.;
   end
   RLcost=2.7445*(R^2.5025); % 2002 dollars
   RBcosttot=(Rotorcost+RMcost+RLcost)*B; % 2002 dollars
   fprintf(fid,'Rotor Blade mass,cost:%10.2f %10.2f\n',(Rmass*B),RBcosttot);
   TCC=TCC+RBcosttot;
   TTM=TTM+Rmass*B;

% Rotor hub mass and cost
   Hmass=0.954*Rmass*B+5680.3; % hub mass
   Hcost=Hmass+5680.3; % 2002 dollars
   fprintf(fid,'Rotor Hub mass,cost:%10.2f %10.2f\n',Hmass,Hcost);
   TCC=TCC+Hcost;
   TTM=TTM+Hmass;

% Pitch mechanism and bearings mass and cost

   PBmass=0.1295*Rmass*B +491.31; % total pitch bearing mass
   TotalPBmass = 1.328*PBmass +555;
   TotPS=0.4801*((2*R)^2.6578); % total pitch system in 2002 dollars
   fprintf(fid,'Pitch Mechanism mass,cost:%10.2f %10.2f\n',PBmass,TotPS);
   TCC=TCC+TotPS;
   TTM=TTM+TotalPBmass;

% Spinner nose cone mass and cost
   if((2*R) > 38.2)
      Conemass=(18.5*2*R) - 520.5; % spinner mass
   else
      Conemass=0.;
   end
```

```
   Conecost=5.57*Conemass; % 2002 dolars
   fprintf(fid,'Spinner Nose mass,cost:%10.2f %10.2f\n',Conemass,Conecost);
   TCC=TCC+Conecost;
   TTM=TTM+Conemass;

% Low speed shaft mass and cost
   FLSSmass=0.0142*((2*R)^2.888);
   FLSScost=0.010*((2*R)^2.877); % 2002 dollars
   fprintf(fid,'Low Speed Shaft mass,cost:%10.2f %10.2f\n',FLSSmass,FLSScost);
   TCC=TCC+FLSScost;
   TTM=TTM+FLSSmass;

% Main bearing mass and cost
   Bmass=(0.000123*2*R - 0.000123)*((2*R)^2.5);
   Bcost=35.2*Bmass; % 2002 dollars
   fprintf(fid,'Main Bearing mass,cost:%10.2f %10.2f\n',Bmass,Bcost);
   TCC=TCC+Bcost;
   TTM=TTM+Bmass;

% Gear box mass and cost (medium-speed, single stage)
   Torque=Prated/omega; % kN-m
   GBmass=88.29*(Torque^0.774);
   GBcost=74.10*Prated; % 2002 dollars
   fprintf(fid,'Gear Box mass,cost:%10.2f  %10.2f\n',GBmass,GBcost);
   TCC=TCC+GBcost;
   TTM=TTM+GBmass;

% Mechanical break/high-speed coupling cost and mass
   BCcost=(1.9894*Prated)-0.1141; % 2002 dollars
   BCmass=0.1*BCcost;
   fprintf(fid,'Mech. Brake/Coupling mass,cost:%10.2f  %10.2f\n',BCmass,BCcost);
   TCC=TCC+BCcost;
   TTM=TTM+BCmass;

% Electric generator mass and cost (medium-speed permanent magnet with single stage drive)
   EGmass=10.51*(Prated)^0.9223;
   EGcost=54.73*Prated; % 2002 dollars
   fprintf(fid,'Electric Generator mass,cost:%10.2f %10.2f\n',EGmass,EGcost);
   TCC=TCC+EGcost;
   TTM=TTM+EGmass;

% Variable speed electronics cost
   VSEcost=79.0*Prated; % 2002 dollars
   fprintf(fid,'Variable Speed Electronics cost:%10.2f\n',VSEcost);
   TCC=TCC+VSEcost;

% Yaw drive and bearing mass and cost
   YBDmass=0.00144*((2*R)^3.314);
   YBDcost=0.0678*((2*R)^2.964); % 2002 dollars
   fprintf(fid,'Yaw Drive and Bearing mass,cost:%10.2f %10.2f\n',YBDmass,YBDcost);
   TCC=TCC+YBDcost;
   TTM=TTM+YBDmass;

% Mainframe mass and cost (mediium-speed permanant magnet generator with single stage drive)
   FMFmass=1.295*((2*R)^1.953);
   FMFcost=303.96*((2*R)^1.067); % 2002 dollars
   fprintf(fid,'Mainframe mass,cost:%10.2f  %10.2f\n',FMFmass,FMFcost);
   TCC=TCC+FMFcost;
   TTM=TTM+FMFmass;

% Internal platform support structure mass and cost
   PFormmass=0.125*FMFmass;
   PFormcost=9.7*PFormmass; % 2002 dollars
   fprintf(fid,'Internal Platform mass,cost:%10.2f %10.2f\n',PFormmass,PFormcost);
   TCC=TCC+PFormcost;
   TTM=TTM+PFormmass;

% Electrical connections cost
   ECcost=40.0*Prated; % 2002 dollars
   fprintf(fid,'Elecctrical Connection cost:%10.2f\n',ECcost);
   TCC=TCC+ECcost;
```

```
% Hydraulic and cooling systems mass and cost
   HCSmass=0.08*Prated;
   HCScost=12.0*Prated; % 2002 dollars
   fprintf(fid,'Hydraulic and Cooling Systems mass,cost:%10.2f  %10.2f\n',HCSmass,HCScost);
   TCC=TCC+HCScost;
   TTM=TTM+HCSmass;

% Nacelle cover cost and mass
   FNCcost=11.537*Prated+3849.7; % 2002 dollars
   %FNCmass=0.1*FNCcost02;
   FNCmass=0.1*FNCcost; % ematlis
   fprintf(fid,'Nacelle Cover mass,cost:%10.2f %10.2f\n',FNCmass,FNCcost);
   TCC=TCC+FNCcost;
   TTM=TTM+FNCmass;

% Control, safety system, condition onitoring cost
   CSSCMcost=35000.0; % 2002 dollars
   fprintf(fid,'Control and Safety System cost:%10.2f\n',CSSCMcost);
   TCC=TCC+CSSCMcost;

% Tower mass and cost
   AreaR=pi*R^2;
   if(AreaR > 3560.0)
        Tmass=(0.3973*AreaR*H)-1414;
   else
        Tmass=0.;
   end
   Tcost=1.5*Tmass; % 2002 dollars
   fprintf(fid,'Tower mass,cost:%10.2f  %10.2f\n',Tmass,Tcost);
   TCC=TCC+Tcost;
   TTM=TTM+Tmass;

% Turbine Capital Cost in 2002 dollars

   fprintf(fid,'\n');
   fprintf(fid,'Turbine Total Mass, Turbine Capital Cost:%10.2f %10.2f\n',TTM,TCC);
   fprintf(fid,'\n');
% For Plotting
   TCCplot(IR)=TCC;

% BALANCE OF STATION

   BOS=0;

% Transportation Cost
   Tcostfactor=(1.581e-5)*(Prated)^2 - 0.0375*Prated + 54.7;
   Transpocost=Tcostfactor*Prated; % 2002 dollars
   fprintf(fid,'Transportation cost:%10.2f\n',Transpocost);
   BOS=BOS+Transpocost;

% Roads and civil work cost
   RCWcostfactor = (2.17e-6)*(Prated)^2 - 0.0145*Prated + 69.54;
   RCWcost=RCWcostfactor*Prated; % 2002 dollars
   fprintf(fid,'Roads and Civil Work cost:%10.2f\n',RCWcost);
   BOS=BOS+RCWcost;

% Assembly and installation
   AIcost= 1.965*(H*2*R)^1.1736; % 2002 dolars
   fprintf(fid,'Assemblty and Installation cost:%10.2f\n',AIcost);
   BOS=BOS+AIcost;

% Electrical interface/connections cost
   EICcostfactor=(3.49e-6)*(Prated)^2 - 0.0221*Prated;
   if(EICcostfactor < 0)
      EICcostfactor=0.;
   end
   EICcost=EICcostfactor*Prated; % 2002 dollars
   fprintf(fid,'Electrical Interface/Connections cost:%10.2f\n',EICcost);
   BOS=BOS+EICcost;
```

```
% Engineering and permit cost
   EPcostfactor=(9.94e-4)*Prated + 20.31;
   EPcost=EPcostfactor*Prated; % 2002 dollars
   fprintf(fid,'Engineering and Permit cost:%10.2f\n',EPcost);
   BOS=BOS+EPcost;

% Foundation cost
   Fcost=303.24*(H*AreaR)^0.4037; % 2002 dollars
   fprintf(fid,'Foundation cost:%10.2f\n',Fcost);
   BOS=BOS+Fcost;

% Balance of Station Cost in 2002 dollars
   fprintf(fid,'\n');
   fprintf(fid,'Balance of Station (BOS) cost:%10.2f\n',BOS);
   fprintf(fid,'\n');
%  For plotting
   BOSplot(IR)=BOS;

% Levelized replacement cost
   LRcostfactor=10.7;
   LRcost=LRcostfactor*Prated; % 2002 dollars
   fprintf(fid,'Levelized Replacement cost:%10.2f\n',LRcost);
   fprintf(fid,'\n');

% Wind turbine capacity factor

   FK=kref*((1.0-0.088*log(zref/10.))/(1.0-0.088*log(H/10.))); % weibul k
   FC=cref*((H/zref)^0.23); % weibul c
   CapFac=exp(-(Vcutin/FC)^FK) - exp(-(Vcutout/FC)^FK);
   fprintf(fid,'Capacity Factor (%%)=%f\n',CapFac*100);
   fprintf(fid,'\n');
%  For plotting
   CapFactor(IR)=CapFac;

% Net annual production
   AEP=CapFac*24.*365.*Prated; % MW-h
   fprintf(fid,'Net Annual Production AEP(MW-h):%10.2f\n',AEP/1000.);
   fprintf(fid,'\n');
%  For plotting
   AEPplot(IR)=AEP;

% Operations and maintenence cost
   OMcost=0.007*AEP; % 2002 dollars
   fprintf(fid,'Operation and Maintenance cost:%10.2f\n',OMcost);
   fprintf(fid,'\n');

% Land lease cost
   LLcost=0.00108*AEP; % 2002 dollars
   fprintf(fid,'Land Lease cost:%10.2f\n',LLcost);
   fprintf(fid,'\n');

% Annual operating expenses (AOE)
   AOE=LLcost/AEP + (OMcost + LRcost)/AEP; % 2002 $/(kW-h)
   fprintf(fid,'Annual Operating Expenses AOE $/kw-h:%10.6f\n',AOE);
   fprintf(fid,'\n');

% Cost of electricity
   FRC=0.1185; % fixed rate charge (11.85%)
   COE=(FRC*(TCC+BOS))/AEP + AOE; % 2002 $/(kW-h)
   fprintf(fid,'Cost of Electricity COE $/kw-h:%10.6f\n',COE);
   fprintf(fid,'\n');
   fprintf(fid,'\n');
%  For plotting
   COEplot(IR)=COE;

end
% End outer R loop

fclose(fid);

% Plotting
```

```
if (PLOTTING==1)

figure(1)
plot(Rs,TCCplot./1000,'sk');
xlabel('R (m)')
ylabel('Turbine Cost (1000 dollars)')
title('\beta=3, \lambda=7, V(rated)=11m/s')
legend('Turbine','location','northwest')
legend boxoff

figure(2)
plot(Rs,BOSplot./1000,'sk');
xlabel('R (m)')
ylabel('BOS Cost (1000 dollars)')
title('\beta=3, \lambda=7, V(rated)=11m/s')
legend('BOS','location','northwest')
legend boxoff

figure(3)
plot(Hub,CapFactor.*100,'sk');
xlabel('Hub Height (m)')
ylabel('Capacity Factor (percent)')
ylim([50 100])
title('\beta=3, \lambda=7, V(rated)=11m/s')
legend('V_{cut-in}=4m/s','location','northwest')
legend boxoff

figure(4)
plot(Rs,AEPplot./1000,'sk');
xlabel('R (m)')
ylabel('AEP (MW-h)')
title('\beta=3, \lambda=7, V(rated)=11m/s')
legend('V_{cut-in}=4m/s','location','northwest')
legend boxoff

figure(5)
plot(Rs,COEplot.*100,'sk');
xlabel('R (m)')
ylabel('COE ($cents/kW-h)')
title('\beta=3, \lambda=7, V(rated)=11m/s')
legend('V_{cut-in}=4m/s','location','northwest')
legend boxoff
end
```

Index

“Sky Serpent concept, 289
“Wind Donut concept, 285
“Wind Lens shrouded rotor concept, 284
“Wind Stalk bladeless concept, 288
“energy kite concept, 289
1100 A.D. Europe, 4
1200 A.D. China, 1
1700 B.C. Persian, 1

A-scale weighting, 199
acoustic noise, 194
actuator disk, 74
aerodynamic loads, 143
aerodynamic sound mechanisms, 202
aerodynamic torque control, 121
aesthetic shaped wind turbines, 290
airborne wind turbines, 289
airfoil geometry, 53, 61
anemometers, 44
angle of attack, 51
angular deformations, 151
angular induction factor, 83
annual energy production, 248
annual operating expenses, 250
assembly and installation cost, 259
Atmospheric boundary layer, 23
average wind speeed, 26
axial induction control, 129
axial induction factor, 81

background sound, 210
balance of station cost, 248
batteries, 217
bending moment of inertia, 270
bending moments, 150
Betz limit, 81, 280
Betz optimum chord, 115
blade deflection, 271
blade element momentum theory, 87, 95
blade element theory, 107
blade segments, 89
blade twist estimate, 109
bladeless wind energy concepts, 287
Blyth Turbine, 5
boundary layer thickness, 23
bouyant hydroelectric storage, 241
box beam main spar, 270
Brush Turbine, 5
Buckingham Pi Theorem, 54

CAES power plant, 232
CAES systems, 229
camber, 53
certification criteria, 157
chemical storage, 217
compressed air storage, 227
conservation of angular momentum, 84
control loads, 148
control volume, 77
control, safety condition monitoring systems cost, 257
Coriolis force, 22
cost function, 235
cost of electricity, 275
cup anemometer, 44

Darrieus Turbine, 10
decibels, 196
design power, 271
design wind speed, 271

differential force, 90
differential power, 90
differential torque, 84, 90
drag, 51
dynamic control, 120
dynamic loads, 146

economic impact, 273
economic summary, 275
electric demand, 216
electric generator cost, 255
electrical connections cost, 257
electrical interface cost, 259
electrical potential storage, 217
electrical torque control, 122
electrochemical storage, 216
energy estimate, 37
engineering and permit cost, 260
evolutionary algorithm, 179
evolutionary genetic algorithm, 178
extreme wind loads, 157

fixed rate charge, 249
flow separation, 59
flywheel storage, 229
freestream velocity, 51

gearbox cost, 254
Genetic Algorithm, 263
geostrophic wind, 22
Glauert analysis, 83
gravitational loads, 144
gravitational storage, 217
Gurney flap, 137
gyroscopic loads, 147

historic wind turbine cost of electricity, 265
hydraulic and cooling systems cost, 257
hydroelectric storage, 226
hydrogen storage, 224

incremental force, 90
incremental power coefficient, 85
infrasound, 199
initial blade sizing, 107
initial capital cost, 249
initial capitalization cost, 248
isobars, 22

Juul Turbine, 6

kinetic energy storage, 217

la Cour Turbine, 6
land lease cost, 250, 260
latent heat storage, 217
leading-edge roughness effects, 65
levelized cost of electricity, 248
levelized replacement cost, 248, 260
lift, 51
lift coefficient, 59
lift control, 135
lift-to-drag ratio, 64
low-speed shaft cost, 254

main bearing cost, 254
mainframe cost, 256
maximum energy capture, 178
maximum power, 79
maximum power coeffcieint, 108
maximum thrust, 79
mechanical break cost, 255
mechanical power, 108
micro wind turbines, 291
modern airfoil section shapes, 68
modulus of elasticity, 270
momentum analysis, 83

NACA airfoil family, 61
nacelle cover cost, 257
net annual production, 275
noise standards, 211
normal force coefficient, 90

operation and maintenance cost, 248, 260
optimization formulation, 179
optimization process, 185
optimum tip-speed ratio, 121
optimum wind farm design, 172

pitch angle, 88
pitch control, 120
pitch controlled, 88
pitch mechanism cost, 253
pitch moment coefficient, 64
pitch regulated, 121, 125
Pitot-static pressure
anemometers, 46
plasma actuator lift control, 138
Post Mill Turbines, 4
power coefficient, 76, 108
power component, 26
power management, 120
power train efficiency, 108
Prandtl tip loss factor, 92
pressure drag, 59
principle bending axes, 149
probable wind speeds, 274
propeller anemometer, 45

Rayleigh distribution, 36
Rayleigh energy estimate, 40
reactive power, 122
replacement cost, 250
Reynolds number, 64
roads, civil work cost, 259
roadway utility wind turbines, 291
rotor blade design evolution, 112
rotor construction, 148
rotor cost, 251
rotor deflections, 152
rotor design, 107
rotor hub cost, 253
rotor mass, 251
rotor radius history, 109
rotor vibrations, 153
roughness tolerant airfoils, 70

Savonius Turbine, 12
shrouded rotors, 283
small utility wind turbines, 291
Smeaton Turbine, 5
Smith-Putman Turbine, 8
Smock Mill Turbines, 4
sonic anemometers, 47
sound directionality, 204
sound emission, 194
sound intensity, 195
sound level weighting, 198
sound power level, 196
sound pressure level, 196
sound propogation, 205
sound waves, 195
spinner nose cone cost, 253
stall, 59
stall regulated, 121, 125
standard deviation, 28
stream-tube, 73
structural design, 143
structural failure causes, 143
supercapacitors, 222
supervisory control, 120
surface roughness categories, 24
synchronous machine, 122

tangential force coefficient, 90
thrust coefficient, 76
tip loss factor, 93
tip-speed-ratio, 84
torque control, 120
tower cost, 257
trade study, 269
trailing edge flap control, 136
transportation cost, 258
turbine control elements, 119
turbine wake model, 165

variable speed control, 126
variable speed electronics cost, 256
variable speed pitch regulated, 126
variable speed stall regulated, 126
velocity profile, 25
vertical wind turbines, 280

wake velocity deficit, 167
Weibull coefficients, 30, 274
Weibull cumulative distribution, 32
Weibull distribution, 30
Weibull energy estimate, 38
Weibull model fits, 33
Weibull standard deviation, 31

wind data acquisition
instrumentation, 48
wind energy architecture, 286
wind energy by the numbers, 18
wind farm design optimization, 170
wind farms, 164
wind turbine sound, 201
wind turbine wakes, 165
wind velocity frequency, 28

yaw drive and bearing
cost, 256